NUMERICAL MATHEMATICS AND SCIENTIFIC COMPUTATION

NUMERICAL MATHEMATICS AND SCIENTIFIC COMPUTATION

*P. Dierckx: *Curve and surface fitings with splines*
*H. Wilkinson: *The algebraic eigenvalue problem*
*I. Duff, A. Erisman, and J. Reid: *Direct methods for sparse matrices*
*M. J. Baines: *Moving finite elements*
*J. D. Pryce: *Numerical solution of Sturm-Liouville problems*
K. Burrage: *Parallel and sequential methods for ordinary differential equations*
Y. Censor and S. A. Zenios: *Parallel optimization: theory, algorithms and applications*
M. Ainsworth, J. Levesley and M. Marletta: *Wavelets, multilevel methods and elliptics PDEs*

Monographs marked with an asterisk (*) appeared in the series "Monographs in Numerical Analysis" which has been folded into, and is continued by, the current series.

Wavelets, Multilevel Methods and Elliptic PDEs

M. AINSWORTH

Department of Mathematics and Computer Science
University of Leicester

J. LEVESLEY

Department of Mathematics and Computer Science
University of Leicester

M. MARLETTA

Department of Mathematics and Computer Science
University of Leicester

W. A. LIGHT

Department of Mathematics and Computer Science
University of Leicester

CLARENDON PRESS • OXFORD

1997

Oxford University Press, Great Clarendon Street, Oxford OX2 6DP
Oxford New York
Athens Auckland Bangkok Bogota Bombay
Buenos Aires Calcutta Cape Town Dar es Salaam
Delhi Florence Hong Kong Istanbul Karachi
Kuala Lumpur Madras Madrid Melbourne
Mexico City Nairobi Paris Singapore
Taipei Tokyo Toronto Warsaw
and associated companies in
Berlin Ibadan

Oxford is a trade mark of Oxford University Press

Published in the United States
by Oxford University Press Inc., New York

A catalogue record for this book is available from the British Library

Library of Congress Cataloging in Publication Data
(Data available)

ISBN 0 19 850190 0

Typeset by the authors
Printed in Great Britain by
Biddles Ltd, Guildford & King's Lynn

Preface

The Seventh EPSRC Numerical Analysis Summer School was held at the University of Leicester during the summer of 1996, from the 8th to the 19th of July. During the first week there were lectures on neural networks, spectral problems for ODEs and PDEs, and wavelets; in the second week, the topics covered were multipole methods, hierarchic modelling in mechanics, and multilevel methods. The lecture notes of five of the six speakers are presented in this volume.

This was the second meeting in the EPSRC Numerical Analysis Summer School series to be held in Leicester: the 1994 meeting was also held here. As for the 1994 meeting, we attempted to mix speakers from areas which are well represented in UK research with others working in areas in which there are few UK researchers. There were approximately 90 participants, drawn mainly from the UK, though we were also pleased by the level of support from Europe and overseas.

All our speakers were charged with the difficult task of delivering lectures at a level which would be comprehensible to first year graduate students while still being sufficiently advanced to touch on current research topics. The high level of audience participation in many of the lectures indicated how successfully the first of these objectives was achieved; the presence of recent (or indeed unpublished) results in many of these lecture notes testifies to the success in attaining the second. It would be unjust to single out any one speaker for particular praise, but we do feel that Rick Beatson deserves special thanks: arriving in Leicester from New Zealand, and expecting to be a normal participant, he suddenly found himself 'offered' the rôle of speaker by the unfortunate withdrawal of Leslie Greengard due to illness. The resulting lectures are presented as joint work.

A long established tradition of the summmer schools is that of afternoon seminars delivered by participants and organized by *local experts*. The local experts for 1996 were Alastair Watson of Dundee and Endre Süli of Oxford. We would like to thank them both for the efficiency with which they brought order to those hectic summer afternoons, for covering up many of our mistakes, and for their great good humour.

Mark Ainsworth, Jeremy Levesley, Will Light, Marco Marletta
University of Leicester, January 1997

Contents

Contributors

Rick Beatson Department of Mathematics and Statistics, University of Canterbury, New Zealand.

Leslie Greengard Courant Institute of Mathematical Sciences, New York University, USA.

Michael Plum Mathematisches Institut I, Universität Karlsruhe, Germany.

Christoph Schwab Seminar für Angewandte Mathematik, ETH Zürich, Switzerland.

Gilbert Strang Department of Mathematics, Massachusetts Institute of Technology, USA.

Jinchao Xu Department of Mathematics, Pennsylvania State University, USA.

Invited Participants

Rick Beatson Department of Mathematics and Statistics, University of Canterbury, New Zealand.

George Cybenko School of Engineering, Dartmouth College, Hanover, New Hampshire, USA.

Leslie Greengard Courant Institute of Mathematical Sciences, New York University, USA.

Michael Plum Mathematisches Institut I, Universität Karlsruhe, Germany.

Christoph Schwab Seminar für Angewandte Mathematik, ETH Zürich, Switzerland.

Gilbert Strang Department of Mathematics, Massachusetts Institute of Technology, USA.

Jinchao Xu Department of Mathematics, Pennsylvania State University, USA.

A short course on fast multipole methods

Rick Beatson

Department of Mathematics and Statistics, University of Canterbury

Leslie Greengard

Courant Institute of Mathematical Sciences, New York University [1]

In this series of lectures, we describe the analytic and computational foundations of fast multipole methods, as well as some of their applications. They are most easily understood, perhaps, in the case of particle simulations, where they reduce the cost of computing all pairwise interactions in a system of N particles from $O(N^2)$ to $O(N)$ or $O(N \log N)$ operations. They are equally useful, however, in solving certain partial differential equations by first recasting them as integral equations. We will draw heavily from the existing literature, especially Greengard [23, 24, 25]; Greengard and Rokhlin [29, 32]; Greengard and Strain [34].

1 Introduction

1.1 Kernels

Many problems in computational physics require the evaluation of all pairwise interactions in large ensembles of particles. The N-body problem of gravitation (or electrostatics), for example, requires the evaluation of

$$\Phi(\mathbf{x}_j) = \sum_{\substack{i=1 \\ i \neq j}}^{N} \frac{m_i}{r_{ij}}. \tag{1.1}$$

for the gravitational potential and

$$\mathbf{E}(\mathbf{x}_j) = \sum_{\substack{i=1 \\ i \neq j}}^{N} m_i \frac{\mathbf{x}_j - \mathbf{x}_i}{r_{ij}^3} \tag{1.2}$$

for the gravitational field. Here, $\mathbf{x}_i$ denotes the location of the ith particle, m_i denotes the mass of the ith particle, and r_{ij} denotes the Euclidean distance between $\mathbf{x}_i$ and $\mathbf{x}_j$. Problems of electrostatics are governed by

[1] The work of Leslie Greengard was supported by the Applied Mathematical Sciences Program of the US Department of Energy, by a NSF Presidential Young Investigator Award and by a Packard Foundation Fellowship.

Coulomb's law, which takes the same form as eqs. (1.1) and (1.2), except that mass is always of one sign, whereas charge is not.

There is no reason, of course, to evaluate the gravitational field at the source locations only. Thus, we will also consider computing

$$\Phi(\mathbf{y}_j) = \sum_{\substack{i=1 \\ i \neq j}}^{N} \frac{m_i}{\|\mathbf{x}_i - \mathbf{y}_j\|} \tag{1.3}$$

for some large number of target points $\mathbf{y}_j$. We will also consider the fields due to continuous distributions of mass,

$$\mathbf{E}(\mathbf{x}) = -\int m(\mathbf{y}) \frac{\mathbf{x} - \mathbf{y}}{|\mathbf{x} - \mathbf{y}|^3} \, d\mathbf{y}. \tag{1.4}$$

A similar integral arises in magnetostatics, expressing the magnetic induction $\mathbf{B}$ due to a steady-state current density $\mathbf{J}$. This is the Biot–Savart law (Jackson 1975)

$$\mathbf{B}(\mathbf{x}) = \frac{1}{c} \int \mathbf{J}(\mathbf{y}) \times \frac{\mathbf{x} - \mathbf{y}}{|\mathbf{x} - \mathbf{y}|^3} \, d\mathbf{y}, \tag{1.5}$$

where c is the speed of light. Acoustic scattering processes give rise to N-body problems of the form

$$\Phi(\mathbf{x}_j) = \sum_{\substack{i=1 \\ i \neq j}}^{N} W_i \frac{e^{ikr_{ij}}}{r_{ij}}, \tag{1.6}$$

but they are beyond the scope of these lectures.

For an N-body problem in diffusion, consider the heat equation

$$u_t = \Delta u$$

with initial temperature distribution

$$u(\mathbf{x}, 0) = w(\mathbf{x}).$$

At any later time $T > 0$, the temperature field is well known to be given by

$$u(\mathbf{x}) = \frac{1}{\sqrt{(4\pi T)^{3/2}}} \int e^{-|\mathbf{x}-\mathbf{y}|^2/4T} \, w(\mathbf{y}) \, d\mathbf{y}. \tag{1.7}$$

A discrete analog of this is the N-body problem

$$u(\mathbf{x}_j) = \frac{1}{\sqrt{(4\pi T)^{3/2}}} \sum_{i=1}^{N} w_i \, e^{-r_{ij}^2/4T}. \tag{1.8}$$

Each of the preceding examples involves an integral of the form

$$u(\mathbf{x}) = \int K(\mathbf{x}, \mathbf{y}) w(\mathbf{y})\, d\mathbf{y} \tag{1.9}$$

or a sum of the form

$$u(\mathbf{x}) = \sum_{i=1}^{N} w_i K(\mathbf{x}, \mathbf{y}_i). \tag{1.10}$$

Direct evaluation of such sums at N target points obviously requires $O(N^2)$ operations, and algorithms which reduce the cost to $O(N^\alpha)$ with $1 \leq \alpha < 2$, $O(N \log N)$, $O(N \log^2 N)$, etc. are referred to as *fast summation methods*. The most well known of these is certainly the Fast Fourier Transform (FFT), which computes

$$u_j = \sum_{k=1}^{N} e^{2\pi i j k/N} w_k$$

for $j = 1, \ldots, N$ in about $5N \log N$ operations. From our point of view, the distinguishing features of the FFT are that it is exact, that it is based on considerations of symmetry (*algebra*), and that it is brittle. By the last point, we mean that it requires a uniform spatial grid to be applicable. Fast multipole methods (FMMs) are different. They are approximate, based on *analytic* considerations, and robust (insensitive to source distribution). In this category we include fast multipole method for the Laplace equation (Rokhlin [43]; Greengard and Rokhlin [29, 30, 32]; Carrier *et al.* [17]; Greengard [23]), the fast Gauss transform (Greengard and Strain [34]), and the fast multipole method for the Helmholtz equation (Rokhlin [44, 45]; Coifman *et al.* [18]). Each one relies on a detailed analysis of the pairwise interaction, and each one permits rigorous *a priori* error bounds.

There are a number of other, good algorithms for accelerating a variety of N-body calculations. We do not seek to review them here, and refer the reader to Alpert and Rokhlin [1]; Anderson [2, 3]; Appel [4]; Barnes and Hut [5]; Beylkin *et al.* [11]; Bradie *et al.* [13]; Brandt [14]; Brandt and Lubrecht [15]; Canning [16]; Greengard [25]; Hockney and Eastwood [36]; Odlyzko and Schönhage [42]; Van Dommelen and Rundensteiner [47]. Some of these are very broad in their applicability, while others, like FMMs, are intended for a specific kernel.

1.2 Degenerate kernels

Consider now a generic summation problem of the form (1.10), where the kernel $K(\mathbf{x}, \mathbf{y})$ can be expressed as a finite series

$$K(\mathbf{x}, \mathbf{y}) = \sum_{k=1}^{p} \phi_k(\mathbf{x}) \psi_k(\mathbf{y}). \tag{1.11}$$

Such kernels are called finite rank or degenerate kernels, and N-body problems governed by them are easily resolved. First, one computes the *moments*

$$A_k = \sum_{i=1}^{N} w_i \psi_k(\mathbf{y}_i).$$

Second, one evaluates $u(\mathbf{x})$ at each desired point via the formula

$$u(\mathbf{x}) = \sum_{k=1}^{p} A_k \phi_k(\mathbf{x}).$$

The amount of work required is of the order $O(Np)$. While the N-body problems of mathematical physics are not of this type, the degenerate case serves as a useful model. Note that, independent of the details of the source distribution, $u(\mathbf{x})$ must always be a linear combination of the functions $\phi_1(\mathbf{x}), \ldots, \phi_p(\mathbf{x})$. In other words, the *dimension* of the function space where $u(\mathbf{x})$ lives is much smaller than the N-dimensional space containing the data $\{\mathbf{y}_i, w_i\}$ and there is a tremendous amount of compression in the transformation from the latter to the former. The notion of compression is fundamental to the fast Gauss transform and to the FMM for the Laplace equation, although not to the FMM for the Helmholtz equation.

2 Hierarchical and fast multipole methods in one dimension

Most tree codes and fast multipole methods are based on a judicious combination of the following key features.

- *A specified acceptable accuracy of computation* ϵ.
- *A hierarchical subdivision of space* into panels, or clusters, of sources.
- *A far field expansion* of the "kernel" $k(\mathbf{x}, \mathbf{y})$ in which the influence of source and evaluation points separates.
- (Optional) *The conversion of far field expansions into local expansions.*

The first feature is very simple but absolutely crucial. Once it is admitted that the result of a calculation is needed only to a certain accuracy, then approximations can be used. This is a key to many fast methods. Actually, analogous remarks about the need to explicitly admit that a certain accuracy is sufficient can be made about other fast methods such as the multigrid method for solving partial differential equations.

Historically, hierarchical and fast multipole schemes were first developed in two and three dimensions, where they are of great practical importance. However, for pedagogical reasons, we will deviate from the historical order of developement and first develop a scheme in a one variable setting. In this simpler setting the essential ideas of hierarchical and fast multipole schemes stand out with great clarity, facilitating an understanding of the more complicated, and much more practically important, two and three dimensional schemes which will be discussed later.

2.1 Fast evaluation of multiquadrics

The one dimensional example we shall consider is the fast evaluation of one dimensional multiquadric radial basis functions. Consider the basic function

$$\phi(x) = \sqrt{x^2 + c^2} \,,$$

where $0 < c \leq h$, and a corresponding spline, or multiquadric radial basis function,

$$s(\cdot) = \sum_{j=1}^{N} d_j \phi(\cdot - x_j) \,.$$

In this section we will develop a fast method of evaluating $s(x)$ when the number of different evaluation points x is large.

The first step is to develop a series expansion of $\phi(x-t)$ valid for "large" values of $|x|$. To this end consider the function

$$g(u) = (u-t)\sqrt{1 + \frac{c^2}{(u-t)^2}},$$

where $u \in \mathbf{C}$ and $\sqrt{\cdot}$ denotes the principal branch of the complex square root. $g(u)$ coincides with $\phi(u-t)$ for all real u greater than t and with $-\phi(u-t)$ for all real u less than t. Now $1 + \{c^2/(u-t)^2\}$ is both real and non-positive only for values of u on the line segment from $t - ic$ to $t + ic$. Hence g is analytic in the region $|u| > \sqrt{t^2 + c^2}$ and can be expanded in a Laurent series about zero valid for $|u| > \sqrt{t^2 + c^2}$. Computing this Laurent series yields the expansion

$$\begin{aligned} \phi\,(x-t) &= \sqrt{(x-t)^2 + c^2} \\ &= \operatorname{sign}(x)\Big\{ x - t + \frac{1}{2}c^2x^{-1} \\ &\qquad + \frac{1}{2}tc^2x^{-2} + \frac{1}{8}(4t^2c^2 - c^4)x^{-3} \\ &\qquad + \frac{1}{8}(4t^3c^2 - 3tc^4)x^{-4} + \cdots + q_p(c,t)x^{-p} + \cdots \Big\} \end{aligned} \tag{2.1}$$

of $\phi\,(x-t)$. The corresponding truncated expansions are endowed with the error bound

$$\left|\phi(x-t) - \text{sign}(x)\left\{ x - t + \frac{1}{2}c^2x^{-1} + \cdots + q_p(c,t)x^{-p}\right\}\right|$$
$$\leq 2(|t|+c)\left(\frac{\sqrt{t^2+c^2}}{|x|}\right)^{p+1}\frac{1}{1-\frac{\sqrt{t^2+c^2}}{|x|}} \quad (2.2)$$

for $|x| > \sqrt{t^2+c^2}$.

Note that each term $q_j(c,t)x^{-j}$ in the series (2.2) separates into the product of a coefficient depending on the source point alone, namely $q_j(c,t)$, and a homogeneous function depending on the evaluation point alone, namely x^{-j}. Thus the truncated far field expansions corresponding to different sources $t = x_j$ in a panel T may be summed. The form of the series stays the same. Having several sources rather than only one merely changes the values of the coefficients of the homogeneous functions $\{x^{-j}\}$.

To be more precise we associate with a panel T that part of the spline s due to sources lying in the panel by setting

$$s_T(\cdot) = \sum_{j:x_j\in T} d_j\phi(\cdot - x_j).$$

Then, presuming for the moment that the panel is centered about the point $x = 0$, we approximate s_T by a far field expansion also centered at $x = 0$,

$$r_T(x) = \text{sign}(x)\left\{a_{-1}x + a_0 + a_1x^{-1} + \cdots a_px^{-p}\right\}. \quad (2.3)$$

This series is obtained from the truncated expansion (2.2) for a single source at t with unit weight by choosing $t = x_j$, multiplying by d_j, and then summing over all sources x_j in the panel.

If the source panel is $T = [-h, h)$, and the geometry of the source and evaluation regions is as in Fig. 1 below, then in the evaluation region

$$\max_{j:x_j\in T}\frac{\sqrt{x_j^2+c^2}}{|x|} \leq \frac{\sqrt{h^2+h^2}}{|x|} \leq \frac{\sqrt{2}}{3} = \frac{1}{2.12\ldots}.$$

Hence, using the estimate of equation (2.2)

$$|s_T(x) - r_T(x)| \leq D_T4h\left(\frac{1}{2.12\ldots}\right)^{p+1}\frac{1}{1-\frac{1}{2.12\ldots}} \quad (2.4)$$

for all x in the evaluation region, where $D_T = \sum_{j:x_j\in T}|d_j|$.

If the source panel is $[t-h, t+h)$ centered at $x = t$ rather than at $x = 0$ then the appropriate form of the far field expansion is also centered

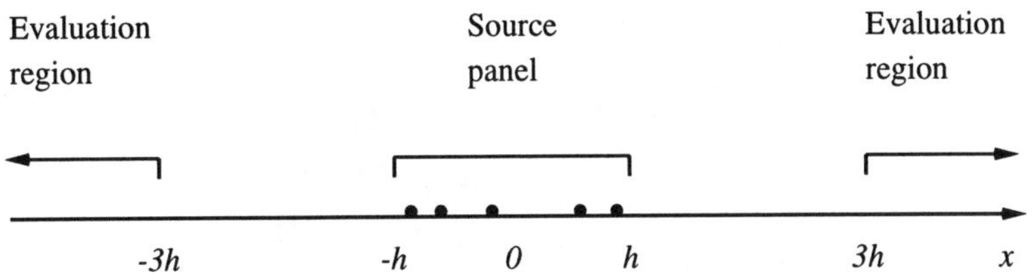

FIG. 1. Source and evaluation regions.

at $x = t$, and making the natural change of variable in (2.3) is

$$r_T(x) = \operatorname{sign}(x-t)\left\{a_{-1}(x-t) + a_0 + a_1(x-t)^{-1} + \cdots a_p(x-t)^{-p}\right\}. \tag{2.5}$$

It enjoys the error bound

$$|s_T(x) - r_T(x)| \leq D_T 4h \left(\frac{1}{2.12\ldots}\right)^{p+1} \frac{1}{1 - \frac{1}{2.12\ldots}} \tag{2.6}$$

whenever $|x - t| \geq 3h$, where t is the center of the source panel T and h is its radius. Thus the far field expansion, r_T, converges at a conveniently fast rate whenever x is separated from the panel T by at least the diameter of T. Such points x will be said to be **well separated** from T. A panel U will be said to be **well separated** from T if all of its points are well separated from T. Approximate evaluation of $s_T(x)$ via the the far field approximation $r_T(x)$ will be much quicker than direct evaluation whenever the number of terms in the series is much smaller than the number of sources in the panel T.

We are now ready to construct a fast evaluation scheme by combining the far field expansions with a hierarchical subdivision of space. The underlying idea is to summarize/approximate the influence of many sources by evaluating a few short series, thus reducing the cost in flops of a single extra evaluation from $\mathcal{O}(N)$ to $\mathcal{O}(\log N)$, or even $\mathcal{O}(1)$.

Assume that the problem has been standardized so that all the source and evaluation points lie in the interval $[0, 1]$. We subdivide $[0, 1]$ into a binary tree of panels as illustrated in Fig. 2 below.

At any point x we will approximate $s(x)$ by summing the direct influence, $s_T(x)$, of the near neighbours of the childless panel containing x, together with the far field approximation $r_T(x)$ to the influence of panels, T, far away from x. Since the underlying motive is to save floating point operations we always use the largest possible, that is the highest level, source panel for the far field series. For example, suppose the tree is as in Fig. 2 and is a uniform binary tree down to level 3. Then to approximately evaluate $s(x)$ for x in the panel $[0, \frac{1}{8}]$ we use

$$\begin{aligned} s(x) \approx\ & s_{[0,\frac{1}{8})}(x) + s_{[\frac{1}{8},\frac{1}{4})}(x) \\ & + r_{[\frac{1}{4},\frac{3}{8})}(x) + r_{[\frac{3}{8},\frac{1}{2})}(x) \\ & + r_{[\frac{1}{2},\frac{3}{4})}(x) + r_{[\frac{3}{4},1]}(x). \end{aligned}$$

Note in particular the use of the two level 2 approximations $r_{[\frac{1}{2},\frac{3}{4})}$ and $r_{[\frac{3}{4},1]}$, rather than the four level 3 approximations $r_{[\frac{1}{2},\frac{5}{8})},\ldots,r_{[\frac{7}{8},1]}$. The use of these larger/higher level panels enables us to approximate the influence of the sources in the interval $[\frac{1}{2},1]$ with two rather than four series evaluations: that is, it halves the flop count for this part of the evaluation task. Focusing on the use of $r_{[\frac{1}{2},\frac{3}{4})}(x)$ this is allowable since the panel $[\frac{1}{2},\frac{3}{4})$ is well separated from the panel $[0,\frac{1}{4})$, the parent of $[0,\frac{1}{8})$. Furthermore, the use of $r_{[\frac{3}{4},1]}(x)$ is allowable since $[\frac{3}{4},1]$ is well separated from $[0,\frac{1}{4})$.

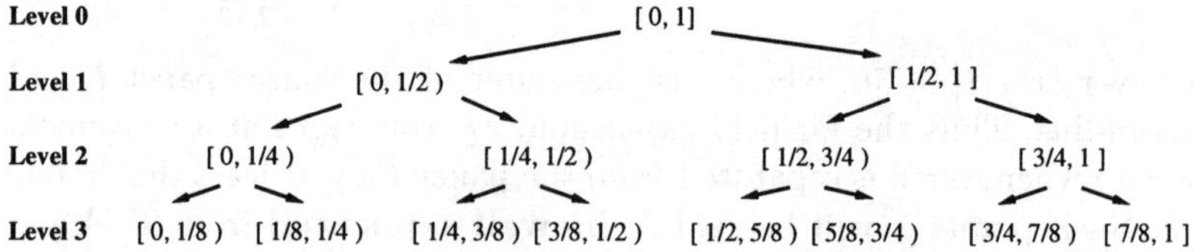

FIG. 2. Binary tree structure induced by a uniform subdivision of the unit interval.

Similarly, to approximately evaluate $s(x)$ in the panel $[\frac{1}{2},\frac{5}{8})$ we would use

$$\begin{aligned} s(x) \approx\ & s_{[\frac{3}{8},\frac{1}{2})}(x) + s_{[\frac{1}{2},\frac{5}{8})}(x) + s_{[\frac{5}{8},\frac{3}{4})}(x) \\ & + r_{[\frac{1}{4},\frac{3}{8})}(x) + r_{[\frac{3}{4},\frac{7}{8})}(x) + r_{[\frac{7}{8},1]}(x) \\ & + r_{[0,\frac{1}{4})}(x). \end{aligned}$$

In view of the error bound (2.6) the overall error in such a procedure will be bounded by ϵ if the error in approximating $\phi(x-t)$ at every level is bounded by $\epsilon/||\mathbf{d}||_1$.

This motivates a simple tree code for evaluating a univariate multiquadric.

A hierarchical code for evaluating a univariate mutiquadric radial basis function: setup part.

Input: the desired accuracy ϵ, $0 < \epsilon < \frac{1}{2}$, the source locations and weights $\{x_j, d_j\}_{j=1}^{N}$.

Step 1: *Choose* $p \approx |\log_{2.12}(\epsilon/||\mathbf{d}||_1)|$.

This choice guarantees that the desired accuracy in approximating $s(x)$ will be attained.

Step 2:

Choose the number of levels $m \approx \log_2 N$.

If the data is approximately uniformly distributed this choice guarantees that each childless panel contains a number of sources which is bounded independently of N.

Step 3:

Work down from the root panel, $[0,1]$, *subdividing each panel at levels* $0, \ldots, m-1$ *in half, and assigning sources* x_j *to the panels containing them.*

The work here is one comparison for every source for each level from 0 through $m-1$ in order to assign the sources associated with each parent panel to the correct child panel.

Step 4:

Work up the tree from level m *to level* 2, *calculating the far field expansions associated with each panel.*

The work here is calculating the $p+2$ coefficients $a_{-1}, a_0, a_1, \ldots a_p$ of the expansion $r_T(\cdot)$, for each panel T.

Once the setup phase has been completed we can use the far field approximations for fast evaluation in the manner of the examples immediately preceding the description of the setup phase. Given x one proceeds down the levels from level 2 to level m. At each level one uses the approximations from far away panels (those from which the panel containing x is well separated) and ignores source panels whose influence has already been incorporated at a higher level. No work is undertaken for near neighbours until the finest level. The method is of a divide and conquer type with near neighbours at level $k-1$ being split to yield smaller near neighbour panels and some source panels whose far field approximations can be used at level k. At the finest level the near neighbours are no longer to be split, and their influence is incorporated directly by adding the "direct" evaluations $s_T(x)$. In this way the largest possible source panels are always used for the far field approximations, and thus the flop count is minimized. A typical situation at an intermediate level k is shown in Fig. 3.

A hierarchical code for evaluating a univariate multiquadric radial basis function: Evaluation part.

The evaluation procedure can be expressed in recursive and non-recursive forms, both of which generalize easily to nonuniform partitions of the problem region.

Recursive form of the evaluation part

Given the root node R of the binary tree of panels and the point x at which

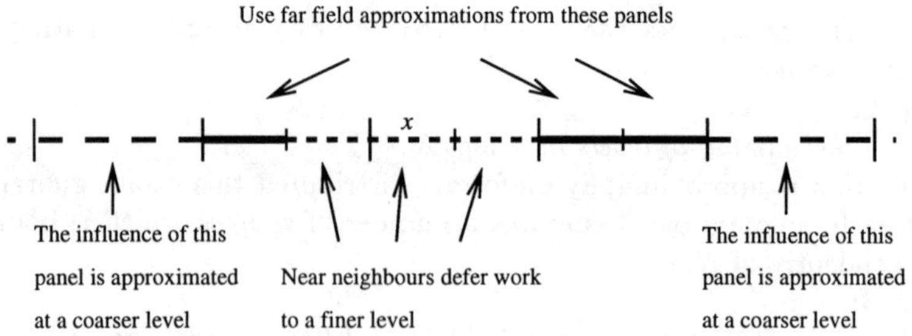

FIG. 3. A typical configuration of source panels at level k when evaluating at x.

$s(x)$ is to be evaluated, a call to routine *eval_recur* specified below with the root node and x returns an approximation to $s(x)$.

In the pseudo-code to follow a C++ like notation is used. Thus the two children of a panel T are denoted by T->child[0] and T->child[1], and remarks are indicated by a leading double slash, //.

```
eval_recur(T,x) {
    if (T is far away from x) then
        // Approximate evaluation gives sufficient accuracy.
        // Perform it and terminate descent of this branch.
        return( r_T(x) )
    else if (T is childless)
        // Cannot descend further on this branch.
        // Evaluate the influence of T directly.
        return ( s_T(x) )
    else
        // Descend the branch to a finer level in the
        // tree where approximations may be allowable.
        return ( eval_recur( T->child[0], x)
                        + eval_recur( T->child[1], x) )
    end if
}
```

We give an equivalent non-recursive expression of the evaluation phase below. This may be more appealing to some readers, and could give rise to more efficient implementations in some environments.

Nonrecursive form of the evaluation part

Given the root node R of the binary tree of panels and the point x at which $s(x)$ is to be evaluated the following pseudo-code returns an approximation to the value of $s(x)$ in the variable a.

Step1:

Initialize a to zero.

Step 2:
For k from 2 to m, add to a the far field approximation $r_T(x)$ to the influence of each panel T well separated from x, which is not a subset of a panel already used at a coarser level (see Fig. 3).
Step 3:
Let Q be the bottom level (level m) panel containing x. Add to a the direct evaluation $s_T(x)$ of the influence of each near neighbour T of Q.

Note that in the case of a uniform subdivision deciding which panel a point lies in requires only a single multiplication by $1/h$. Also, if we work from level m backwards up to level 2 then the panel at level $k-1$ containing x is the parent of the panel at level k containing x, so that even this single multiplication is avoidable except at the bottom-most level.

We are now ready to estimate the computational cost of a single extra evaluation, ignoring the cost of setup. The far field part of the work involves the evaluation of at most 3 far field expansions of $p+2$ terms at each level from 2 to m. Hence the flop count arising from the evaluation of far field expansions is $\mathcal{O}(mp)$. The direct part of the evaluation is the direct evaluation of the influence of at most 3 bottom level panels. Since each bottom level panel contains only $\mathcal{O}(1)$ sources the flop count arising from the direct part of the evaluation is $\mathcal{O}(1)$. Hence the incremental cost in flops of an additional evaluation is

$$\mathcal{O}(mp) \approx \mathcal{O}(|\log(\epsilon/\|\mathbf{d}\|_1)| \log N),$$

which is an order of magnitude faster than the $\mathcal{O}(N)$ flop count for a single direct evaluation of $s(x)$.

2.2 Enhancements

The algorithm outlined above can be improved in several ways.

First, it is clearly inefficient to split uniformly down to level m irrespective of the distribution of sources. The panel splitting should be done in a more adaptive manner. For example subdividing a panel only when it contains a certain minimum number of sources. Also, once the sources have been allocated to children there is no reason to keep the child panels of uniform size. They can be shrunk to the minimum size necessary to contain all their sources. In this way the tree would be more refined in regions where there is a high density of sources, and finish after relatively few levels in a region where the density of sources is low.

Second, the far field expansions are of the form

$$r_T(x) = \operatorname{sign}(x-t)\left(a_{-1}(x-t) + a_0 + a_1(x-t)^{-1} + \cdots + a_p(x-t)^{-p}\right).$$

Such a function is smooth away from t. Hence it can be well approximated by a truncated Taylor series in any panel well separated from t. Employing this idea instead of summing the values of Laurent series one can approximate them with Taylor polynomials, and sum these polynomials, i.e. add corresponding coefficients. The result in the end is a polynomial associated with each panel, T, from level 2 to the bottom level, which approximates the influence of all sources in panels far away from T. Then approximate evaluation of $s(x)$ involves the evaluation of one polynomial of order approximately p, plus the calculation of the direct influence of at most three panels each containing $\mathcal{O}(1)$ sources. Therefore the incremental cost of a single extra evaluation is reduced from $\mathcal{O}(|\log(\epsilon/\|\mathbf{d}\|_1)|\log N)$ flops to $\mathcal{O}(|\log(\epsilon/\|\mathbf{d}\|_1)|)$ flops. There is a price to pay of course. This price is the extra work of converting Laurent expansions into local polynomial expansions, summing such polynomial expansion coefficients, and shifting such expansions to be centered at the center of the child panel. However, even for relatively moderate values of N, the fast multipole scheme described in this paragraph is quicker than the hierarchical scheme previously described. A variety of hierarchical and fast multipole methods for evaluating radial basis functions are described in the papers [6, 7, 8, 9].

3 The fast Gauss transform

Given a set of points $\mathbf{x}_i = (x_i, y_i)$ and a set of source strengths w_i, the function

$$U(\mathbf{x}) = \sum_{i=1}^{N} w_i e^{-|\mathbf{x}-\mathbf{x}_i|^2/4T} \tag{3.1}$$

will be referred to as the (discrete) Gauss transform. This is an infinitely differentiable function and the derivatives decay rapidly in space for any $T > 0$. To be a bit more precise, we introduce the Hermite functions $h_n(x)$, defined by

$$h_n(x) = (-1)^n \frac{d^n}{dx^n}(e^{-x^2}).$$

Consider now a two-dimensional problem, where a source $\mathbf{x}_i$ is located in a square with center $\mathbf{c} = (c_1, c_2)$ and side length $\sqrt{T}$. Then the heat kernel

$$e^{-|\mathbf{x}-\mathbf{x}_i|^2/4T}$$

can be expressed as an Hermite series

$$e^{-|\mathbf{x}-\mathbf{x}_i|^2/4T} = \sum_{n_1,n_2=0}^{\infty} \Phi_{n_1n_2}(\mathbf{x}-\mathbf{c})\Psi_{n_1n_2}(\mathbf{x}_i-\mathbf{c}), \tag{3.2}$$

where, for $\mathbf{x} = (x, y)$,

$$\Psi_{n_1n_2}(\mathbf{x}) = \frac{1}{n_1!n_2!}\left(\frac{x}{\sqrt{4T}}\right)^{n_1}\left(\frac{y}{\sqrt{4T}}\right)^{n_2},$$

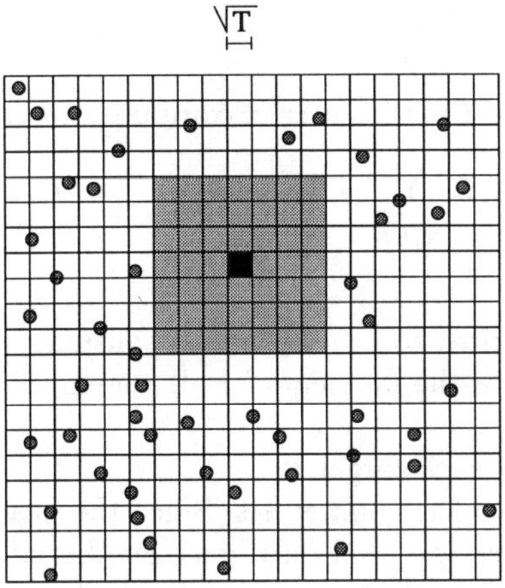

FIG. 4. The fast Gauss transform mesh.

and

$$\Phi_{n_1n_2}(\mathbf{x}) = h_{n_1}\left(\frac{x}{\sqrt{4T}}\right) h_{n_2}\left(\frac{y}{\sqrt{4T}}\right).$$

One can also derive a precise estimate of the error in truncating the series (3.2) after a finite number of terms [34]:

$$\left| e^{-|\mathbf{x}-\mathbf{x}_i|^2/4T} - \sum_{n_1,n_2=0}^{p} \Phi_{n_1n_2}(\mathbf{x})\Psi_{n_1n_2}(\mathbf{x}_i) \right| \le \left(\frac{1}{p!}\right)\left(\frac{1}{8}\right)^p. \tag{3.3}$$

Such a rapidly decaying error makes the heat kernel very nearly degenerate. Suppose, for example, that one is satisfied with four digits of accuracy. Then $p = 4$ is sufficient to guarantee it. For eight digits, choose $p = 6$; for fourteen digits, choose $p = 10$; and so on. It turns out that (3.2) together with its error bound (3.3) is the only analytical tool needed to construct a fast Gauss transform. It remains only to organize the computation so that the finite series can be used effectively.

We begin by assuming that all sources and targets lie in a square B_0 of unit area on which we superimpose a uniform mesh which subdivides B_0 into finer squares of side length $\sqrt{T}$ (Fig. 4). [If $T \ge 1$, B_0 needs no further refinement.]

A simple fast Gauss transform

Step 1:
Sort the N sources into the fine squares of the uniform mesh.
Step 2:
Choose p sufficiently large that the error estimate (3.3) is less than the

desired precision ϵ.
Step 3:
For each fine square B with center c, compute the moments

$$A_{n_1 n_2} = \sum_{\mathbf{x}_j \in B} w_j \Psi_{n_1 n_2}(\mathbf{x}_j), \quad \text{for } n_1, n_2 \le p.$$

Operation count:
Each source contributes to exactly one expansion, so that the amount of work required to form the moments for all nonempty boxes is proportional to Np^2.
Step 4: (Repeat for each target $\mathbf{x}$)
Find the box B in which point $\mathbf{x}$ lies. Since the heat kernel decays exponentially fast in space, we need only consider the influence of certain near neighbours in our decomposition of B_0 (the shaded region in Fig. 4). To be more precise, ignoring all but the nearest $(2n+1)^2$ boxes incurs an error of the order $e^{-n^2/4}$. For $n = 6$, this is approximately 10^{-4}. The influence of the sources contained in each of these nearby boxes can be obtained by evaluating the expression

$$\sum_{n_1,n_2=0}^{p} A^j_{n_1 n_2} \Phi_{n_1 n_2}(\mathbf{x} - \mathbf{c}_j),$$

where $\{A^j_{n_1 n_2}\}$ are the precomputed moments for neighbour j and $\mathbf{c}_j$ is its center.
Operation count:
The amount of work required is of the order $(2n+1)^2 p^2$ for each target, which is a constant. The total amount of work, therefore, is proportional to $N + M$ where N is the number of sources and M is the number of targets.

It is worth considering two extreme regimes. For very small times, say $T = 10^{-10}$, the effect of each heat source is vanishingly small beyond a distance of approximately 10^{-5}. Assuming the sources and targets are widely distributed, the question of rapidly computing sums of the form (3.1) becomes substantially one of sorting and finding near neighbours. For large times, say $T = 1$, then all sources have nonnegligible interactions, but only one box is constructed, one set of moments is computed, and one expansion is evaluated for each target. As time marches forward, there is less and less information content in the temperature field $U(\mathbf{x})$ as would be expected from a diffusion process.

For a more complex but more efficient version of the method, we refer the reader to the original paper by Greengard and Strain [34]. For 100,000 sources and targets, the algorithm is about three orders of magnitude faster than direct summation. Finally, note that the algorithm performs *worst* when sources are uniformly distributed. That case provides an upper bound on the number of expansions which need to be evaluated at each target position. In other words, the more clustering present in the source distribution, the better. This behavior is atypical of standard numerical methods based, for example, on finite difference equation solvers.

4 The FMM in two dimensions

The evaluation of gravitational or Coulombic interactions is somewhat more complicated than the evaluation of interactions governed by the heat kernel. The force is long-ranged and nonsmooth (at least locally). It requires a new set of tools for organizing the computation. Appel [4], Barnes and Hut [5], and others working in the astrophysics community developed what have come to be known as "tree codes" in order to overcome the computational obstacle presented by this N-body problem. They are based on the observation that at some distance from the sources, the gravitational field is smooth and should be representable in some compressed form.

This lecture begins with the two-dimensional case, since both the analytic and computational aspects of the problem are easier to understand. We will use the language of electrostatics, and assume it known that a two-dimensional point charge located at $\mathbf{x}_0 = (x_0, y_0) \in \mathbf{R}^2$ gives rise to potential and electrostatic fields at any location $\mathbf{x} = (x, y) \neq \mathbf{x}_0$ of the form

$$\phi_{\mathbf{x}_0}(x, y) = -\log(||\mathbf{x} - \mathbf{x}_0||) \tag{4.1}$$

and

$$E_{\mathbf{x}_0}(x, y) = \frac{(\mathbf{x} - \mathbf{x}_0)}{||\mathbf{x} - \mathbf{x}_0||^2} \tag{4.2}$$

respectively. Away from source points, the potential ϕ is harmonic, that is, it satisfies the Laplace equation

$$\nabla^2 \phi = \frac{\partial^2 \phi}{\partial x^2} + \frac{\partial^2 \phi}{\partial y^2} = 0 \ . \tag{4.3}$$

Since for every harmonic function u, there exists an analytic function w for which $u = Re(w)$, we will use complex analysis to simplify notation. Equating (x, y) with the complex point z, we note that

$$\phi_{\mathbf{x}_0}(\mathbf{x}) = Re(-\log(z - z_0)), \tag{4.4}$$

and will refer to the analytic function $\log(z)$ as the potential due to a charge. We will continue to use analytic expressions for the potentials due to more complicated charge distributions, even though we are only interested in their real part. For future reference, we note that we can extract the electrostatic field from the complex potential.

Lemma 4.1 *If $u = Re(w)$ describes the potential field at (x, y), then the corresponding force field is given by*

$$\nabla u = (u_x, u_y) = (Re(w'), -Im(w')), \tag{4.5}$$

where w' is the derivative of w.

Consider now a point charge of strength q, located at z_0. A straightforward calculation shows that for any z with $|z| > |z_0|$,

$$\phi_{z_0}(z) = q\log(z - z_0) = q\left(\log(z) - \sum_{k=1}^{\infty}\frac{1}{k}\left(\frac{z_0}{z}\right)^k\right). \tag{4.6}$$

This provides us with a means of computing the multipole expansion due to a collection of charges (Rokhlin [43]; Greengard and Rokhlin [29]).

Lemma 4.2 (Multipole expansion) *Suppose that m charges of strengths $\{q_i, i = 1, ..., m\}$ are located at points $\{z_i, i = 1, ..., m\}$, with $|z_i| < r$. Then for any z with $|z| > r$, the potential $\phi(z)$ induced by the charges is given by*

$$\phi(z) = Q\log(z) + \sum_{k=1}^{\infty}\frac{a_k}{z^k}, \tag{4.7}$$

where

$$Q = \sum_{i=1}^{m} q_i \quad \textit{and} \quad a_k = \sum_{i=1}^{m}\frac{-q_i z_i^k}{k}. \tag{4.8}$$

Furthermore, for any $p \geq 1$,

$$\begin{aligned}\left|\phi(z) - Q\log(z) - \sum_{k=1}^{p}\frac{a_k}{z^k}\right| &\leq \frac{1}{p+1}\alpha\left|\frac{r}{z}\right|^{p+1} \\ &\leq \left(\frac{A}{p+1}\right)\left(\frac{1}{c-1}\right)\left(\frac{1}{c}\right)^p,\end{aligned} \tag{4.9}$$

where

$$c = \left|\frac{z}{r}\right|, \quad A = \sum_{i=1}^{m}|q_i|, \ \textit{and} \quad \alpha = \frac{A}{1 - |\frac{r}{z}|}. \tag{4.10}$$

Proof The form of the multipole expansion follows immediately from the lemma above by summing over the m expansions corresponding to the m sources z_i. To see the error bound observe that

$$\left|\phi(z) - Q\log(z) - \sum_{k=1}^{p}\frac{a_k}{z^k}\right| \leq \left|\sum_{k=p+1}^{\infty}\frac{a_k}{z^k}\right|.$$

Substituting expression (4.8) for a_k into the above

$$\left|\sum_{k=p+1}^{\infty} \frac{a_k}{z^k}\right| \leq A \sum_{k=p+1}^{\infty} \frac{r^k}{k|z|^k} \leq \frac{A}{p+1} \sum_{k=p+1}^{\infty} \left|\frac{r}{z}\right|^k$$
$$= \frac{\alpha}{p+1} \left|\frac{r}{z}\right|^{p+1} = \left(\frac{A}{p+1}\right)\left(\frac{1}{c-1}\right)\left(\frac{1}{c}\right)^p.$$

which is the required result. □

Note that if $c \geq 2$, then

$$\left|\phi(z) - Q\log(z) - \sum_{k=1}^{p} \frac{a_k}{z^k}\right| \leq A\left(\frac{1}{2}\right)^p. \tag{4.11}$$

The reader may recall that in the fast Gauss transform, the error bound (3.3) was independent of target location. Here, the amount of *compression* achieved depends both on the desired precision and the distance of the target from the sources.

The ability to form multipole expansions and compute error bounds via (4.9) is all that is required for an $O(N \log N)$ fast summation algorithm. It must be combined, however, with a recursive "divide and conquer" strategy.

4.1 The $N \log N$ algorithm

To simplify the number of issues addressed, let us assume for the moment that the particles are fairly homogeneously distributed in a square. In order to make systematic use of multipole expansions, we introduce a hierarchy of boxes which refine the computational domain into smaller and smaller regions. At refinement level 0, we have the entire computational domain. Refinement level $l+1$ is obtained recursively from level l by subdivision of each box into four equal parts. This yields a natural tree structure, where the four boxes at level $l+1$ obtained by subdivision of a box at level l are considered its children.

Definition 4.3 *Two boxes are said to be* **near neighbours** *if they are at the same refinement level and share a boundary point. (A box is a near neighbour of itself.)*

Definition 4.4 *Two boxes are said to be* **well separated** *if they are at the same refinement level and are not near neighbours.*

Definition 4.5 *With each box i is associated an* **interaction list**, *consisting of the children of the near neighbours of i's parent which are well separated from box i (Fig. 4).*

The basic idea is to consider clusters of particles at successive levels of spatial refinement, and to compute interactions between distant clusters by means of multipole expansions when possible. It is clear that at levels

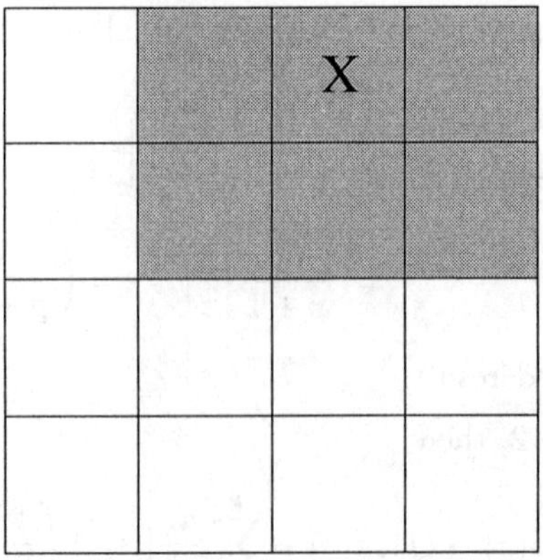

FIG. 5. The first step of the algorithm. Interactions between particles in box X and the white boxes can be computed via multipole expansions. Interactions with near neighbours (grey) are not computed.

0 and 1, there are no pairs of boxes which are well separated. At level 2, on the other hand, sixteen boxes have been created and there are a number of well separated pairs. Multipole expansions can then be used to compute interactions between these well separated pairs (Fig. 5) with rigorous bounds on the error. In fact, the bound (4.11) is valid so that given a precision ϵ, we need to use $p = \log_2(1/\epsilon)$ terms.

It remains to compute the interactions between particles contained in each box with those contained in the box's near neighbours, and this is where recursion enters the picture. After each level 2 box is refined to create level 3, we seek to determine which other boxes can be interacted with by means of multipole expansions. But notice that those boxes outside the region of the parent's nearest neighbours are already accounted for (at level 2) and that interactions with current near neighbours cannot accurately be computed by means of an expansion. The remaining boxes correspond exactly to the interaction list defined above (Fig. 6).

The nature of the recursion is now clear. At every level, the multipole expansion is formed for each box due to the particles it contains. The resulting expansion is then evaluated for each particle in the region covered by its interaction list (Fig. 4).

We halt the recursive process after roughly $\log N$ levels of refinement. The amount of work done at each level is of the order $O(N)$. To see this, note first that approximately $N\,p$ operations are needed to create all expansions, since each particle contributes to p expansion coefficients. Secondly, from the point of view of a single particle, there are at most 27 boxes (the maximum size of the interaction list) whose expansions are computed, so that $27\,N\,p$ operations are needed for all evaluations.

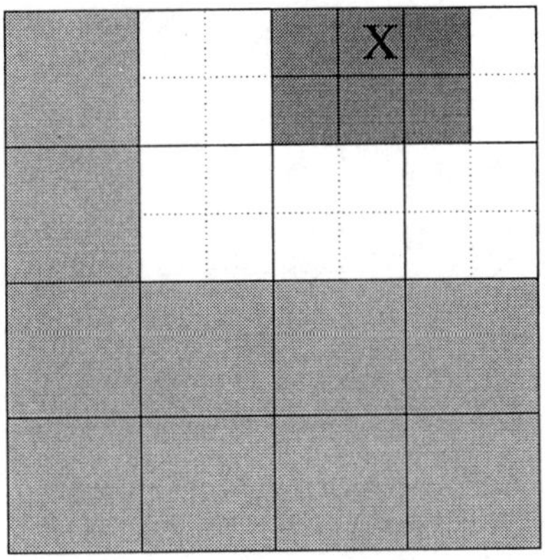

FIG. 6. The second step of the algorithm. After refinement, note that the particles in the box marked X have already interacted with the most distant particles (light grey). They are now well separated from the particles in the white boxes, so that these interactions can be computed via multipole expansions. The near neighbour interactions (dark grey) are not computed.

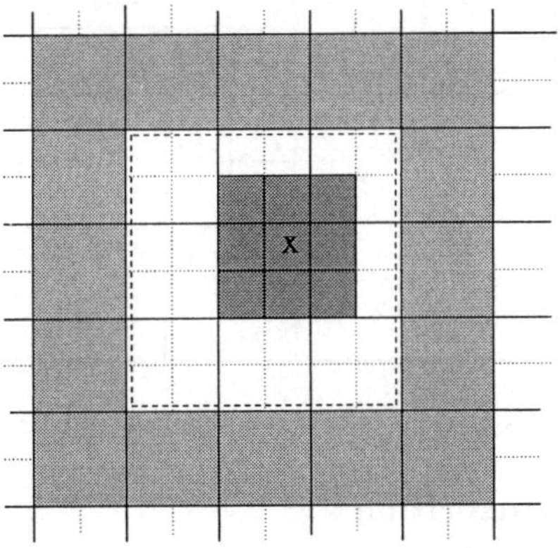

FIG. 7. Subsequent steps of the algorithm. The interaction list for box X is indicated in white.

At the finest level, we have created roughly $4^{\log_4 N} = N$ boxes and it remains only to compute interactions between nearest neighbours. By the assumption of homogeneity, there are $O(1)$ particles per box, so that this last step requires about $8\,N$ operations (Fig. 5). The total cost is approximately

$$28Np\log N + 8N. \tag{4.12}$$

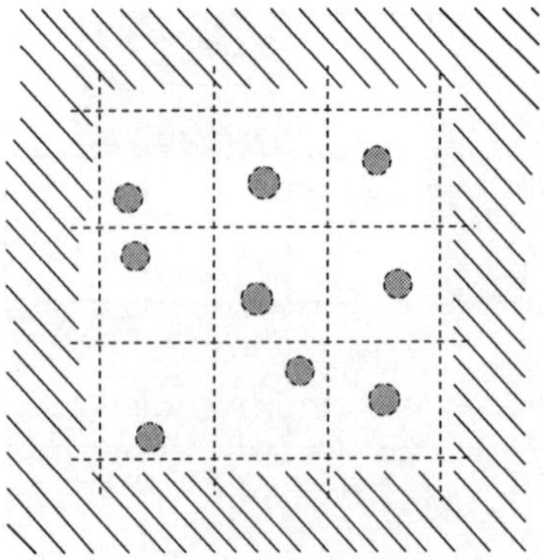

FIG. 8. At the finest level, interactions with near neighbours are computed directly.

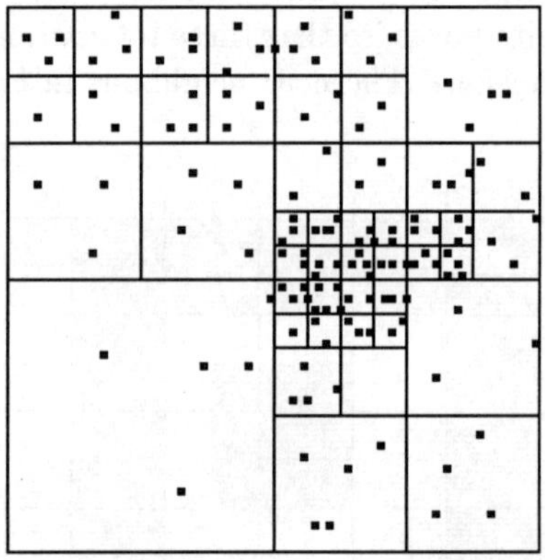

FIG. 9. An adaptive data structure

4.2 The adaptive algorithm

When the distribution of particles is nonuniform, a somewhat different strategy must be employed. One option is that during the refinement process, each box is examined to determine whether it contains any particles. If so, it is subdivided further. If not, it is pruned from the tree structure and ignored at subsequent levels (Fig. 9). The complexity of this adaptive algorithm is harder to state precisely, since it depends on the total number of refinement levels which is not determined a priori. For distributions of practical interest, this turns out to be proportional to $\log N$. Note that if the interparticle spacing collapses as N^{-p}, where p is independent of N, then $p \cdot \log N$ levels are needed. It is, therefore, quite reasonable to refer to the adaptive algorithm as also being of the order $N \log N$.

The algorithm just described is similar to that of Van Dommelen and

Rundensteiner [47].

4.3 The FMM

In order to develop an $O(N)$ method, we need several further analytic results concerning multipole expansions. Lemma 4.6 provides a formula for shifting the center of a multipole expansion, Lemma 4.7 describes how to convert such an expansion into a local (Taylor) expansion in a circular region of analyticity, and Lemma 4.8 furnishes a mechanism for shifting the center of a Taylor expansion within a region of analyticity. We state the second result without complete proof. A complete proof can be found in Greengard and Rokhlin [29] and Greengard [23].

Lemma 4.6 (Translation of a multipole expansion) *Suppose that*

$$\phi(z) = a_0 \log(z - z_0) + \sum_{k=1}^{\infty} \frac{a_k}{(z - z_0)^k} \tag{4.13}$$

is a multipole expansion of the potential due to a set of m charges of strengths $q_1, q_2, \ldots, q_m$, all of which are located inside the circle D of radius R with center at z_0. Then for z outside the circle D_1 of radius $(R + |z_0|)$ and center at the origin,

$$\phi(z) = a_0 \log(z) + \sum_{l=1}^{\infty} \frac{b_l}{z^l}, \tag{4.14}$$

where

$$b_l = -\frac{a_0 z_0^l}{l} + \sum_{k=1}^{l} a_k z_0^{l-k} \binom{l-1}{k-1}, \tag{4.15}$$

with $\binom{l}{k}$ the binomial coefficients. Furthermore, for any $p \geq 1$,

$$\left| \phi(z) - a_0 \log(z) - \sum_{l=1}^{p} \frac{b_l}{z^l} \right| \leq \left(\frac{A}{1 - \left| \frac{|z_0| + R}{z} \right|} \right) \left| \frac{|z_0| + R}{z} \right|^{p+1} \tag{4.16}$$

with A defined in (4.10).

Proof The form of the translated multipole expansion follows from the two equations

$$\log(z - z_0) = \log\left(z\left(1 - \frac{z_0}{z}\right)\right) = \log(z) - \sum_{\ell=1}^{\infty} \frac{1}{\ell} \left(\frac{z_0}{z}\right)^{\ell},$$

and

$$(z - z_0)^{-k} = \sum_{\ell=k}^{\infty} \binom{\ell-1}{k-1} \frac{z_0^{\ell-k}}{z^{\ell}},$$

both valid for $|z| > |z_0|$, by substituting, summing, and truncating. The error bound follows from the uniqueness of the multipole expansion, which implies that the expansion about the origin obtained indirectly above must be identical with the one obtained directly. Therefore its truncations enjoy the error bound of the direct expansion, namely (4.9). □

Lemma 4.7 (Conversion of a multipole expansion into a local expansion) *Suppose that m charges of strengths $q_1, q_2, ..., q_m$ are located inside the circle D_1 with radius R and center at z_0, and that $|z_0| > (c+1)R$ with $c > 1$. Then the corresponding multipole expansion (4.13) converges inside the circle D_2 of radius R centered about the origin. Inside D_2, the potential due to the charges is described by a power series:*

$$\phi(z) = \sum_{l=0}^{\infty} b_l \cdot z^l, \tag{4.17}$$

where

$$b_0 = a_0 \log(-z_0) + \sum_{k=1}^{\infty} \frac{a_k}{z_0^k}(-1)^k, \tag{4.18}$$

and

$$b_l = -\frac{a_0}{l \cdot z_0^l} + \frac{1}{z_0^l} \sum_{k=1}^{\infty} \frac{a_k}{z_0^k} \binom{l+k-1}{k-1} (-1)^k , \qquad \text{for } l \geq 1. \tag{4.19}$$

Furthermore, an error bound for the truncated series is given by

$$\left| \phi(z) - \sum_{l=0}^{p} b_l \cdot z^l \right| < \frac{A(4e(p+c)(c+1)+c^2)}{c(c-1)} \left(\frac{1}{c}\right)^{p+1}, \tag{4.20}$$

where A is defined in (4.10) and e is the base of natural logarithms.

The form of the local series follows from substituting the expressions

$$\begin{aligned} \log(z - z_0) &= \log\left(-z_0\left(1 - \frac{z}{z_0}\right)\right) \\ &= \log(-z_0) - \sum_{\ell=1}^{\infty} \frac{1}{\ell} \left(\frac{z}{z_0}\right)^{\ell}, \end{aligned}$$

and

$$\begin{aligned} (z - z_0)^{-k} &= \left(\frac{1}{-z_0}\right)^k \left(\frac{1}{1 - \frac{z}{z_0}}\right)^k \\ &= \left(\frac{1}{-z_0}\right)^k \sum_{\ell=0}^{\infty} \binom{\ell+k-1}{k-1} \left(\frac{z}{z_0}\right)^{\ell}, \end{aligned}$$

both valid for $|z| \leq R$, into the multipole expansion (4.13). The proof of the error bound may be found in Greengard and Rokhlin [29] or Greengard [23].

Lemma 4.8 (Translation of a local expansion) *Translation of a complex polynomial centered about* z_0

$$\sum_{k=0}^{p} a_k (z - z_0)^k, \tag{4.21}$$

into a complex polynomial centered about 0

$$\sum_{k=0}^{n} b_k z^k, \tag{4.22}$$

can be achieved by the complete Horner scheme

for j *from* 0 *to* $p-1$ *do*
  for k *from* $p-j-1$ *to* $p-1$ *do*
    $a_k := a_k - z_0 \, a_{k+1}$
  end
end

which given the vector of coefficients **a** *overwrites it with* **b**.

Proof This scheme is essentially nested multiplication and can be derived from the relation

$$\begin{aligned} & \left(a_m z^j + a_{m-1} z^{j-1} + \cdots + a_{m-j}\right)(z - z_0) + a_{m-j-1} \\ = \; & a_m z^{j+1} + \tilde{a}_{m-1} z^j + \cdots + \tilde{a}_{m-j-1}, \end{aligned}$$

which corresponds to an intermediate stage in converting expression (4.21) into expression (4.22) by performing nested multiplication. □

Observation 1: Suppose that we have created the multipole expansion for each of the four children of some box in the mesh hierarchy. We would like to form a single expansion for the parent box without re-examining each particle. Lemma 4.6 provides just such a mechanism at the cost of $4p^2$ operations, since each shift requires p^2 work.

Observation 2: Suppose that we have a local expansion for a box b which describes the field induced by all particles outside b's nearest neighbours. We would like to transmit this information to b's children. Lemma 4.8 provides just such a mechanism at the cost of $4p^2$ operations, since each shift requires p^2 work.

Note now that the $N \log N$ scheme was of that complexity because all particles were "accessed" at every level of refinement. The analytic manipulations just described allow us to avoid this, and their incorporation into the $N \log N$ scheme results in the FMM. Here, we simply indicate the modifications which lead to its implementation.

Initialization

Choose a number of levels so that there are, on average, s particles per box at the finest level. (The number of boxes is then approximately N/s.)

Upward Pass

In the $N \log N$ scheme, we proceeded from the coarsest to the finest level, forming multipole expansions for every box. In the FMM, we begin at the finest level, and create multipole expansions from the source positions and strengths. The expansions for all boxes at all higher levels are then formed by the merging procedure delineated in Observation 1.

Downward Pass

In the $N \log N$ scheme, whenever a box b was under consideration, we used its multipole expansion to compute interactions with all particles contained in the boxes of b's interaction list. In the FMM, we *convert* the multipole expansion into a local expansion about the centers of all boxes in b's interaction list, using Lemma 4.7.

After these calculations are completed, we are left with a local expansion in each box at each level. Beginning at the coarsest level, these local expansions are *shifted* to the children's level and added to the children's local expansions, as described in Observation 2. After this recursive process reaches the finest refinement level, a local expansion will have been created for each box which describes the field due to all particles outside the box's near neighbours. It is only this expansion which is evaluated. The near neighbour interactions, as before, are computed directly.

The total operation count is approximately

$$N\,p + 29\left(\frac{N}{s}\right)\,p^2 + N\,p + 9N\,s.$$

The terms correspond to formation of the multipole expansions, shifting the expansions, evaluation of the local expansions and computation of the near neighbour interactions.

Choosing $s = p$, this yields

$$40N\,p,$$

which compares favorably with the estimate (4.12) even for modest N. This is not the end of the story, however. The richer analytic structure of the FMM permits a large number of modifications and optimizations which are not available to other hierarchical schemes. These have to do with the use of symmetry relations to reduce the number of shifts (Greengard and Rokhlin [30]; Wang and LeSar [49]) as well as schemes which reduce the cost of translation itself (Greengard and Rokhlin [31]; Hrycak and Rokhlin [37]; Elliott and Board [21]). We will discuss these variants in greater detail when we come to the three-dimensional algorithm.

As for the adaptive FMM, a rudimentary complexity analysis is given by Carrier *et al.* [17]. A more complete analysis is given by Nabors *et al.* [41].

Typical timings for high precision FMM calculations are shown in the following table, taken from Hrycak and Rokhlin [37].

N	T_{HR}	T_{CGR}	T_{dir}	E_{HR}	E_{CGR}	E_{dir}
400	0.5	1.4	1.0	$3.5\,10^{-13}$	$3.5\,10^{-13}$	$3.0\,10^{-13}$
1600	2.3	5.2	17.1	$1.1\,10^{-12}$	$1.1\,10^{-12}$	$1.1\,10^{-12}$
6400	8.7	25.8	285.2	$4.9\,10^{-13}$	$4.9\,10^{-13}$	$2.7\,10^{-12}$
25600	34.7	110.6	(4400)	$2.5\,10^{-12}$	$2.5\,10^{-12}$	—

Here T_{HR} denotes the timing for the Hrycak and Rokhlin method, T_{CGR} denotes the timing for the Carrier *et al.* method and T_{dir} the timing for direct evaluation; the E's are the corresponding errors.

5 The FMM in three dimensions

In three dimensions, as in two dimensions, functions which satisfy the Laplace equation

$$\nabla^2\Phi = \frac{\partial^2\Phi}{\partial x^2} + \frac{\partial^2\Phi}{\partial y^2} + \frac{\partial^2\Phi}{\partial z^2} = 0 \tag{5.1}$$

are referred to as harmonic functions. The theory of such functions is generally called potential theory, and is beautifully described in the classic text by Kellogg [39]. A shorter description is available in the text by Wallace [48] and the first part of this lecture is abstracted from Greengard [23].

Assume now that the task at hand is to compute the Coulomb interactions in a large particle system. If a point charge of strength q is located at $P_0 = (x_0, y_0, z_0)$, then the potential and electrostatic field due to this charge at a distinct point $P = (x, y, z)$ are given by

$$\Phi = \frac{1}{\tilde{r}} \tag{5.2}$$

and

$$\vec{E} = -\nabla\Phi = (\frac{x - x_0}{\tilde{r}^3}, \frac{y - y_0}{\tilde{r}^3}, \frac{z - z_0}{\tilde{r}^3})\,, \tag{5.3}$$

respectively, where $\tilde{r}$ denotes the distance between points P_0 and P.

We would like to derive a series expansion for the potential at P in terms of its distance r from the origin. For this, let the spherical coordinates of P be (r, θ, ϕ) and of P_0 be (ρ, α, β). Letting γ be the angle between the vectors P and P_0, we have from the law of cosines

$$\tilde{r}^2 = r^2 + \rho^2 - 2r\rho\cos\gamma, \tag{5.4}$$

with

$$\cos\gamma = \cos\theta\cos\alpha + \sin\theta\sin\alpha\cos(\phi - \beta). \tag{5.5}$$

Thus,

$$\frac{1}{\tilde{r}} = \frac{1}{r\sqrt{1 - 2\frac{\rho}{r}\cos\gamma + \frac{\rho^2}{r^2}}} = \frac{1}{r\sqrt{1 - 2u\mu + \mu^2}}\,, \tag{5.6}$$

having set

$$\mu = \frac{\rho}{r} \quad \text{and} \quad u = \cos\gamma\,. \tag{5.7}$$

For $\mu < 1$, we may expand the inverse square root in powers of μ, resulting in a series of the form

$$\frac{1}{\sqrt{1 - 2u\mu + \mu^2}} = \sum_{n=0}^{\infty} P_n(u)\mu^n \tag{5.8}$$

where

$$P_0(u) = 1, \quad P_1(u) = u, \quad P_2(u) = \frac{3}{2}(u^2 - \frac{1}{3}), \cdots \tag{5.9}$$

and, in general, $P_n(u)$ is the Legendre polynomial of degree n. Our expression for the field now takes the form

$$\frac{1}{\tilde{r}} = \sum_{n=0}^{\infty} \frac{\rho^n}{r^{n+1}} P_n(u). \tag{5.10}$$

The formula (5.10), unfortunately, does not meet our needs. The parameter u depends on both the source and the target locations. We cannot use it to expand the field due to a large number of sources. A more general representation will require the introduction of spherical harmonics.

For this, we first write the Laplace equation in spherical coordinates:

$$\frac{1}{r^2}\frac{\partial}{\partial r}\left(r^2\frac{\partial\Phi}{\partial r}\right) + \frac{1}{r^2\sin\theta}\frac{\partial}{\partial\theta}\left(\sin\theta\frac{\partial\Phi}{\partial\theta}\right) + \frac{1}{r^2\sin^2\theta}\frac{\partial^2\Phi}{\partial\phi^2} = 0. \tag{5.11}$$

The standard solution of this equation by separation of variables results in an expression for the field as a series, the terms of which are known as spherical harmonics:

$$\Phi = \sum_{n=0}^{\infty} \sum_{m=-n}^{n} \left(L_n^m r^n + \frac{M_n^m}{r^{n+1}} \right) Y_n^m(\theta, \phi). \tag{5.12}$$

The terms $Y_n^m(\theta, \phi) r^n$ are referred to as spherical harmonics of degree n, the terms $Y_n^m(\theta, \phi)/r^{n+1}$ are called spherical harmonics of degree $-n-1$, and the coefficients L_n^m and M_n^m are known as the moments of the expansion.

The spherical harmonics can be expressed in terms of partial derivatives of $1/r$, but we will simply define them via the relation

$$Y_n^m(\theta, \phi) = \sqrt{\frac{(n-|m|)!}{(n+|m|)!}} \cdot P_n^{|m|}(\cos\theta) e^{im\phi}, \tag{5.13}$$

omitting a normalization factor of $\sqrt{(2n+1)/4\pi}$.

The special functions P_n^m are called associated Legendre functions and can be defined by the Rodrigues' formula

$$P_n^m(x) = (-1)^m (1-x^2)^{m/2} \frac{d^m}{dx^m} P_n(x).$$

5.1 The multipole expansion

We will need a well known result from the theory of spherical harmonics.

Theorem 5.1 (Addition theorem for Legendre polynomials) *Let P and Q be points with spherical coordinates (r, θ, ϕ) and (ρ, α, β), respectively, and let γ be the angle subtended between them. Then*

$$P_n(\cos\gamma) = \sum_{m=-n}^{n} Y_n^{-m}(\alpha, \beta) \cdot Y_n^m(\theta, \phi) \ . \tag{5.14}$$

Theorem 5.2 (Multipole expansion) *Suppose that k charges of strengths $\{q_i,\ i = 1, ..., k\}$ are located at the points $\{Q_i = (\rho_i, \alpha_i, \beta_i),\ i = 1, ..., k\}$, with $|\rho_i| < a$. Then for any $P = (r, \theta, \phi) \in \mathbf{R}^3$ with $r > a$, the potential $\phi(P)$ is given by*

$$\phi(P) = \sum_{n=0}^{\infty} \sum_{m=-n}^{n} \frac{M_n^m}{r^{n+1}} \cdot Y_n^m(\theta, \phi) \ , \tag{5.15}$$

where

$$M_n^m = \sum_{i=1}^{k} q_i \cdot \rho_i^n \cdot Y_n^{-m}(\alpha_i, \beta_i). \tag{5.16}$$

Furthermore, for any $p \geq 1$,

$$\left| \phi(P) - \sum_{n=0}^{p} \sum_{m=-n}^{n} \frac{M_n^m}{r^{n+1}} \cdot Y_n^m(\theta,\phi) \right| \leq \frac{A}{r-a} \left(\frac{a}{r}\right)^{p+1} , \tag{5.17}$$

where

$$A = \sum_{i=1}^{k} |q_i| . \tag{5.18}$$

5.2 The $N \log N$ scheme

As in the two-dimensional case, Theorem 5.2 is all that is required to construct an $N \log N$ scheme of arbitrary precision. Since we discussed that case in some detail in a previous lecture, we simply indicate the principal differences.

- The particles are contained within a cube which is subdivided in a recursive manner. In the nonadaptive case, each box has *eight* children, 27 nearest neighbours and an interaction list of dimension 189.
- A multipole expansion is made up of p^2 terms rather than p terms. Furthermore, a simple geometric calculation shows that for a target in a well-separated cube, the rate of decay of the error with p looks like $(\sqrt{3}/3)^p$ rather than $(1/2)^p$ or $(\sqrt{2}/3)^p$.

Thus, the cost of the $N \log N$ scheme is approximately

$$189 N p^2 \log_8 N + 27N, \tag{5.19}$$

where $p = \log_{\sqrt{3}}(1/\epsilon)$.

5.3 The $O(N)$ scheme

In order to accelerate the preceding scheme, we need some additional analytic machinery to shift multipole expansions and obtain local representations.

Theorem 5.3 (Translation of a multipole expansion) *Suppose that l charges of strengths $q_1, q_2, \cdots, q_l$ are located inside the sphere D of radius a with center at $Q = (\rho, \alpha, \beta)$, and that for points $P = (r, \theta, \phi)$ outside D, the potential due to these charges is given by the multipole expansion*

$$\Phi(P) = \sum_{n=0}^{\infty} \sum_{m=-n}^{n} \frac{O_n^m}{r'^{\,n+1}} \cdot Y_n^m(\theta', \phi') , \tag{5.20}$$

where $P - Q = (r', \theta', \phi')$. *Then for any point* $P = (r, \theta, \phi)$ *outside the sphere* D_1 *of radius* $(a + \rho)$,

$$\Phi(P) = \sum_{j=0}^{\infty} \sum_{k=-j}^{j} \frac{M_j^k}{r^{j+1}} \cdot Y_j^k(\theta, \phi) \ , \tag{5.21}$$

where

$$M_j^k = \sum_{n=0}^{j} \sum_{m=-n}^{n} \frac{O_{j-n}^{k-m} \cdot i^{|k|-|m|-|k-m|} \cdot A_n^m \cdot A_{j-n}^{k-m} \cdot \rho^n \cdot Y_n^{-m}(\alpha, \beta)}{A_j^k} \ , \tag{5.22}$$

with A_n^m *defined by*

$$A_n^m = \frac{(-1)^n}{\sqrt{(n-m)!(n+m)!}}. \tag{5.23}$$

Furthermore, for any $p \geq 1$,

$$\left| \Phi(P) - \sum_{j=0}^{p} \sum_{k=-j}^{j} \frac{M_j^k}{r^{j+1}} \cdot Y_j^k(\theta, \phi) \right| \leq \left(\frac{\sum_{i=1}^{l} |q_i|}{r - (a + \rho)} \right) \left(\frac{a + \rho}{r} \right)^{p+1} . \tag{5.24}$$

Theorem 5.4 (Conversion of a multipole expansion into a local expansion) *Suppose that* l *charges of strengths* $q_1, q_2, \cdots, q_l$ *are located inside the sphere* D_Q *of radius* a *with center at* $Q = (\rho, \alpha, \beta)$, *and that* $\rho > (c+1)a$ *with* $c > 1$. *Then the corresponding multipole expansion (5.20) converges inside the sphere* D_0 *of radius* a *centered at the origin. Inside* D_0, *the potential due to the charges* $q_1, q_2, \cdots, q_l$ *is described by a local expansion:*

$$\Phi(P) = \sum_{j=0}^{\infty} \sum_{k=-j}^{j} L_j^k \cdot Y_j^k(\theta, \phi) \cdot r^j \ , \tag{5.25}$$

where

$$L_j^k = \sum_{n=0}^{\infty} \sum_{m=-n}^{n} \frac{O_n^m \cdot i^{|k-m|-|k|-|m|} \cdot A_n^m \cdot A_j^k \cdot Y_{j+n}^{m-k}(\alpha, \beta)}{(-1)^n A_{j+n}^{m-k} \cdot \rho^{j+n+1}} \ , \tag{5.26}$$

with A_r^s *defined by (5.23). Furthermore, for any* $p \geq 1$,

$$\left| \Phi(P) - \sum_{j=0}^{p} \sum_{k=-j}^{j} L_j^k \cdot Y_j^k(\theta, \phi) \cdot r^{j+1} \right| \leq \left(\frac{\sum_{i=1}^{l} |q_i|}{ca - a} \right) \left(\frac{1}{c} \right)^{p+1} . \tag{5.27}$$

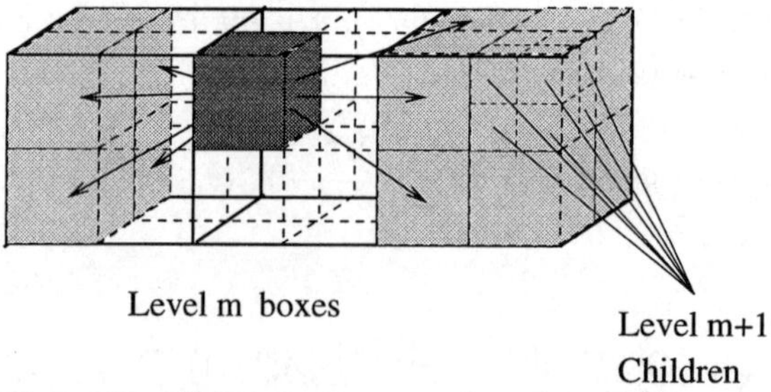

FIG. 10. In the FMM, a box transmits its far field expansion to the boxes in its interaction list. Those boxes then pass the information down to their children.

Theorem 5.5 **(Translation of a local expansion)** *Let $Q = (\rho, \alpha, \beta)$ be the origin of a local expansion*

$$\Phi(P) = \sum_{n=0}^{p} \sum_{m=-n}^{n} O_n^m \cdot Y_n^m(\theta', \phi') \cdot r'^{\,n} , \tag{5.28}$$

where $P = (r, \theta, \phi)$ and $P - Q = (r', \theta', \phi')$. Then

$$\Phi(P) = \sum_{j=0}^{p} \sum_{k=-j}^{j} L_j^k \cdot Y_j^k(\theta, \phi) \cdot r^j , \tag{5.29}$$

where

$$L_j^k = \sum_{n=j}^{p} \sum_{m=-n}^{n} \frac{O_n^m \cdot i^{|m|-|m-k|-|k|} \cdot A_{n-j}^{m-k} \cdot A_j^k \cdot Y_{n-j}^{m-k}(\alpha, \beta) \cdot \rho^{n-j}}{(-1)^{n+j} \cdot A_n^m} , \tag{5.30}$$

with A_r^s defined by eqn. (5.23).

We can now construct the FMM by analogy with the two-dimensional case.

Upward Pass

- Form multipole expansions at finest level (from source positions and strengths).
- Form multipole expansions at coarser levels by merging, according to Theorem 5.3.

Downward Pass

- Account for interactions at each level by Theorem 5.4.
- Transmit information to finer levels by Theorem 5.5.

The total operation count is

$$189\left(\frac{N}{s}\right)p^4 + 2Np^2 + 27Ns.$$

Choosing $s = 2p^2$, the operation count becomes approximately

$$150Np^2. \tag{5.31}$$

This would appear to beat the estimate (5.19) for any N, but there is a catch. The number of terms p needed for a fixed precision in the $N \log N$ scheme is smaller than the number of terms needed in the FMM described above. To see why, consider two interacting cubes A and B of unit volume, with sources in A and targets in B. The worst-case multipole error decays like $(\sqrt{3}/3)^p$, since $\sqrt{3}/2$ is the radius of the smallest sphere enclosing cube A and $3/2$ is the shortest distance to a target in B. The conversion of a multipole expansion in A to a local expansion in B, however, satisfies an error bound which depends on the sphere enclosing B as well as the sphere enclosing A and has a worst case error estimate of the order $(3/4)^p$.

In the original FMM (Greengard and Rokhlin [30]; Greengard [23]), it was suggested that one redefine the nearest neighbour list to include "second nearest neighbours." The error can then be shown to decay approximately like $(0.4)^p$. However, the number of near neighbours increases to 125 and the size of the interaction list increases to 875. In the latest generation of FMMs (Hrycak and Rokhlin [37]; Greengard and Rokhlin [33]), it turns out that one can recover the same accuracy as the $N \log N$ schemes without this modification. The relevant ideas will be sketched below.

Returning now to the formula (5.31), it is clear that the major obstacle to achieving reasonable efficiency at high precision is the cost of the multipole to local translations ($189p^4$ operations per box).

There are a number of schemes for reducing that cost. The first we describe is based on rotating the coordinate system (Fig. 11). By inspection of Theorem 5.4, it is easily shown that translation in the z direction only requires p^3 operations. A fast translation scheme could then be obtained as follows.

(1) Rotate the coordinate system (p^3 operations) so that the vector connecting the source box and the target box lies along the z axis.
(2) Shift the expansion along the z axis (p^3 operations).
(3) Rotate back to the original coordinate system (p^3 operations).

The total operation count becomes

$$189\left(\frac{N}{s}\right)3p^3 + 2Np^2 + 27Ns.$$

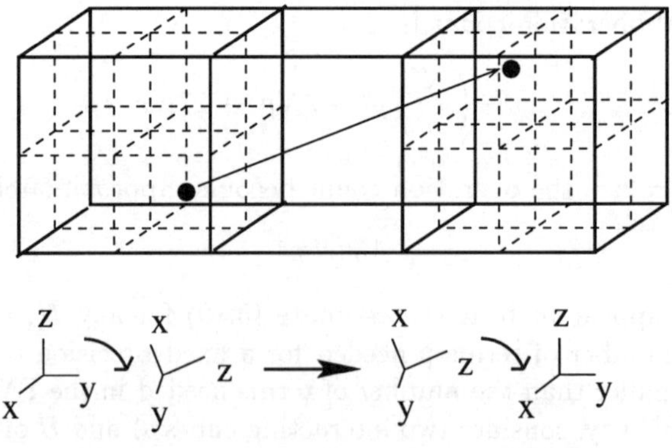

FIG. 11. Fast translation via rotation.

Choosing $s = 3p^{3/2}$, the operation count becomes approximately

$$270N\,p^{3/2} + 2N\,p^2. \tag{5.32}$$

Over the last few years, a number of diagonal translation schemes were developed, requiring $O(p^2)$ work (Greengard and Rokhlin [31]; Berman [10]; Elliott and Board [21]). Unfortunately, these schemes are all subject to certain numerical instabilities. They can be overcome, but at additional cost.

5.4 Exponential expansions

The latest generation of FMMs is based on combining multipole expansions with plane wave expansions. The two-dimensional theory is described in Hrycak and Rokhlin [37] and we will just sketch the three-dimensional version here (Greengard and Rokhlin [33]). The starting point is the integral representation

$$\frac{1}{r} = \frac{1}{2\pi}\int_0^\infty e^{-\lambda z}\int_0^{2\pi} e^{i\lambda(x\cos\alpha + y\sin\alpha)}\,d\alpha\,d\lambda.$$

To get a discrete representation, one must use some quadrature formula. The inner integral, with respect to α, is easily handled by the trapezoidal rule (which achieves spectral accuracy for periodic functions), but the outer integral requires more care. Laguerre quadrature is an appropriate choice here, but even better performance can be obtained using generalized Gaussian quadrature rules (Yarvin [50]). The central idea, however, is that the multipole expansion can be converted to an expansion in exponentials.

$$\sum_{n=0}^{p}\sum_{m=-n}^{n}\frac{M_n^m Y_n^m(\theta,\phi)}{r^{n+1}} \approx \sum_{j=1}^{P_l}\sum_{k=1}^{K_j} e^{-\lambda_j(z - ix\cos\theta_k - iy\sin\theta_k)}S(j,k).$$

Unlike multipole expansions, however, these plane wave expansions are dir-

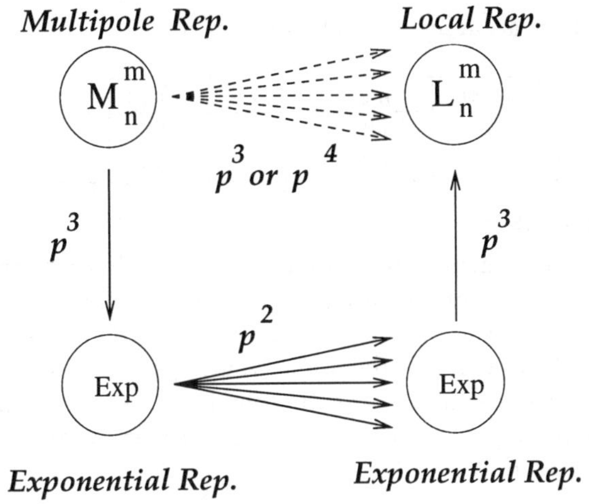

FIG. 12. Multipole translation can be replaced by exponential translation.

ection dependent. The one written above is for the $+z$ direction, and a *different* one is needed for emanation through each of the six faces of the cube.

The reason for preferring exponentials is that translation corresponds to multiplication and requires only p^2 work. Conversion from the multipole representation to the six outgoing exponentials requires $3p^3$ work. Thus, we can replace $189p^4$ or $189p^3$ operations per box with $3p^3 + 189p^2$ (Fig. 12).

The total operation count is now

$$189\,\frac{N}{s}\,p^2 + 2\,N\,p^2 + 27N\,s + 3\frac{N}{s}\,p^3.$$

Setting $s = 2p$, the total operation count is about

$$200\,N\,p + 3.5N\,p^2.$$

The new FMM has only been tested on relatively uniform distributions using a nonadaptive implementation. For three digit accuracy, $p = 8$ and on a Sun SPARCStation 2, we have

N	Levels	T_{Rot}	$T_{Fourier}$	T_{Exp}	T_{dir}
1000	3	13	12	8	14
8000	4	160	140	67	900
64000	5	1700	1400	610	(58000)

T_{Rot} is the time for the FMM using rotations, $T_{Fourier}$ is the time required for the diagonal translation scheme based on the convolution structure, and T_{Exp} is the time required for the plane wave scheme sketched above.

At ten digits of accuracy, $p = 30$ and

N	Levels	T_{Rot}	$T_{Fourier}$	T_{Exp}	T_{dir}
5000	3	1030	—	230	350
40000	4	12300	—	1930	(22,400)

6 Conclusions

FMMs are efficient schemes for computing N-body interactions. Applications to boundary value problems for the Laplace and biharmonic equations can be found, for example, in Greenbaum *et al.* [22]; Greengard *et al.* [26]; Greengard and Moura [28]; Nabors and White [40]; Rokhlin [43]. Applications to the Poisson equation can be found in Greengard and Lee [27]; Russo and Strain [46]. For applications to molecular dynamics, see Board *et al.* [12]; Ding *et al.* [20].

Bibliography

1. B. K. Alpert and V. Rokhlin. *A Fast Algorithm for the Evaluation of Legendre Expansions.* SIAM J. Sci. Stat. Comput. **12**, 158–179 (1991).
2. C. R. Anderson. *A Method of Local Corrections for Computing the Velocity Field Due to a Distribution of Vortex Blobs.* J. Comput. Phys. **62**, 111–123 (1986).
3. C. R. Anderson. *An Implementation of the Fast Multipole Method Without Multipoles.* SIAM J. Sci. Stat. Comput. **13**, 923–947 (1992).
4. A. W. Appel. *An Efficient Program for Many-body Simulation.* Siam. J. Sci. Stat. Comput. **6**, 85–103 (1985).
5. J. Barnes and P. Hut. *A Hierarchical $O(N \log N)$ Force-Calculation Algorithm.* Nature **324**, 446–449 (1986).
6. R. K. Beatson and W. A. Light (1996). *Fast evaluation of Radial Basis Functions: Methods for 2-dimensional Polyharmonic Splines.* To appear in IMA J. Numer. Anal.
7. R. K. Beatson and G. N. Newsam. *Fast Evaluation of Radial Basis Functions, I.* Computers Math. Applic. **24**, 7–19 (1992).
8. R. K. Beatson and G. N. Newsam (1996). *Fast Evaluation of Radial Basis functions: Moment Based Methods.* To appear in SIAM J. Sci. Comput.
9. R. K. Beatson, M. J. D. Powell and A. M. Tan (1996). *Fast Evaluation of Radial Basis Functions: Methods for 3-dimensional Polyharmonic Splines.* In preparation.
10. L. Berman. *Grid-Multipole Calculations.* SIAM J. Sci. Comput. **16**, 1082–1091 (1995).
11. G. Beylkin, R. Coifman and V. Rokhlin. *Fast Wavelet Transforms and Numerical Algorithms I.* Comm. Pure and Appl. Math. **44**, 141–183

(1991).

12. J.A. Board, J. W. Causey, J. F. Leathrum, A. Windemuth, and K. Schulten. *Accelerated Molecular Dynamics Simulation with the Parallel Fast Multipole Method.* Chem. Phys. Let. **198**, 89–94 (1992).
13. B. Bradie, R. Coifman, and A. Grossmann. *Fast Numerical Computations of Oscillatory Integrals Related to Acoustic Scattering, I.* Appl. and Comput. Harmonic Analysis **1**, 94–99 (1993).
14. A. Brandt. *Multilevel Computations of Integral Transforms and Particle Interactions with Oscillatory Kernels.* Comp. Phys. Comm. **65**, 24–38 (1991).
15. A. Brandt and A. A. Lubrecht. *Multilevel Matrix Multiplication and Fast Solution of Integral Equations.* J. Comput. Phys. **90**, 348–370 (1990).
16. F. X. Canning. *Transformations that Produce a Sparse Moment Method Matrix.* J. Electromagnetic Waves and Appl. **4**, 893–913 (1990).
17. J. Carrier, L. Greengard, and V. Rokhlin. *A Fast Adaptive Multipole Algorithm for Particle Simulations.* SIAM J. Sci. and Stat. Comput. **9**, 669–686 (1988).
18. R. Coifman, V. Rokhlin, and S. Wandzura. *The Fast Multipole Method for the Wave Equation: A Pedestrian Prescription.* IEEE Antennas and Propagation Magazine **35**, 7–12 (1993).
19. J. Crank. The Mathematics of Diffusion. Clarendon Press, Oxford (1964).
20. H.-Q. Ding, N. Karasawa, and W. A. Goddard III. *Atomic Level Simulations on a Million Particles: The Cell Multipole Method for Coulomb and London Nonbond Interactions.* J. Chem. Phys. **97**, 4309–4315 (1992).
21. W. D. Elliott and J.A. Board. *Fast Fourier Transform Accelerated Fast Multipole Algorithm.* SIAM J. Sci. Comput. **17**, 398–415 (1996).
22. A. Greenbaum, L. Greengard, and G. B. McFadden. *Laplace's Equation and the Dirichlet–Neumann Map in Multiply Connected Domains.* J. Comp. Phys. **105**, 267–278 (1993).
23. L. Greengard. The Rapid Evaluation of Potential Fields in Particle Systems. MIT Press, Cambridge (1988).
24. L. Greengard. *The Numerical Solution of the N-body Problem.* Computers in Physics **4**, p. 142–152 (1990).
25. L. Greengard. *Fast Algorithms for Classical Physics.* Science **265**, 909–914 (1994).
26. L. Greengard, M. C. Kropinski and A. Mayo. *Integral Equation Methods for Stokes Flow and Isotropic Elasticity.* J. Comput. Phys. **125**,

403–414 (1996).

27. L. Greengard and J.-Y. Lee. *A Direct Adaptive Poisson Solver of Arbitrary Order Accuracy.* J. Comput. Phys. **125**, 415–424 (1996).
28. L. Greengard and M. Moura. *On the Numerical Evaluation of Electrostatic Fields in Composite Materials.* Acta Numerica 379–410 (1994).
29. L. Greengard and V. Rokhlin. *A Fast Algorithm for Particle Simulations.* J. Comput. Phys. **73**, 325–348 (1987).
30. L. Greengard and V. Rokhlin. *Rapid Evaluation of Potential Fields in Three Dimensions.* Vortex Methods, C. Anderson and C. Greengard eds. Lecture Notes in Mathematics, vol. 1360, Springer–Verlag (1988).
31. L. Greengard and V. Rokhlin. *On the Efficient Implementation of the Fast Multipole Algorithm.* Department of Computer Science Research Report 602, Yale University (1988).
32. L. Greengard and V. Rokhlin. *On the Evaluation of Electrostatic Interactions in Molecular Modeling.* Chemica Scripta **29A**, 139–144 (1989).
33. L. Greengard and V. Rokhlin (1996). *An Improved Fast Multipole Algorithm in Thre Dimensions*, to appear.
34. L. Greengard and J. Strain. *The Fast Gauss Transform.* SIAM J. Sci. Stat. Comput. **12**, 79–94 (1991).
35. M. Gu and S. C. Eisenstat. *A Divide-And-Conquer Algorithm for the Symmetric Tridiagonal Eigenproblem.* Department of Computer Science Research Report 932, Yale University (1992).
36. R. W. Hockney and J. W. Eastwood. Computer Simulation Using Particles. McGraw–Hill, New York (1981).
37. T. Hrycak and V. Rokhlin. *An Improved Fast Multipole Algorithm for Potential Fields.* Department of Computer Science Research Report 1089, Yale University (1995).
38. J. D. Jackson. Classical Electrodynamics. Wiley, New York (1975).
39. O. D. Kellogg. Foundations of Potential Theory. Dover, New York (1953).
40. K. Nabors and J. White. *FastCap: A Multipole-Accelerated 3-D Capacitance Extraction Program.* IEEE Transactions on Computer-Aided Design **10**, 1447–1459 (1991).
41. K. Nabors, F. T. Korsmeyer, F. T. Leighton, and J. White. *Preconditioned, Adaptive, Multipole-Accelerated Iterative Methods for Three-Dimensional First-Kind Integral Equations of Potential Theory.* SIAM J. Sci. Stat. Comput. **15**, 713–735 (1994).
42. A. M. Odlyzko and A. Schönhage. *Fast Algorithms for Multiple Evaluations of the Riemann Zeta Function.* Trans. Am. Math. Soc. **309**, 797–809 (1988).
43. V. Rokhlin. *Rapid Solution of Integral Equations of Classical Potential*

Theory. J. Comput. Phys. **60**, 187–207 (1985).

44. V. Rokhlin. *Rapid Solution of Integral Equations of Scattering Theory in Two Dimensions.* J. Comput. Phys. **86**, 414–439 (1990).
45. V. Rokhlin. *Diagonal Forms of Translation Operators for the Helmholtz Equation in Three Dimensions.* Appl. and Comput. Harmonic Analysis **1**, 82–93 (1993).
46. G. Russo and J.A. Strain. *Fast Triangulated Vortex Methods for the 2D Euler Equations.* J. Comput. Phys. **111**, 291–323 (1994).
47. L. Van Dommelen and E. A. Rundensteiner. *Fast, Adaptive Summation of Point Forces in the Two-Dimensional Poisson Equation.* J. Comput. Phys. **83**, 126–147 (1989).
48. P. R. Wallace. Mathematical Analysis of Physical Problems. Dover, New York (1984).
49. H. Y. Wang and R. LeSar. *An Efficient Fast-multipole Algorithm based on an Expansion in the Solid Harmonics.* J. Chem. Phys. **104**, 4173–417 (1996).
50. N. Yarvin (1996). *Generalized Gaussian Quadratures and Fast Summation Schemes*, Ph.D. Dissertation, Yale University, in preparation.

Eigenvalue problems for differential equations

Michael Plum

Mathematisches Institut I, Universität Karlsruhe, Germany

1 Eigenvalue problems in physics and engineering

In this first chapter, we will report on examples showing how eigenvalue problems can arise from practical physical or engineering problems. A whole class of examples is provided by the wave equation via the separation of variables technique. Therefore, we will start with some practical problems modelled by the wave equation, in Section 1.1. The separation of variables technique and its application to the examples will be discussed in Section 1.2. Finally, in Section 1.3, we will present further practical examples of eigenvalue problems, with origins other than the wave equation.

1.1 The wave equation and related problems

Here we will be concerned with some practical physical problems which are mathematically modelled by the linear wave-type equation

$$\begin{aligned}
&\frac{\partial^2 u}{\partial t^2}(x,t) + L[u](x,t) = f(x,t) && (x \in \Omega,\ t > 0),\\
&B[u](x,t) = g(x,t) && (x \in \partial\Omega,\ t > 0),\\
&u(x,0) = u_0(x) && (x \in \overline{\Omega}),\\
&\frac{\partial u}{\partial t}(x,0) = u_1(x) && (x \in \overline{\Omega}),
\end{aligned} \tag{1.1}$$

with $\Omega \subset \mathbb{R}^n$ denoting some domain with boundary $\partial\Omega$ which is Lipschitz continuous (i.e., $\partial\Omega$ is everywhere locally representable as graph of a Lipschitz continuous function), and with L denoting a uniformly elliptic operator (acting only on the space variable x), e.g.

$$L = -\Delta \equiv -\sum_{i=1}^{n} \frac{\partial^2}{\partial x_i^2} \quad \text{or} \quad L = \Delta\Delta.$$

B is a boundary operator, e.g.

$$B[u] = u\big|_{\partial\Omega} \quad \text{or} \quad B[u] = \frac{\partial u}{\partial \nu} \equiv \nu^t \nabla u \ \text{ if } \ L = -\Delta$$

(with $\nu : \partial\Omega \to \mathbb{R}^n$ denoting the outer unit normal field), or

$$B[u] = \left(u\big|_{\partial\Omega}, \frac{\partial u}{\partial \nu}\right)^{\mathsf{T}} \quad \text{if } L = \Delta\Delta.$$

f, g, u_0, u_1 are given functions.

Example A (vibrating string)

Consider a string of length l with mass density $\rho = dm/dx$. The expression "string" means that its thickness, its weight, and bending forces are negligible. We suppose that a tension force T acts on the endpoints $x = 0$ and $x = l$ of the string, where it is moreover fixed. We consider vibrations (in the (x, u)-plane) of the string with small amplitude $u(x, t)$.

"Freezing" the string at some fixed time t, let $F(x) = (F_1(x), F_2(x))$ denote the tangential force at x pointing in the positive x-direction. Then, for "small" $\delta > 0$, the total force acting on the piece of string between x and $x + \delta$ is

$$F(x+\delta) - F(x) = F'(x)\delta + O(\delta^2).$$

On the other hand, since the amplitude u is "small", the "mass · acceleration" term of this piece reads

$$\left(0, \int_x^{x+\delta} \rho(\tilde{x}) \frac{\partial^2 u}{\partial t^2}(\tilde{x}, t)\, d\tilde{x}\right) = \left(0, \rho(x)\frac{\partial^2 u}{\partial t^2}(x,t)\cdot\delta\right) + O(\delta^2).$$

Consequently, since $\delta > 0$ is arbitrary, Newton's law yields

$$F'(x) = \left(0, \rho(x)\frac{\partial^2 u}{\partial t^2}(x,t)\right).$$

Thus, F_1 is constant, so that $F_1 \equiv T$ since the tension between the endpoints is T, and

$$F_2'(x) = \rho(x)\frac{\partial^2 u}{\partial t^2}(x,t).$$

Since F acts tangentially, we also have $\partial u/\partial x = F_2/F_1 = F_2/T$, and thus,

$$\frac{\partial^2 u}{\partial x^2} = \frac{F_2'(x)}{T} = \frac{\rho(x)}{T}\frac{\partial^2 u}{\partial t^2}.$$

Introducing the local speed of sound $c(x) := \sqrt{\frac{T}{\rho(x)}}$ we therefore obtain the wave equation

$$\frac{\partial^2 u}{\partial t^2} - c^2\frac{\partial^2 u}{\partial x^2} = 0 \quad (0 < x < l,\ t > 0) \tag{1.2}$$

to be satisfied by the amplitude u. If ρ (and thus, c) is constant, we can carry out the time scaling $t \to ct$ to obtain the scaled wave equation

$$\frac{\partial^2 u}{\partial t^2} - \frac{\partial^2 u}{\partial x^2} = 0 \quad (0 < x < l,\ t > 0). \tag{1.3}$$

Due to the fact that the string is fixed at the endpoints, we have the Dirichlet boundary conditions

$$u(0,t) = u(l,t) = 0 \quad (t > 0). \tag{1.4}$$

(1.2), (1.4) together with the initial conditions

$$u(x,0) = u_0(x), \quad \frac{\partial u}{\partial t}(x,0) = u_1(x) \quad (0 \leq x \leq l)$$

fixing the initial state of the string, form the complete initial-boundary value problem for the wave equation (1.1).

Example B (vibrating membrane)

The two-dimensional analogue of the vibrating string is the vibrating membrane with (constant) surface mass density ρ, and again with negligible thickness, weight, and bending forces. The membrane is (in its rest position) given as a bounded domain $\Omega \subset \mathbb{R}^2$. We assume that a uniform tension force T acts on the boundary $\partial\Omega$, where the membrane is moreover fixed, and that the membrane reacts isotropically to the tension, i.e. at each of its points the tension force is the same in all directions. Assuming again that the vibration amplitude is "small", we now obtain, after a derivation which is technically a bit more involved than in the string example, the two-dimensional wave equation

$$\frac{\partial^2 u}{\partial t^2} - \Delta u = 0 \qquad (x \in \Omega,\ t > 0). \tag{1.5}$$

Since the membrane is fixed at its boundary, we obtain the Dirichlet boundary condition

$$u(x,t) = 0 \qquad (x \in \partial\Omega,\ t > 0). \tag{1.6}$$

In addition, as before we have to prescribe the initial state of the membrane, i.e., $u(x,0)$ and $\frac{\partial u}{\partial t}(x,0)$ $(x \in \overline{\Omega})$.

Example C (vibrating plate)

Compared with a membrane, a plate has the property that bending forces are present. With a suitable scaling, we obtain here the wave-type equation

$$\frac{\partial^2 u}{\partial t^2} + \Delta\Delta u = 0 \qquad (x \in \Omega,\ t > 0). \tag{1.7}$$

Several kinds of boundary conditions can occur here, depending on the physical circumstances. If, for example, the plate is fixed and clamped at its boundary, we have

$$u = 0, \quad \frac{\partial u}{\partial \nu} = 0 \quad (x \in \partial\Omega,\, t > 0). \tag{1.8}$$

If the plate is fixed in a hinge at its boundary (so that no bending forces are present there), we obtain

$$u = 0, \quad \Delta u = 0 \qquad (x \in \partial\Omega,\, t > 0). \tag{1.9}$$

Again, $u(x,0)$ and $\frac{\partial u}{\partial t}(x,0)$ have to be prescribed for $x \in \overline{\Omega}$ as initial conditions.

Example D (Maxwell's equations)

The fundamental problem of electrodynamics, namely the computation of electromagnetic fields from given electric charges and currents, is governed by Maxwell's equations (here, in vacuum and in scaled form)

$$\operatorname{div} E = \rho, \quad \operatorname{curl} E = -\frac{\partial B}{\partial t}, \quad \operatorname{div} B = 0, \quad \operatorname{curl} B = \frac{\partial E}{\partial t} + j,$$

with E and B denoting the required electric and magnetic field, respectively, and with ρ and j representing the given electric charge density and current density, respectively. The equations are required to hold for all $x \in \mathbb{R}^3$ and $t > 0$.

The third equation $\operatorname{div} B = 0$ ensures the existence of some *magnetic vector potential* A satisfying

$$B = \operatorname{curl} A. \tag{1.10}$$

Inserting (1.10) into the second equation, we obtain $\operatorname{curl}\left(E + \frac{\partial A}{\partial t}\right) = 0$, which implies the existence of an *electric scalar potential* ϕ satisfying

$$E + \frac{\partial A}{\partial t} = -\nabla\phi. \tag{1.11}$$

Obviously, (1.10) and (1.11) are invariant under the gauge transformation $A \to A + \nabla\psi$, $\phi \to \phi - \frac{\partial\psi}{\partial t}$, for arbitrary scalar gauge potential ψ. Thus, with A_0 and ϕ_0 denoting arbitrary potentials such that (1.10), (1.11) hold, we can choose a gauge potential ψ satisfying the inhomogeneous wave equation

$$\frac{\partial^2\psi}{\partial t^2} - \Delta\psi = \frac{\partial\phi_0}{\partial t} + \operatorname{div} A_0,$$

the so-called *Lorentz gauge*, to obtain

$$\frac{\partial\phi}{\partial t} + \operatorname{div} A = 0 \tag{1.12}$$

for the gauged potentials $A = A_0 + \nabla\psi$, $\phi = \phi_0 - \frac{\partial\psi}{\partial t}$. Inserting (1.11) into the first Maxwell equation $\operatorname{div} E = \rho$ we obtain $-\frac{\partial}{\partial t}\operatorname{div} A - \Delta\phi = \rho$, so that (1.12) provides the inhomogeneous wave equation

$$\frac{\partial^2 \phi}{\partial t^2} - \Delta\phi = \rho \tag{1.13}$$

for the electric potential ϕ. Inserting (1.10), (1.11) into the fourth Maxwell equation and using the identity curl curl $\equiv \nabla\text{div} - \Delta$ and (1.12), we obtain the inhomogeneous wave equation

$$\frac{\partial^2 A}{\partial t^2} - \Delta A = j \tag{1.14}$$

for the magnetic potential A also.

1.2 Separation of variables

To obtain some particular solutions of the homogeneous equation and boundary condition

$$\begin{aligned} &\frac{\partial^2 u}{\partial t^2} + L[u] = 0 \quad (x \in \Omega,\ t > 0), \\ &B[u] = 0 \qquad\qquad\quad (x \in \partial\Omega,\ t > 0) \end{aligned} \tag{1.15}$$

in (1.1), we try to find u in the form

$$u(x,t) = v(x)w(t) \qquad (x \in \Omega,\ t > 0), \tag{1.16}$$

and assume for the moment that neither v nor w has zeroes. Inserting (1.16) into (1.15) and dividing by $v(x)w(t)$ we obtain

$$\frac{w''(t)}{w(t)} + \frac{L[v](x)}{v(x)} = 0 \qquad (x \in \Omega,\ t > 0), \tag{1.17}$$

$$B[v](x) = 0 \qquad (x \in \partial\Omega,\ t > 0). \tag{1.18}$$

Since the first term in (1.17) depends only on t, and the second only on x, both must be constant:

$$\frac{w''(t)}{w(t)} = -\lambda \quad (t > 0), \qquad \frac{L[v](x)}{v(x)} = \lambda \quad (x \in \Omega) \tag{1.19}$$

for some $\lambda \in \mathbb{C}$. Thus, with $\omega \in \mathbb{C}$ satisfying $\omega^2 = \lambda$, the first equation in (1.19) has (for $\lambda \neq 0$) the general solution

$$w(t) = \alpha \cos(\omega t) + \beta \sin(\omega t) \qquad (t > 0) \tag{1.20}$$

(where $\alpha, \beta \in \mathbb{C}$), while the second equation in (1.19), together with (1.18), yields the eigenvalue problem

$$\begin{aligned} L[v] &= \lambda v \quad (x \in \Omega), \\ B[v] &= 0 \quad\ \ (x \in \partial\Omega). \end{aligned} \tag{1.21}$$

If (v_k, λ_k) are eigenpairs of problem (1.21) and $\omega_k^2 = \lambda_k \neq 0$, we obtain from (1.20), (1.16), and the linearity of problem (1.15), that for all $\alpha_k, \beta_k \in \mathbb{C}$, a solution of (1.15) is given by

$$u(x,t) := \sum_k [\alpha_k \cos(\omega_k t) + \beta_k \sin(\omega_k t)] v_k(x) \quad (x \in \overline{\Omega},\ t \geq 0), \tag{1.22}$$

where, in the case of an infinite series, its convergence properties must of course be investigated.

To satisfy the *initial* conditions $u(x,0) = u_0(x)$, $\frac{\partial u}{\partial t}(x,0) = u_1(x)$ $(x \in \overline{\Omega})$ with the solutions in (1.22), we must therefore find α_k, β_k satisfying

$$\sum_k \alpha_k v_k(x) = u_0(x), \quad \sum_k \beta_k \omega_k v_k(x) = u_1(x) \quad (x \in \overline{\Omega}). \tag{1.23}$$

This poses the important question of whether (and in what sense) a given function can be expanded into a series of eigenfunctions of problem (1.21), or, formulated differently, the question of *completeness* of the eigenfunctions. We will address this question many times in this article.

It should be noted that the fully inhomogeneous problem (1.1) can easily be reduced to a homogeneous differential equation with homogeneous boundary conditions (i.e., to the problem treated above) if a particular solution $\hat{u}$ of this inhomogeneous problem (without initial condition!) is known. To obtain $\hat{u}$ one may use Duhamel's method, which is closely related to the variation of constants method for ordinary differential equations.

Example A (continued)

In the case of the one-dimensional wave equation (1.3), (1.4) governing the vibrating string, the related eigenvalue problem (1.21) reads

$$-v'' = \lambda v \quad (0 < x < l), \qquad v(0) = v(l) = 0,$$

and straightforward calculations show that all eigenpairs (v_k, λ_k) are given by

$$\lambda_k = \frac{k^2\pi^2}{l^2}, \quad v_k(x) = \sin\left(\frac{k\pi x}{l}\right) \qquad (0 \leq x \leq l,\ k \in \mathbb{N}).$$

The mathematical theory which we will present in the next chapter will show that these v_k are indeed complete: for each $u \in L_2(0,l)$ (i.e., each measurable, quadratically integrable $u : (0,l) \to \mathbb{C}$), there exist $c_k \in \mathbb{C}$ $(k \in \mathbb{N})$ such that

$$u = \sum_{k=1}^{\infty} c_k v_k, \tag{1.24}$$

with convergence to be understood in the L_2-sense:

$$\lim_{N\to\infty}\int_0^l \left|u(x)-\sum_{k=1}^{N} c_k v_k(x)\right|^2 dx = 0.$$

The coefficients c_k in (1.24) can be computed by multiplying (1.24) by v_j and integrating over $(0,l)$. Since $\int_0^l v_k v_j\,dx = \frac{1}{2} l\delta_{kj}$, we obtain

$$c_j = \frac{2}{l}\int_0^l u(x)v_j(x)\,dx \qquad (j\in\mathbb{N}). \tag{1.25}$$

Eqn. (1.24), with these c_j (and with $v_k(x)=\sin\left(\frac{k\pi x}{l}\right)$) is often called the generalized *Fourier expansion* of u.

For the purposes discussed above (in particular, for obtaining the identities (1.23) and the initial conditions for u given by (1.22) in the *pointwise* sense), one would like to have pointwise or even uniform convergence in (1.24). For general – even continuous – functions u, this is false: consider, for instance, a continuous function u satisfying $u(0)\neq 0$; then, the convergence in (1.24) cannot take place at the point $x=0$. However, for

$$u\in C_2[0,l] \text{ satisfying } u(0)=u(l)=0, \tag{1.26}$$

one obtains uniform convergence of the Fourier series as follows: with c_k $(k\in\mathbb{N})$ given by (1.25), let

$$u_N := \sum_{k=1}^{N} c_k v_k, \qquad \hat{c}_k := \frac{2}{l}\int_0^l u''(x)v_k(x)\,dx.$$

Partial integration and use of (1.26) provides $\hat{c}_k = -c_k\left(\frac{k\pi}{l}\right)^2$. Thus, since (1.24) also holds with u'' and $\hat{c}_k$ in place of u and c_k, we obtain

$$u'' = \sum_{k=1}^{\infty}\hat{c}_k v_k = \sum_{k=1}^{\infty} c_k\left(-\frac{k^2\pi^2}{l^2}v_k\right) = \sum_{k=1}^{\infty} c_k v_k'' = \lim_{N\to\infty} u_N'',$$

with convergence to be understood in the L_2-sense. Finally, partial integration provides

$$\int_0^l |u'-u_N'|^2\,dx = -\int_0^l (u-u_N)\overline{(u''-u_N'')}\,dx$$
$$\leq \left[\int_0^l |u-u_N|^2\,dx\right]^{1/2}\cdot\left[\int_0^l |u''-u_N''|^2\,dx\right]^{1/2} \to 0 \quad (N\to\infty).$$

Altogether, $u_N\to u$, $u_N'\to u'$, $u_N''\to u''$ in the L_2-sense. Thus, using Sobolev's Embedding Theorem (see [2, pp. 97, 98]) we obtain $u_N\to u$,

$u'_N \to u'$ uniformly, i.e.,

$$u = \sum_{k=1}^{\infty} c_k v_k, \quad u' = \sum_{k=1}^{\infty} c_k v'_k \text{ uniformly on } [0, l]$$

for u satisfying (1.26).

Using (1.22), (1.23), (1.24), (1.25), we obtain the solution of the one-dimensional wave equation (1.3), (1.4) with initial condition $u(x,0) = u_0(x)$, $\frac{\partial u}{\partial t}(x,0) = u_1(x)$ $(0 \le x \le l)$ to be

$$\begin{aligned} u(\cdot,t) &= \frac{2}{l}\sum_{k=1}^{\infty}\left[\left(\int_0^l u_0(y)v_k(y)\,dy\right)\cos\left(\frac{k\pi t}{l}\right)\right. \\ &\quad \left. + \frac{l}{k\pi}\left(\int_0^l u_1(y)v_k(y)\,dy\right)\sin\left(\frac{k\pi t}{l}\right)\right] v_k(\cdot). \end{aligned} \tag{1.27}$$

Eqn. (1.27) is just the harmonic decomposition of the string vibrations known from music theory.

Example B (continued)

For the vibrating membrane problem (1.5), (1.6), the related eigenvalue problem (1.21) reads

$$-\Delta v = \lambda v \quad (x \in \Omega), \qquad v = 0 \quad (x \in \partial\Omega). \tag{1.28}$$

Closed form solutions of this eigenvalue problem are known only for some particular domains Ω, e.g., for a rectangle

$$\Omega = (0, l_1) \times \cdots \times (0, l_n) \subset \mathbb{R}^n,$$

all eigenpairs are given by

$$\lambda_{(k_1,\ldots,k_n)} = \pi^2 \sum_{j=1}^{n} \left(\frac{k_j}{l_j}\right)^2, \quad v_{(k_1,\ldots,k_n)}(x) = \prod_{j=1}^{n} \sin\left(\frac{k_j \pi x_j}{l_j}\right)$$

$(x \in \overline{\Omega},\ (k_1, \ldots, k_n) \in \mathbb{N}^n)$, and for a circular disk

$$\Omega = \{x \in \mathbb{R}^2 : |x|^2 < R^2\} \subset \mathbb{R}^2,$$

the eigenpairs are, in polar coordinates (r, φ),

$$\begin{aligned} \lambda_{(k,j)} = \frac{q_{k,j}^2}{R^2}, \quad v_{(k,j)}^{(1)}(r,\varphi) &= J_k\left(\frac{q_{k,j}}{R} r\right)\cos(k\varphi), \\ v_{(k,j)}^{(2)}(r,\varphi) &= J_k\left(\frac{q_{k,j}}{R} r\right)\sin(k\varphi) \end{aligned}$$

($r \in [0, R]$, $\varphi \in [0, 2\pi]$, $k \in \mathbb{N}_0$, $j \in \mathbb{N}$), with J_k denoting the k-th Bessel function and $q_{k,j}$ ($j \in \mathbb{N}$) its zeroes; $v^{(2)}_{(k,j)}$ has to be cancelled if $k = 0$ (see [13, p. 260]).

Even if closed form solutions for problem (1.28) are available only in particular cases, the theory to be presented in the next chapter shows that, if the boundary $\partial\Omega$ is sufficiently regular (and – as already required earlier – Ω is bounded), there exists a sequence (v_k, λ_k) of eigenpairs with $\lambda_k > 0$, and the eigenfunctions are complete:

$$u = \sum_{k=1}^{\infty} c_k v_k, \qquad c_k = \int_{\Omega} u v_k \, dx \Big/ \int_{\Omega} |v_k|^2 \, dx$$

for each $u \in L_2(\Omega)$ (i.e., each measurable, quadratically integrable $u : \Omega \to \mathbb{C}$), with convergence of the series to be understood in the L_2-sense, and uniformly if $n \leq 3$ and $u \in C_2(\overline{\Omega})$, $u = 0$ on $\partial\Omega$.

As in the one-dimensional case, we obtain the solution of the wave equation (1.5), (1.6) with initial condition $u(x, 0) = u_0(x)$, $\frac{\partial u}{\partial t}(x, 0) = u_1(x)$ ($x \in \overline{\Omega}$) as

$$\begin{aligned} u(\cdot, t) \;=\; \sum_{k=1}^{\infty} \left[\int_{\Omega} |v_k|^2 \, dy \right]^{-1} & \left[\left(\int_{\Omega} u_0 v_k \, dy \right) \cos\left(\sqrt{\lambda_k}\, t\right) \right. \\ & \left. + \frac{1}{\sqrt{\lambda_k}} \left(\int_{\Omega} u_1 v_k \, dy \right) \sin\left(\sqrt{\lambda_k}\, t\right) \right] v_k(\cdot). \end{aligned}$$

Example C (continued)

For the vibrating plate problem (1.7), (1.8), the related eigenvalue problem (1.21) is

$$\Delta\Delta u = \lambda u \quad (x \in \Omega), \qquad u = \frac{\partial u}{\partial \nu} = 0 \quad (x \in \partial\Omega)$$

(correspondingly for problem (1.7), (1.9)). In this case too, the mathematical theory provides a sequence of eigenpairs with the eigenfunctions being complete, if the boundary $\partial\Omega$ is sufficiently regular (and Ω is bounded).

Example D (continued)

For the wave equations (1.13), (1.14) for the potentials ϕ and A, together with the boundary conditions

$$\phi \to 0, \quad A \to 0 \quad \text{for} \quad |x| \to \infty$$

fixing the zero level of the potentials (which makes sense if the electric charges and currents are localized in some bounded set), the eigenvalue problem (1.21) related to the corresponding homogeneous wave equation reads

$$-\Delta v = \lambda v \quad (x \in \mathbb{R}^3), \qquad v \to 0 \text{ for } |x| \to \infty. \tag{1.29}$$

Concerning the question of completeness of eigenfunctions, problem (1.29) shows a behaviour which is very different from the other examples discussed so far:

$$\text{there exist no eigenvalues of problem (1.29) at all!} \tag{1.30}$$

We prove this statement for the one-dimensional analogue, where $-\Delta u$ is replaced by $-u''$: The solutions of the differential equation $-v'' = \lambda v$ (on $\mathbb{R}$) are $v(x) = \alpha e^{i\omega x} + \beta e^{-i\omega x}$ (if $\lambda \neq 0$, $\omega^2 = \lambda$) resp. $v(x) = \alpha x + \beta$ (if $\lambda = 0$), with $\alpha, \beta \in \mathbb{C}$, but none of these (except $v \equiv 0$) tends to zero for $x \to \infty$ and for $x \to -\infty$.

Of course, (1.30) implies in particular that no expansion of a given function into a series of eigenfunctions (generalized Fourier expansion) is possible. But nevertheless, some kind of expansion exists here, too, namely in the form of a *Fourier integral* over the "almost" eigenfunctions $e^{i\omega x}$, $e^{-i\omega x}$ ($\omega \in \mathbb{R}$ resp. $\lambda \geq 0$):

If $u : \mathbb{R} \to \mathbb{R}$ is measurable with $\int_{-\infty}^{\infty} |u|\, dx < \infty$, then

$$u(x) = \int_{-\infty}^{\infty} c(\omega) e^{i\omega x}\, d\omega, \quad \text{where } c(\omega) = \frac{1}{2\pi} \int_{-\infty}^{\infty} u(y) e^{-i\omega y}\, dy.$$

Thus, under corresponding assumptions on the initial functions u_0 and u_1, the solution of the initial value problem for the one-dimensional (homogeneous) wave equation is

$$u(x,t) = \int_{-\infty}^{\infty} \left[c_0(\omega) \cos(\omega t) + c_1(\omega) \frac{\sin(\omega t)}{\omega} \right] e^{i\omega x}\, d\omega, \tag{1.31}$$

where

$$c_j(\omega) = \frac{1}{2\pi} \int_{-\infty}^{\infty} u_j(y) e^{-i\omega y}\, dy \qquad (j = 0, 1).$$

Defining $d_l(\omega) := \frac{1}{2}\left[c_0(\omega) - i\,\frac{c_1(\omega)}{\omega}\right]$, $d_r(\omega) := \frac{1}{2}\left[c_0(\omega) + i\,\frac{c_1(\omega)}{\omega}\right]$, we obtain

$$u(x,t) = \int_{-\infty}^{\infty} \left[d_l(\omega) e^{i\omega(x+t)} + d_r(\omega) e^{i\omega(x-t)} \right] d\omega,$$

i.e., a decomposition into *simple waves* running to the left ($e^{i\omega(x+t)}$) and to the right ($e^{i\omega(x-t)}$), respectively.

Finally, we mention the multidimensional generalization of (1.31), which decomposes the solution of the wave equation in higher dimensions into *plane waves*:

$$u(x,t) = \int_{\mathbb{R}^n} \left[c_0(\omega) \cos(|\omega| t) + c_1(\omega) \frac{\sin(|\omega| t)}{|\omega|} \right] e^{i\omega \cdot x}\, d\omega_1 \cdots d\omega_n,$$

where $c_j(\omega) = \dfrac{1}{(2\pi)^n} \displaystyle\int_{\mathbb{R}^n} u_j(y) e^{-i\omega\cdot y}\, dy_1 \cdots dy_n \ (j = 0, 1)$.

1.3 Some other eigenvalue problems

In this section, we mention some practical examples of eigenvalue problems with origins other than wave-type equations.

Example E (Schrödinger's equation)
The quantum mechanical behaviour of particles is governed by Schrödinger's equation

$$\frac{1}{i}\frac{\partial u}{\partial t} - \Delta u + V(x) u = 0 \qquad (t > 0,\ x \in \mathbb{R}^n), \quad u \to 0 \ \text{for} \ |x| \to \infty,$$

with $V : \mathbb{R}^n \to \mathbb{R}$ denoting the *potential* acting on the particle (system). The solution u is the *wave* function of the particle, fixed by an additional initial condition $u(x, 0) = u_0(x)$ $(x \in \mathbb{R}^n)$.

This problem, too, admits the separation of variables technique discussed for the wave equation in the previous section. This leads to the eigenvalue problem

$$-\Delta v + V(x) v = \lambda v \quad (x \in \mathbb{R}^n), \qquad v \to 0 \ \text{for} \ |x| \to \infty, \tag{1.32}$$

the eigenvalues of which are the possible *energy levels* of the particle (system). Some simple but nevertheless important particular examples of potentials are:

i) $V \equiv 0$, describing the *free* particle. As stated in (1.30) already, problem (1.32) has *no eigenvalues* in this case, but all $\lambda \geq 0$ are some kind of "almost" eigenvalues with the "almost" eigenfunctions $e^{i\omega\cdot x}$ (where $\omega \in \mathbb{R}^3$, $|\omega|^2 = \lambda$) which do not satisfy the boundary condition in (1.32), but nevertheless are *bounded* for $|x| \to \infty$.

A precise mathematical definition of these "almost" eigenvalues, the *continuous spectrum*, will be given in Section 2.4 of the next chapter.

The physical meaning of the above considerations is that free particles do not have "bound states" (i.e., eigenvalues allowing only certain energy levels), but can take nonnegative energy in a "blurred" way. The corresponding wave functions $\exp[i(\omega \cdot x - |\omega|^2 t)]$ are *plane waves*.

ii) $V(x) = -\dfrac{1}{|x|}$ $(x \in \mathbb{R}^3 \setminus \{0\})$, the Coulomb potential, describing the electron in the *hydrogen atom* (with infinite mass nucleus). The eigenvalue problem (1.32) can be shown (by separation into polar coordinate variables) to have an infinite number of negative eigenvalues, corresponding to discrete energy levels in the atomic bonding, and – as in the free particle case – the interval $[0, \infty)$ as continuous spectrum, corresponding to a free electron which has left the atom.

iii) $V(x) = -\frac{1}{|x^{(1)}|} - \frac{1}{|x^{(2)}|} + \frac{1}{|x^{(1)} - x^{(2)}|}$ $(x = (x^{(1)}, x^{(2)})^\top \in \mathbb{R}^6$, $x^{(1)}, x^{(2)} \neq 0$, $x^{(1)} \neq x^{(2)})$, describing the two electrons in the helium atom. Here, closed form solutions of problem (1.32) no longer exist, and numerical methods are required for the computation of the energy levels. Nevertheless, the general situation is similar to the hydrogen atom.

Example F (bending of a beam)
Consider a beam of length l which is fixed and horizontally clamped at its right endpoint. A horizontal force P is acting from the left at its left endpoint. In the bended position (if bending takes place), choose a coordinate system with origin in the beam's left endpoint, and with horizontal x-axis. Thus, for "small" deviations $u(x)$ $(0 \leq x \leq l)$ from the rest position $u \equiv 0$, we have the boundary condition

$$u(0) = u'(l) = 0. \tag{1.33}$$

The moment $M(x)$ acting at the point $(x, u(x))$ of the beam is $P \cdot u(x)$. On the other hand, $M(x)$ is proportional to the curvature $\kappa(x) = -u''(x)/[1 + u'(x)^2]^{3/2}$ of the beam at $(x, u(x))$; the proportionality factor $\alpha > 0$ is the *stiffness* of the beam. Thus, with $\lambda = P/\alpha$, we have the nonlinear problem

$$\frac{-u''}{[1 + (u')^2]^{3/2}} = \lambda u \qquad (0 < x < l). \tag{1.34}$$

For small deviations, we may *linearize* (1.34) to obtain, together with (1.33), the eigenvalue problem

$$-u'' = \lambda u \quad (0 < x < l), \qquad u(0) = u'(l) = 0, \tag{1.35}$$

with eigenpairs

$$\lambda_k = \left(k - \frac{1}{2}\right)^2 \frac{\pi^2}{l^2}, \quad u_k(x) = \sin\left[\left(k - \frac{1}{2}\right)\frac{\pi x}{l}\right] \qquad (0 \leq x \leq l,\ k \in \mathbb{N}).$$

Thus, for each $k \in \mathbb{N}$, we obtain a new bending mode when the force P equals $P_k := \alpha\lambda_k$. For forces $P \in (P_k, P_{k+1})$ resp. for $\lambda \in (\lambda_k, \lambda_{k+1})$, problem (1.35) has the trivial solution $u \equiv 0$ only, which seems to indicate that no bending can take place (which contradicts physical evidence). However, this is an effect of the linearization. The nonlinear problem (1.34), (1.33) has indeed k nontrivial solutions for each $\lambda > \lambda_k$. For $\lambda \leq \lambda_1$ resp. $P \leq P_1$, both the nonlinear and the linearized problem have the trivial solution only: no bending can take place if the force P is below the critical value $P_1 = \alpha\pi^2/(4l^2)$.

Example G (Sturm–Liouville problems)
Let $a \in \mathbb{R} \cup \{-\infty\}$, $b \in \mathbb{R} \cup \{+\infty\}$, $a < b$, and let $p, q, w : (a, b) \to \mathbb{R}$

denote measurable functions such that $p > 0$, $w > 0$ on (a, b) and p^{-1}, q, w are absolutely integrable on compact subintervals of (a, b). Consider the *Sturm–Liouville problem*

$$-(pu')' + qu = \lambda wu \quad \text{on} \quad (a, b) \tag{1.36}$$

occurring, e.g., as the radial part of elliptic eigenvalue problems such as the Schrödinger equation (1.32).

Depending on the data a, b, p, q, w, the eigenvalue problem (1.36) is either "well posed" as it stands, or additional boundary conditions must be required. The corresponding case distinction is made in the famous limit point/limit circle theory by H. Weyl which we will not discuss here; see [35] for details. The set of eigenfunctions may or may not be complete.

We wish to remark that higher order and system analogues of problem (1.36) are also of interest; see [21].

2 Mathematical theory

In this second chapter, we report on the theoretical background needed for the treatment of eigenvalue problems. In particular, we will systematically address the question of completeness of eigenfunctions.

2.1 Compact symmetric operators in Hilbert space

Let $(H, \langle\cdot,\cdot\rangle)$ denote a (complex) separable Hilbert space, and let $K : H \to H$ denote a linear operator which is *symmetric* (i.e., $\langle Ku, v\rangle = \langle u, Kv\rangle$ for $u, v \in H$), so that all its eigenvalues are real and eigenelements corresponding to different eigenvalues are orthogonal, and which is moreover *compact* (i.e., for each bounded sequence (u_k) in H, (Ku_k) has a convergent subsequence), so that the set of its eigenvalues has no accumulation point except (possibly) zero, and each nonzero eigenvalue λ has finite multiplicity.

Theorem 2.1 *Let* $\dim H = \infty$ *and suppose that the image space* $K(H) = \{Ku : u \in H\}$ *is dense in* H *(i.e., each* $v \in H$ *is an accumulation point of* $K(H)$*). Then, there exists a sequence* $(u_k, \lambda_k)_{k\in\mathbb{N}}$ *of eigenpairs of* K *with the following properties:*

a) $(|\lambda_k|)_{k\in\mathbb{N}}$ *is monotonically nonincreasing and converges to zero; moreover,* $\lambda_k \neq 0$ *for all* k*;*

b) $(u_k)_{k\in\mathbb{N}}$ *is orthonormal:* $\langle u_k, u_j\rangle = \delta_{kj}$ $(k, j \in \mathbb{N})$*;*

c) $(u_k)_{k\in\mathbb{N}}$ *is complete:* $u = \sum_{k=1}^{\infty} \langle u, u_k\rangle u_k$ *for all* $u \in H$*;*

d) $|\lambda_k| = \max\left\{\dfrac{|\langle Ku, u\rangle|}{\langle u, u\rangle} : u \in H \setminus \{0\},\ \langle u, u_j\rangle = 0\ (j = 1, \ldots, k-1)\right\}$
for all $k \in \mathbb{N}$.

The proof of this theorem can be found in every functional analysis textbook.

In the next two sections, we will apply Theorem 2.1 to certain classes of eigenvalue problems for differential equations.

2.2 Eigenvalue problems for regular symmetric ordinary differential operators

Consider the differential operator (of order m)

$$L[u](x) := \sum_{\nu=0}^{m} p_\nu(x) u^{(\nu)}(x) \qquad (a \le x \le b,\ u \in C_m[a,b])$$

with continuous, complex-valued coefficients $p_0, \ldots, p_m$ such that $p_m(x) \ne 0$ $(a \le x \le b)$, and the m boundary operators (of order $\le m-1$)

$$B_\mu[u] := \sum_{\nu=0}^{m-1} \left[\alpha_{\mu\nu} u^{(\nu)}(a) + \beta_{\mu\nu} u^{(\nu)}(b)\right] \qquad (\mu = 0, \ldots, m-1)$$

with complex coefficients $\alpha_{\mu\nu}$, $\beta_{\mu\nu}$, and suppose that, for all $u, v \in C_m[a,b]$ satisfying the boundary conditions $B_\mu[u] = B_\mu[v] = 0$ $(\mu = 0, \ldots, m-1)$,

$$\int_a^b L[u]\bar{v}\,dx = \int_a^b u\overline{L[v]}\,dx. \tag{2.1}$$

In addition, let w be a continuous, real-valued function on $[a,b]$ satisfying $w(x) > 0$ $(a \le x \le b)$. Consider the eigenvalue problem

$$L[u] = \lambda w u \quad (a < x < b), \qquad B_\mu[u] = 0 \quad (\mu = 0, \ldots, m-1). \tag{2.2}$$

Examples. i) Consider problem (1.36), with the additional assumptions that $-\infty < a < b < +\infty$, that p, q, w are continuous on $[a,b]$, and that $p > 0$, $w > 0$ on $[a,b]$, and with boundary operators

$$B_0[u] = -\alpha_0 u'(a) + \gamma_0 u(a), \qquad B_1[u] = \alpha_1 u'(b) + \gamma_1 u(b)$$

(with real α_j, γ_j satisfying $\alpha_j^2 + \gamma_j^2 > 0$ for $j = 0, 1$), or the *periodic* boundary operators

$$B_0[u] = u(b) - u(a), \qquad B_1[u] = (pu')(b) - (pu')(a).$$

Then, $[-pu'\bar{v} + pu\bar{v}']_a^b = 0$ for all $u, v \in C_2[a,b]$ satisfying $B_0[u] = B_1[u] = 0$, so that partial integration provides (2.1).

ii) Let $L[u] := \frac{1}{i} u'$, the *momentum operator* from quantum mechanics. Using the periodic boundary operator $B_0[u] := u(b) - u(a)$, we find that (2.1) holds true for all $u, v \in C_1[a,b]$ satisfying $B_0[u] = B_0[v] = 0$, again by partial integration.

The aim of this section is to prove

Theorem 2.2 *There exists a sequence $(u_k, \lambda_k)_{k \in \mathbb{N}}$ of eigenpairs of prob-*

lem (2.2) such that $u_k \in C_m[a,b]$, $B_\mu[u_k] = 0$ $(\mu = 0, \dots, m-1;\ k \in \mathbb{N})$, *and* $(u_k)_{k\in\mathbb{N}}$ *is orthonormal and complete in* $L_2(a,b)$ *(the Hilbert space of square integrable measurable functions) endowed with the inner product*

$$\langle u, v\rangle := \int_a^b w u \bar{v}\, dx,$$

i.e., for all $k, j \in \mathbb{N}$ *and all* $u \in L_2(a,b)$

$$\langle u_k, u_j\rangle = \delta_{kj}, \qquad u = \sum_{\nu=1}^{\infty} \langle u, u_\nu\rangle u_\nu, \tag{2.3}$$

with the series converging in the L_2*-sense. Moreover, the eigenvalue sequence* $(\lambda_k)_{k\in\mathbb{N}}$ *is real and has no finite accumulation point, and all eigenvalues have finite multiplicity.*

Proof. We choose some $\kappa \in \mathbb{R}$ which is no eigenvalue of problem (2.2). This is possible since (2.1) implies that eigenfunctions corresponding to different eigenvalues are $\langle\cdot,\cdot\rangle$-orthogonal, so that the assumption that all real numbers are eigenvalues provides a supercountable orthonormal set (of eigenfunctions), which cannot exist in a separable Hilbert space.

Thus, the boundary value problem

$$L[u] - \kappa w u = r \text{ on } (a,b), \qquad B_\mu[u] = 0 \quad (\mu = 0, \dots, m-1) \tag{2.4}$$

has only the trivial solution $u \equiv 0$ if $r \equiv 0$, so that it is uniquely solvable (with $u \in C_m[a,b]$) for each $r \in C[a,b]$, according to the alternative theorem for ordinary boundary value problems, and the solution can be represented as

$$u(x) = \int_a^b G(x,t) r(t)\, dt \qquad (a \le x \le b),$$

with $G : [a,b] \times [a,b] \to \mathbb{C}$ denoting the *Green's function* for problem (2.4). Now we choose $H := L_2(a,b)$ and define the linear operator

$$K : H \to H, \quad (Kv)(x) := \int_a^b G(x,t) w(t) v(t)\, dt \qquad (a \le x \le b)$$

and show that

i) K maps H into $C[a,b]$, $K : H \to (C[a,b], \|\cdot\|_\infty)$ is compact, and $K : H \to H$ is compact,

ii) K maps $C[a,b]$ onto $\{u \in C_m[a,b] : B_\mu[u] = 0\ (\mu = 0, \dots, m-1)\} =: D$,

iii) K is $\langle\cdot,\cdot\rangle$-symmetric.

Then, since D is dense in H, so is $K(H)$; Theorem 2.1 therefore provides a sequence $(u_k, \mu_k)_{k\in\mathbb{N}}$ of eigenpairs of K with the properties stated there.

Thus, for all $k \in \mathbb{N}$, $u_k = \mu_k^{-1}(Ku_k) \in C[a,b]$ due to i), and consequently, $u_k = \mu_k^{-1}(Ku_k) \in D$ due to ii), so that the definition of K yields

$$L[u_k] - \kappa w u_k = \mu_k^{-1} w u_k \qquad (a < x < b),$$

which provides all our assertions, with $\lambda_k := \kappa + \mu_k^{-1}$.

Ad i) Let $M \subset H$ be a set bounded by some constant C. We show that $K(M)$ is a relatively compact subset of $C[a,b]$, which then proves the first two statements in i). For this purpose, we have to show, according to the famous Arzelà–Ascoli Theorem, that $K(M)$ is uniformly bounded and equicontinuous. But this is a consequence of Schwarz's inequality and of the uniform continuity of the Green's function: for $v \in M$ and $x, y \in [a,b]$,

$$\begin{aligned} |(Kv)(x)| &\leq \int_0^1 \left[|G(x,t)|\sqrt{w(t)}\right] \cdot \left[\sqrt{w(t)}\,|v(t)|\right] dt \\ &\leq \left[\int_0^1 |G(x,t)|^2 w(t)\,dt\right]^{1/2} \left[\int_0^1 w(t)|v(t)|^2\,dt\right]^{1/2} \\ &\leq C \cdot \max_{x\in[0,1]} \left[\int_0^1 |G(x,t)|^2 w(t)\,dt\right]^{1/2}, \end{aligned}$$

$$\begin{aligned} |(Kv)(x) - (Kv)(y)| &\leq \int_0^1 \left[|G(x,t) - G(y,t)|\sqrt{w(t)}\right] \cdot \left[\sqrt{w(t)}\,|v(t)|\right] dt \\ &\leq C \cdot \left[\int_0^1 |G(x,t) - G(y,t)|^2 w(t)\,dt\right]^{1/2} < \varepsilon \end{aligned}$$

for $|x-y| < \delta(\varepsilon)$. To show finally that $K: H \to H$ is compact, let $(v_k)_{k\in\mathbb{N}}$ denote a bounded sequence in H. Then $(Kv_k)_{k\in\mathbb{N}}$ has a $\|\cdot\|_\infty$-convergent subsequence due to what has just been shown. Since $\|u\| := \sqrt{\langle u,u\rangle} \leq \sqrt{\int_a^b w\,dt} \cdot \|u\|_\infty$ for all $u \in C[a,b]$, the same subsequence converges in H, too.

Ad ii) This is a direct consequence of the properties of the Green's function, together with the continuity of w and the coefficients of L, and the positivity of w.

Ad iii) Obviously, (2.1) also holds with L replaced by $L - \kappa w$, for all $u, v \in D$. Thus, $w^{-1}(L - \kappa w) : D \to C[a,b]$ is $\langle\cdot,\cdot\rangle$-symmetric. Consequently, also $K\big|_{C[a,b]} : C[a,b] \to D$, which is just the inverse of $w^{-1}(L - \kappa w)$, is $\langle\cdot,\cdot\rangle$-symmetric. Since $C[a,b]$ is dense in H and K is continuous, K is $\langle\cdot,\cdot\rangle$-symmetric. ∎

2.3 Eigenvalue problems for second-order symmetric elliptic differential operators

We now consider the second-order elliptic differential operator

$$L[u](x) := -\sum_{\mu,\nu=1}^{n} \frac{\partial}{\partial x_\mu}\left(p_{\mu\nu}\frac{\partial u}{\partial x_\nu}\right)(x) + q(x)u(x) \qquad (x \in \Omega) \tag{2.5}$$

with $p_{\mu\nu} = \bar{p}_{\nu\mu} \in C_1(\overline{\Omega})$, $q = \bar{q} \in C(\overline{\Omega})$, $\sum_{\mu,\nu=1}^{n} p_{\mu\nu}(x)\xi_\mu\bar{\xi}_\nu \geq p_0|\xi|^2$ $(x \in \overline{\Omega},\ \xi \in \mathbb{C}^n)$ for some constant $p_0 > 0$. Here, $\Omega \subset \mathbb{R}^n$ is a bounded domain with sufficiently regular boundary $\partial\Omega$ (e.g., it suffices if $\partial\Omega$ is a C_2-manifold or if $n = 2$ and Ω is a convex polygon).

We restrict ourselves to *second*-order operators (2.5) for simplicity of presentation. Most of the arguments can be carried over to higher order operators such as $\Delta\Delta$ (see Example C). For the same reason, we only consider *Dirichlet* boundary conditions, $u = 0$ on $\partial\Omega$.

As in the previous section, let $w \in C(\overline{\Omega})$ denote a real-valued function satisfying $w(x) > 0$ for $x \in \overline{\Omega}$, and consider the eigenvalue problem

$$L[u] = \lambda w u \ \text{ on } \Omega, \qquad u = 0 \ \text{ on } \partial\Omega. \tag{2.6}$$

In contrast to the ordinary differential equation (2.2) we cannot expect to find eigenfunctions of (2.6) in $C_2(\overline{\Omega})$, without additional smoothness assumptions. It is more natural to look for eigenfunctions in

$$D := H_2(\Omega) \cap \overset{\circ}{H}_1(\Omega),$$

where $H_k(\Omega)$ is the Sobolev space of k-times weakly differentiable functions with weak derivatives (up to order k) in $L_2(\Omega)$ (the space of measurable, square integrable functions on Ω). Endowed with the inner product $\langle\cdot,\cdot\rangle_{H_k}$, defined as the sum of the L_2-products of all derivatives up to order k, $H_k(\Omega)$ is a Hilbert space. $\overset{\circ}{H}_1(\Omega)$ is the closure of $\{u \in C_\infty(\overline{\Omega}) :$ u has compact support in $\Omega\}$ in $H_1(\Omega)$. See [2] for (many) more details on Sobolev spaces.

In analogy to the previous section, our aim is here to prove

Theorem 2.3 *There exists a sequence $(u_k, \lambda_k)_{k\in\mathbb{N}}$ of eigenelements of problem (2.6) such that $u_k \in H_2(\Omega) \cap \overset{\circ}{H}_1(\Omega)$ $(k \in \mathbb{N})$, and $(u_k)_{k\in\mathbb{N}}$ is orthonormal and complete in $L_2(\Omega)$ endowed with the inner product*

$$\langle u, v\rangle := \int_\Omega w u \bar{v}\, dx,$$

i.e., for all $k, j \in \mathbb{N}$ and all $u \in L_2(\Omega)$,

$$\langle u_k, u_j\rangle = \delta_{kj}, \qquad u = \sum_{\nu=1}^{\infty} \langle u, u_\nu\rangle u_\nu, \tag{2.7}$$

with the series converging in the L_2-sense, i.e., with respect to $\|\cdot\| := \sqrt{\langle\cdot,\cdot\rangle}$. Moreover, the eigenvalue sequence $(\lambda_k)_{k\in\mathbb{N}}$ is real and converges to $+\infty$, and can therefore be assumed to be ordered by magnitude.

Proof. From (2.5) and $p_{jk} = \bar{p}_{kj}$, $q = \bar{q}$ we find, by partial integration, i.e. Green's formula,

$$\int_\Omega L[u]\bar{v}\,dx = \int_\Omega u\overline{L[v]}\,dx \quad \text{for} \quad u,v \in H_2(\Omega)\cap \overset{\circ}{H}_1(\Omega) \tag{2.8}$$

(compare (2.1)). Thus, with the same arguments as in the proof of Theorem 2.2, we can choose some $\kappa \in \mathbb{R}$ which is not an eigenvalue of problem (2.6), i.e., the boundary value problem

$$L[u] - \kappa w u = r \ \text{ on } \ \Omega, \qquad u \in H_2(\Omega)\cap \overset{\circ}{H}_1(\Omega) \tag{2.9}$$

has only the trivial solution $u \equiv 0$ if $r \equiv 0$. According to the theory of elliptic boundary value problems (see, e.g., [15]), problem (2.9) is therefore uniquely solvable (with $u \in H_2(\Omega)\cap\overset{\circ}{H}_1(\Omega)$) for each $r \in L_2(\Omega)$, and the solution operator

$$S : \left\{ \begin{array}{c} L_2(\Omega) \to H_2(\Omega) \\ r \mapsto u \end{array} \right\}$$

is bounded. Thus, since w is continuous, also $T : L_2(\Omega) \to H_2(\Omega)$, $Tv := S(wv)$ is bounded. Furthermore, the Sobolev–Kondrachev–Rellich embedding theorem (see [2, p. 144]) shows that the embedding

$$E : \left\{ \begin{array}{c} H_2(\Omega) \to L_2(\Omega) \\ u \mapsto u \end{array} \right\}$$

is compact. Consequently,

$$K := E \circ T : L_2(\Omega) \to L_2(\Omega)$$

is compact. Moreover, $w^{-1}(L - \kappa w) : H_2(\Omega)\cap\overset{\circ}{H}_1(\Omega) \to L_2(\Omega)$ is $\langle\cdot,\cdot\rangle$-symmetric due to (2.8), so that also its inverse T, and therefore K, is $\langle\cdot,\cdot\rangle$-symmetric.

Since moreover $K(L_2(\Omega)) = H_2(\Omega)\cap\overset{\circ}{H}_1(\Omega)$ is dense in $L_2(\Omega)$, Theorem 2.1 provides a sequence $(u_k,\mu_k)_{k\in\mathbb{N}}$ of eigenpairs of K with the properties stated there. Thus, for all $k \in \mathbb{N}$, $u_k = \mu_k^{-1}(Ku_k) = \mu_k^{-1}(Tu_k) = \mu_k^{-1}(S(wu_k))$ which imples $u_k \in H_2(\Omega)\cap\overset{\circ}{H}_1(\Omega)$ and

$$L[u_k] - \kappa w u_k = \mu_k^{-1} w u_k \ \text{ on } \ \Omega,$$

which again provides all our assertions with $\lambda_k := \kappa + \mu_k^{-1}$, except that (in contrast to Theorem 2.2) we are now left to show that $-\infty$ is no accumulation point of $(\lambda_k)_{k\in\mathbb{N}}$. But this is an easy consequence of (2.5) and the assumptions thereafter: partial integration provides, for all $k \in \mathbb{N}$,

$$\begin{aligned}\lambda_k &= \int_\Omega L[u_k]\bar{u}_k\,dx = \int_\Omega \left[\sum_{\mu,\nu=1}^n p_{\mu\nu}\frac{\partial u_k}{\partial x_\nu}\frac{\partial \bar{u}_k}{\partial x_\mu} + q|u_k|^2\right] dx \\ &\geq \int_\Omega \left[p_0|\nabla u_k|^2 + \frac{q}{w}w|u_k|^2\right] dx \geq \min_{x\in\overline{\Omega}}\left\{\frac{q(x)}{w(x)}\right\}.\end{aligned}$$

Since $(\lambda_k)_{k\in\mathbb{N}}$ has moreover no *finite* accumulation point, it can be made monotonically nondecreasing by re-ordering $(u_k, \lambda_k)_{k\in\mathbb{N}}$. ∎

2.4 Spectral decomposition in the case of non-complete eigenfunctions

Recall that the eigenvalue problem

$$-\Delta v = \lambda v \;\text{ on }\; \mathbb{R}^n, \qquad v \to 0 \;\text{ for }\; |x| \to \infty$$

has no eigenvalues at all. Nevertheless, a spectral decomposition in *integral* form (Fourier integral) could be achieved; see Example D. Other problems do have eigenvalues, but "not enough" in the sense that the eigenfunctions are not complete (see Example E; hydrogen atom). The reason is that these problems have, besides (possibly) eigenvalues, *continuous spectrum* which has to be taken into account for spectral decompositions.

Let $(H, \langle\cdot,\cdot\rangle)$ denote a separable Hilbert space, $D(A) \subset H$ a dense subspace, and $A : D(A) \to H$ a symmetric operator. We now ask for more than symmetry: we require that A be *selfadjoint*, which means that the following implication holds for each $v \in H$: if some $v^* \in H$ exists such that $\langle Au, v\rangle = \langle u, v^*\rangle$ for all $u \in D(A)$, then $v \in D(A)$.

Selfadjointness means (besides symmetry) that the domain of definition $D(A)$ is, in a certain sense, "large enough" (or even "maximal") so that it contains all elements needed as a "basis" for spectral decompositions.

A number $\lambda \in \mathbb{C}$ is said to belong to the *continuous spectrum* $\sigma_c(A)$ of the selfadjoint operator A if $(A - \lambda) : D(A) \to H$ is one-to-one (so that λ is not an eigenvalue), but not onto.

It can be shown that $\sigma_c(A) \subset \mathbb{R}$ and that, for $\lambda \in \sigma_c(A)$, $(A-\lambda)(D(A))$ is dense in H, and $(A - \lambda)^{-1}$ is unbounded. Moreover, each $\lambda \in \mathbb{C}$ which is neither an eigenvalue of A nor an element of $\sigma_c(A)$ belongs to the resolvent set, i.e., $(A - \lambda) : D(A) \to H$ is one-to-one and onto, and $(A - \lambda)^{-1}$ is bounded.

Let us return for a moment to the case where an orthonormal and complete system $(u_k)_{k\in\mathbb{N}}$ of eigenfunctions $u_k \in D(A)$ of A exists. Then define, for each $\lambda \in \mathbb{R}$,

$$E_\lambda : H \to H, \qquad E_\lambda u := \sum_{\substack{k\in\mathbb{N}\\ \lambda_k \le \lambda}} \langle u, u_k\rangle u_k, \tag{2.10}$$

so that the family $(E_\lambda)_{\lambda\in\mathbb{R}}$ has the following properties:

α) E_λ is symmetric and a projection ($E_\lambda^2 = E_\lambda$),

β) $E_\lambda E_\mu = E_\mu E_\lambda = E_\mu$ for $\mu \le \lambda$,

γ) for each $u \in H$, $E_{\lambda+\varepsilon} u \to E_\lambda u$ for $\varepsilon \to 0+$,

δ) for each $u \in H$, $E_\lambda u \to 0$ for $\lambda \to -\infty$, $E_\lambda u \to u$ for $\lambda \to +\infty$.

Moreover, for all $\lambda \in \mathbb{R}$ and $u \in H$,

$$(E_\lambda - E_{\lambda-0})u = \sum_{\substack{k\in\mathbb{N}\\ \lambda_k = \lambda}} \langle u, u_k\rangle u_k \begin{cases} = 0 \text{ if } \lambda \text{ is not an eigenvalue,} \\ \text{is the projection of } u \text{ onto} \\ \text{the eigenspace if } \lambda \text{ is an} \\ \text{eigenvalue.} \end{cases} \tag{2.11}$$

Consequently,

the mapping $\lambda \mapsto E_\lambda$ is constant between the eigenvalues, and jumps by the projection onto the eigenspaces at the eigenvalues. (2.12)

Furthermore, for all $\lambda \in \mathbb{R}$ and $u \in D(A)$,

$$\lambda(E_\lambda - E_{\lambda-0})u = \sum_{\substack{k\in\mathbb{N}\\ \lambda_k = \lambda}} \langle u, \lambda u_k\rangle u_k = \sum_{\substack{k\in\mathbb{N}\\ \lambda_k = \lambda}} \langle Au, u_k\rangle u_k. \tag{2.13}$$

From (2.11) and (2.13) we obtain, for all $a, b \in \mathbb{R}$, $a < b$,

$$\begin{aligned} \left(\int_a^b dE_\lambda\right) u &= \sum_{\substack{k\in\mathbb{N}\\ a\le\lambda_k\le b}} \langle u, u_k\rangle u_k \qquad (u \in H), \\ \left(\int_a^b \lambda\, dE_\lambda\right) u &= \sum_{\substack{k\in\mathbb{N}\\ a\le\lambda_k\le b}} \langle Au, u_k\rangle u_k \qquad (u \in D(A)), \end{aligned} \tag{2.14}$$

where the *Riemann–Stieltjes integral* $\int_a^b f(\lambda)\, dE_\lambda$ is defined as follows: for each subdivision $\mathscr{Y} = (\mu_0, \ldots, \mu_N)$ such that $\mu_0 < a < \mu_1 < \mu_2 < \cdots < \mu_N = b$, and each allocation $\Lambda := (\lambda^{(1)}, \ldots, \lambda^{(N)})$ of $\mathscr{Y}$ satisfying $\mu_{j-1} \le \lambda^{(j)} \le \mu_j$ $(j = 1, \ldots, N)$ and $\lambda^{(1)} \ge a$, let

$$S(\mathscr{Y}, \Lambda) := \sum_{j=1}^N f(\lambda^{(j)})(E_{\mu_j} - E_{\mu_{j-1}}),$$

and $|\mathcal{Y}| := \max_{j=1,\ldots,N}(\mu_j - \mu_{j-1})$. Provided that some linear operator $B : H \to H$ exists such that $B = \lim_{|\mathcal{Y}|\to 0} S(\mathcal{Y}, \Lambda)$ in the sense that, for each $\varepsilon > 0$, some $\delta > 0$ exists such that $\|B - S(\mathcal{Y}, \Lambda)\| < \varepsilon$ for each subdivision $\mathcal{Y}$ satisfying $|\mathcal{Y}| < \delta$ and each allocation Λ of $\mathcal{Y}$, we define

$$\int_a^b f(\lambda)\, dE_\lambda := B.$$

Letting $a \to -\infty$, $b \to +\infty$ in (2.14), we obtain, due to the completeness of $(u_k)_{k\in\mathbb{N}}$, that

$$\int_{-\infty}^{\infty} dE_\lambda = \mathrm{id}_H, \qquad \int_{-\infty}^{\infty} \lambda\, dE_\lambda = A, \tag{2.15}$$

where id_H denotes the identity on H. In this formulation, the spectral decomposition can be generalized to selfadjoint operators without a complete system of eigenfunctions:

Theorem 2.4 *Let $A : D(A) \to H$ be selfadjoint. Then there exists a spectral family $(E_\lambda)_{\lambda\in\mathbb{R}}$ with the properties α) to δ) above, such that (2.15) holds and moreover,*

$$D(A) = \left\{ u \in H : \int_{-\infty}^{\infty} \lambda^2\, d(\|E_\lambda u\|^2) < \infty \right\}.$$

For a proof of this theorem, see [32, p. 358 ff]. For general selfadjoint operators A, property (2.12) no longer holds true. More precisely, it can be shown that $\lambda_0 \in \mathbb{R}$ belongs to the continuous spectrum of A if and only if $\lambda \mapsto E_\lambda$ is continuous at λ_0 but not constant in each neighbourhood of λ_0. As before, $\lambda \mapsto E_\lambda$ jumps at the eigenvalues, and it is locally constant in the resolvent set.

3 Numerical approximation methods for eigenpairs

As we have seen already in the examples in Chapter 1, many (or even most) eigenvalue problems cannot be solved in closed form, so that numerical methods for their approximate solution are required. Here, for simplicity of presentation, we restrict ourselves to problems with a complete orthonormal system of eigenfunctions, i.e., we exclude problems with continuous spectrum. In fact, some of the numerical methods described here can also be applied in the presence of continuous spectrum, if the eigenvalues to be approximated are "away" from it. The continuous spectrum itself is hard to approximate, since every numerical discretization method will dissolve it into a "crowd" of eigenvalues. In practice, one often computes approximations on several levels of discretization accuracy, and looks on the one hand for eigenvalues which are "stable" (i.e., do not change much if the discretization is improved) and on the other hand for "crowds" of eigenvalues which stay "stable" as a whole, without the single members of the crowd being stable. Then, the "stable" eigenvalues are regarded as approximations to exact eigenvalues, while the "crowds" are assumed to approximate continuous spectrum.

Let $(H, \langle \cdot, \cdot \rangle)$ be a separable Hilbert space, $\dim H = \infty$, $D(A) \subset H$ a dense subspace, and $A : D(A) \to H$ a linear, symmetric operator. We assume that A has an orthonormal and complete system (u_k) of eigenfunctions, and that the eigenvalue sequence (λ_k) has no finite accumulation point, so that essentially two cases can occur:

a) (λ_k) is bounded below, so that it can be assumed to be ordered by magnitude: $\lambda_1 \leq \lambda_2 \leq \cdots \leq \lambda_k \leq \lambda_{k+1} \to +\infty$. The problems investigated in Section 2.3, are examples for this situation.
The case where (λ_k) is bounded *above* (and thus, not below) can be reduced to this case by considering $-A$ in place of A.

b) (λ_k) is neither bounded below nor above. It is then more natural to index it by $\mathbb{Z}$ rather than $\mathbb{N}$, so that it can again be assumed to be ordered by magnitude:

$$-\infty \leftarrow \lambda_{-k-1} \leq \lambda_{-k} \leq \cdots \leq \lambda_{-1} \leq \lambda_0 \leq \lambda_1 \leq \cdots \leq \lambda_k \leq \lambda_{k+1} \to +\infty.$$

In view of the numerical methods to be presented, one must be well aware of the fact that this indexing is not unique! Every shifted indexing ($k \to k + l$ for some fixed $l \in \mathbb{Z}$) is as "natural". Expressed a bit differently, *each* eigenvalue may be called λ_0 (which fixes then the remaining indices).
An example for the situation in case b) is the momentum operator from quantum mechanics with periodic boundary conditions: $Au = \frac{1}{i} u'$, $D(A) = \{u \in C_2[0, 2\pi] : u(0) = u(2\pi)\}$, with the eigenvalues $\lambda_k = k$ and eigenfunctions $u_k(x) = \exp(ikx)$ $(k \in \mathbb{Z})$.

Before discussing numerical methods we have to ask what we *expect* from a numerical approximation method. Since infinitely many eigenpairs cannot be approximated, we can go for

- a *single* eigenpair (u_K, λ_K)
  - with fixed index K (which causes problems in case b), due to the non-unique indexing);
  - with λ_K closest to some given $\mu \in \mathbb{R}$ (e.g., $\mu = 0$); see the inverse power method explained in Section 3.2;
- a *group* $(u_K, \lambda_K), \ldots, (u_{K+N-1}, \lambda_{K+N-1})$ of eigenpairs with fixed N and
  - with fixed starting index K (which again causes problems in case

b)); in case a) one usually takes $K = 1$; see the Rayleigh–Ritz method explained in Section 3.1;

- with $\lambda_K, \ldots, \lambda_{K+N-1}$ closest (in an appropriate sense) to some given $\mu \in \mathbb{R}$ (e.g., $\mu = 0$); see the inverse subspace iteration method explained in Section 3.3.

3.1 The Rayleigh–Ritz method

Here, we assume that the underlying situation is the one in case a). The Rayleigh–Ritz method is well known as a very accurate method for approximating the first N eigenvalues $\lambda_1, \ldots, \lambda_N$. For this purpose, one chooses $M \geq N$ linearly independent trial functions $\tilde{u}_1, \ldots, \tilde{u}_M \in D(A)$ (for instance, finite element basis functions in the case of a differential equation problem) and looks for approximate eigenpairs (U_k, Λ_k) in the form

$$U_k = \sum_{i=1}^{M} x_i^{(k)} \tilde{u}_i, \quad \text{with } x_i^{(k)} \in \mathbb{C},$$

by requiring that $AU_k - \Lambda_k U_k$ is orthogonal to $\text{span}\{\tilde{u}_1, \ldots, \tilde{u}_M\}$ (which will hopefully make it "small"), i.e., that

$$\sum_{i=1}^{M} \langle A\tilde{u}_i, \tilde{u}_j \rangle x_i^{(k)} = \Lambda_k \sum_{i=1}^{M} \langle \tilde{u}_i, \tilde{u}_j \rangle x_i^{(k)} \qquad (j, k = 1, \ldots, M).$$

Thus, one forms the $M \times M$-matrices

$$A_1 := (\langle A\tilde{u}_i, \tilde{u}_j \rangle)_{i,j=1,\ldots,M}, \quad A_2 := (\langle \tilde{u}_i, \tilde{u}_j \rangle)_{i,j=1,\ldots,M}, \tag{3.1}$$

and solves (approximately) the symmetric matrix eigenvalue problem

$$A_1 x = \Lambda A_2 x, \qquad x \in \mathbb{R}^M \setminus \{0\}, \tag{3.2}$$

e.g., by QR, Lanczos, or inverse power methods. With $\Lambda_1, \ldots, \Lambda_M$ denoting its eigenvalues, ordered by magnitude, one usually finds that $\Lambda_1, \ldots, \Lambda_N$ approximate $\lambda_1, \ldots, \lambda_N$ very well, if N is not too large (say, $N \leq M/2$). Moreover, with $x^{(1)}, \ldots, x^{(M)} \in \mathbb{R}^M$ denoting eigenvectors satisfying

$$(x^{(i)})^\top A_2 x^{(j)} = \delta_{ij}, \quad (x^{(i)})^\top A_1 x^{(j)} = \Lambda_i \delta_{ij},$$

the elements $U_k := \sum_{i=1}^{M} x_i^{(k)} \tilde{u}_i$ $(k = 1, \ldots, N)$ approximate $u_1, \ldots, u_N$ (possibly up to orthogonal transformation within each of the eigenspaces). Actually, the quality of the approximations depends of course on the choice of the trial functions $\tilde{u}_1, \ldots, \tilde{u}_M$: the true eigenelements $u_1, \ldots, u_N$ should be "not far" from $\text{span}\{\tilde{u}_1, \ldots, \tilde{u}_M\}$. This requirement may cause difficulties in situations where the true eigenelements have some kinds of singularities which are hard to model by trial functions (e.g., for singular Sturm–Liouville problems in the oscillatory case, where the eigenfunctions oscillate infinitely often close to one endpoint).

3.2 The inverse power method

Here, we want to approximate the eigenpair (u_K, λ_K) with λ_K closest to some given $\mu \in \mathbb{R}$, no matter if case a) or case b) is underlying. We assume that $D(A) \subset (A-\mu)(D(A))$, which is true for most differential equation problems if μ is no eigenvalue (see condition i) below).

We choose some starting approximation $U^0 \in D(A)$ for u_K, $\|U^0\| = 1$, and assume that the following conditions hold true:

i) $0 < |\lambda_K - \mu| < |\lambda_k - \mu|$ for all $k \neq K$,
ii) $\langle U^0, u_K \rangle \neq 0$.

Both conditions are hard to control safely a priori. Condition i) requires μ to be a rough (but not too rough) approximation for λ_K already; moreover, it requires λ_K to be *simple*. Condition ii) is "very likely" to be satisfied if U^0 is chosen randomly; the numerical errors made in the following iterative procedure will moreover usually *help* to satisfy condition ii). In practice, one will often not care too much about conditions i) and ii), and see if the procedure converges "by numerical evidence", even if a convergence *proof* needs both conditions.

Starting with U^0, we construct a sequence $(U^\nu)_{\nu \in \mathbb{N}}$ recursively as follows:

α) compute $V^{\nu+1} \in D(A)$ satisfying $(A-\mu)V^{\nu+1} = U^\nu$;
β) define $U^{\nu+1} := V^{\nu+1}/\|V^{\nu+1}\|$.

Observe that α) can indeed be carried out since $D(A) \subset (A-\mu)(D(A))$. If A is a differential operator, this requires the solution of a linear boundary value problem (for each ν), which can practically be solved only approximately, e.g., by a finite element procedure.

Theorem 3.1 *There exists some $c \in \mathbb{C}$, $|c| = 1$, such that either $(U^\nu)_{\nu \in \mathbb{N}}$ or $((-1)^\nu U^\nu)_{\nu \in \mathbb{N}}$ converges to cu_K.*

Proof. Defining $\psi^{(\nu)} := \sum_k \left(\frac{1}{\lambda_k - \mu}\right)^\nu \langle U^0, u_k \rangle u_k$ we first show that

$$U^\nu = \frac{\psi^{(\nu)}}{\|\psi^{(\nu)}\|} \qquad (\nu \in \mathbb{N}_0). \tag{3.3}$$

For $\nu = 0$, (3.3) holds true since (u_k) is complete and $\|U^0\| = 1$.

If (3.3) holds for some $\nu \in \mathbb{N}$, we find

$$\langle U^\nu, u_k \rangle = (\lambda_k - \mu)^{-\nu} \langle U^0, u_k \rangle / \|\psi^{(\nu)}\|$$

and moreover, with $V^{\nu+1}$ computed in α),

$$(\lambda_k - \mu)\langle V^{\nu+1}, u_k \rangle = \langle V^{\nu+1}, (A-\mu)u_k \rangle = \langle (A-\mu)V^{\nu+1}, u_k \rangle = \langle U^\nu, u_k \rangle,$$

so that $\langle V^{\nu+1}, u_k\rangle = (\lambda_k - \mu)^{-(\nu+1)}\langle U^0, u_k\rangle / \|\psi^{(\nu)}\|$ and thus,

$$V^{\nu+1} = \sum_k \langle V^{\nu+1}, u_k\rangle u_k = \frac{1}{\|\psi^{(\nu)}\|}\sum_k \left(\frac{1}{\lambda_k - \mu}\right)^{\nu+1} \langle U^0, u_k\rangle u_k = \frac{\psi^{(\nu+1)}}{\|\psi^{(\nu)}\|}.$$

Consequently, $U^{\nu+1} = V^{\nu+1}/\|V^{\nu+1}\| = \psi^{(\nu+1)}/\|\psi^{(\nu+1)}\|$, i.e., (3.3) holds for $\nu + 1$. Moreover,

$$\begin{aligned}
\|(\lambda_K - \mu)^\nu \psi^{(\nu)} - \langle U^0, u_K\rangle u_K\|^2 &= \left\| \sum_{k\neq K} \left(\frac{\lambda_K - \mu}{\lambda_k - \mu}\right)^\nu \langle U^0, u_k\rangle u_k \right\|^2 \\
&= \sum_{k\neq K} \left(\frac{\lambda_K - \mu}{\lambda_k - \mu}\right)^{2\nu} |\langle U^0, u_k\rangle|^2 \\
&\leq \max_{k\neq K} \left(\frac{\lambda_K - \mu}{\lambda_k - \mu}\right)^{2\nu} \to 0 \quad (\nu \to \infty)
\end{aligned}$$

due to assumption i). Consequently,

$$(\lambda_K - \mu)^\nu \psi^{(\nu)} \to \langle U^0, u_K\rangle u_K \quad \text{for} \quad \nu \to \infty. \tag{3.4}$$

Using (3.3), (3.4), and assumption ii), we obtain

$$\left(\frac{\lambda_K - \mu}{|\lambda_K - \mu|}\right)^\nu U^\nu \to \frac{\langle U^0, u_K\rangle u_K}{\|\langle U^0, u_K\rangle u_K\|} = \frac{\langle U^0, u_K\rangle}{|\langle U^0, u_K\rangle|} u_K. \qquad \blacksquare$$

In practice, the above iteration α), β) is stopped when either $\|U^{\nu+1} - U^\nu\|$ or $\|U^{\nu+1} + U^\nu\|$ is below some tolerance. Then $U_K := U^{\nu+1}$ is regarded as an approximation for u_K, and an eigenvalue approximation Λ_K is computed, e.g., by use of the Rayleigh quotient: $\Lambda_K := \langle AU_K, U_K\rangle$.

3.3 Inverse subspace iteration

We now want to approximate the eigenpairs

$$(u_K, \lambda_K), \ldots, (u_{K+N-1}, \lambda_{K+N-1}),$$

with $\lambda_K, \ldots, \lambda_{K+N-1}$ closest to some given $\mu \in \mathbb{R}$, in one go. Again, we allow both cases a) and b), and we assume that $D(A) \subset (A - \mu)(D(A))$.

We now choose starting approximations $U_1^0, \ldots, U_N^0 \in D(A)$ for $u_K, \ldots, u_{K+N-1}$ satisfying $\langle U_i^0, U_j^0\rangle = \delta_{ij}$ $(i, j = 1, \ldots, N)$. The conditions i) and ii) in Section 3.2 are now replaced by

i) $0 < |\lambda_j - \mu| < |\lambda_k - \mu|$ for all $j \in \{K, \ldots, K + N - 1\}$, $k \notin \{K, \ldots, K + N - 1\}$,

ii) $\sum_{k=K}^{K+N-1} \langle U_j^0, u_k\rangle u_k$ $(j = 1, \ldots, N)$ are linearly independent.

The remarks made in the previous section after i) and ii) hold here as well, except that now multiple eigenvalues are also allowed.

Starting with $(U_1^0, \dots, U_N^0)$, we now construct a sequence $(U_1^\nu, \dots, U_N^\nu)$, $\nu \in \mathbb{N}$, recursively as follows:

α) compute $V_1^{\nu+1}, \dots, V_N^{\nu+1}$ satisfying $(A-\mu)V_j^{\nu+1} = U_j^\nu$ $(j = 1, \dots, N)$;

β) orthonormalize $V_1^{\nu+1}, \dots, V_N^{\nu+1}$ to obtain $U_1^{\nu+1}, \dots, U_N^{\nu+1}$.

In the case of a differential operator A, part α) requires the solution of N linear boundary value problems (in each iteration step), which in practice is carried out approximately. The orthonormalization required in β) can simply be carried out by the Schmidt process, but a better result is achieved if, at least once in every few (say, five) iteration steps, the orthonormalization is carried out by the Rayleigh–Ritz method, with $V_1^{\nu+1}, \dots, V_N^{\nu+1}$ as trial functions.

In analogy to Theorem 3.1, one can obtain convergence results here, too: the subspaces $\operatorname{span}\{U_1^\nu, \dots, U_N^\nu\}$ converge in an appropriate sense to $\operatorname{span}\{u_K, \dots, u_{K+N-1}\}$. In practice, one will stop the iteration when the distance between $\operatorname{span}\{U_1^\nu, \dots, U_N^\nu\}$ and $\operatorname{span}\{U_1^{\nu+1}, \dots, U_N^{\nu+1}\}$ (measured, e.g., by the distance between their unit spheres) is below some tolerance; however, the final orthonormalization step β) should be a Rayleigh–Ritz process (rather than a simple Schmidt process), which then provides approximate eigenpairs $(U_{K+j-1}, \Lambda_{K+j-1})$ $(j = 1, \dots, N)$.

4 Eigenvalue enclosures

Let $(H, \langle \cdot, \cdot \rangle)$ denote a pre-Hilbert-space, $D(A) \subset H$ a dense linear subspace, and $A : D(A) \to H$ a linear symmetric operator. Again, we consider the eigenvalue problem

$$Au = \lambda u, \quad u \in D(A) \setminus \{0\}. \tag{4.1}$$

For simplicity and unity of presentation, we make the (relaxable) assumptions that $\dim H = \infty$, that H possesses an orthonormal basis $(u_n)_{n \in \mathbb{N}}$ of eigenelements $u_n \in D(A)$ of A, and that the corresponding eigenvalue sequence $(\lambda_n)_{n \in \mathbb{N}}$ converges to $+\infty$, so that it may be assumed to be ordered by magnitude, i.e., case a) is underlying (see Chapter 3).

Our aim is the computation of upper and lower *bounds* (i.e., *enclosure intervals*) for $\lambda_1, \dots, \lambda_N$, with $N \in \mathbb{N}$ denoting some given number.

Possibly the simplest method for computing eigenvalue enclosures is due to D. Weinstein (see [11, Corollary 6.20]); we will describe it in Section 4.1. It is based on an *approximate* eigenpair $(\tilde{u}, \tilde{\lambda}) \in D(A) \times \mathbb{R}$ and on its *defect* $\delta := \|A\tilde{u} - \tilde{\lambda}\tilde{u}\| / \|\tilde{u}\|$, and states that $[\tilde{\lambda} - \delta, \tilde{\lambda} + \delta]$ contains an eigenvalue. However, it does not provide information about *which* eigenvalue that is, i.e., which index it has. Thus, our task of computing enclosure intervals for the *first* N eigenvalues cannot be solved by this method alone.

This situation is no better (indeed, even worse) in Kato's method [23], which we will describe in Section 4.2. This method is superior to D. Weinstein's method in the *accuracy* of the eigenvalue bounds – the diameter of the enclosure interval is, roughly speaking, proportional to δ^2 instead of δ – but it too is unable to provide the index of the enclosed eigenvalue; it even *requires* the index as pre-information.

For both D. Weinstein's and Kato's methods, we will also mention generalized versions providing intervals enclosing *several* eigenvalues (which are needed in the case of multiple or clustered eigenvalues), but this does not solve the index problem. It finally turns out that a lower bound for λ_{N+1} is needed as pre-information to make sure that the enclosed eigenvalues have the "right" indices.

A method which does not need any spectral pre-information and which also yields very accurate – but only *upper* – eigenvalue bounds (including the index information) is the Rayleigh–Ritz method which we discussed already as an *approximation* method; we will return to it in Section 4.3 (see, e.g., [31, Theorem 40.1 and Remarks 40.1, 40.2, 39.10]). It generalizes the fact that the Rayleigh quotient of any trial function (or approximate eigenfunction) $\tilde{u} \in D(A) \setminus \{0\}$ is an upper bound for the first eigenvalue, to the statement that some kinds of "generalized Rayleigh quotients" (given as the eigenvalues of some matrix eigenvalue problem), formed with N linearly independent trial functions (or approximate eigenfunctions), are upper bounds for the first N eigenvalues.

The computation of *lower* bounds of comparable quality is a much harder task. A method which is *optimal* in a certain sense was proposed by Lehmann [27]. Like the Rayleigh–Ritz method, it is based on a variational principle, it uses N linearly independent trial functions (approximate eigenfunctions), and it requires the computation of bounds to some matrix eigenvalues. However, it provides *lower* eigenvalue bounds, and it needs, as D. Weinstein's and Kato's methods, a lower bound for λ_{N+1} as pre-information. Besides this essential problem, its practical realization often involves another difficulty, not so much for problems of type (4.1), but for generalized eigenvalue problems involving a second operator B on the right-hand side. The latter difficulty was resolved by Goerisch (and his co-workers) in a famous way by his "XbT-concept" (see [16–19]). We will describe Lehmann's method in Section 4.4.

The remaining fundamental problem of obtaining – a priori – a lower bound for λ_{N+1} can sometimes be solved by means of a *comparison problem*: an operator A_0 has to be constructed such that sufficiently accurate (lower) bounds for its eigenvalues λ_n^0 are available (e.g., by solution in closed form), and moreover, the inequalities $\lambda_n^0 \leq \lambda_n$ $(n \in \mathbb{N})$ hold true (which might follow, e.g., from variational characterizations). For many problems, however, the lower bound for λ_{N+1} obtained in this way is too rough. To cover such cases we propose a *homotopy method* connecting A_0

and $A =: A_1$ by a family $(A_s)_{0\le s\le 1}$ of operators such that, for all $n \in \mathbb{N}$, the eigenvalue λ_n^s increases in s. Following the homotopy in small steps and choosing, at each step, the problem obtained from the previous step as a comparison problem (in the sense explained above) we succeed in computing the desired bounds. The homotopy method will be explained in more detail in Section 4.5.

Another method (or class of methods) using a simple "base problem" and a connection between the base and given problem, is the method of *intermediate problems* which has been developed by A. Weinstein [33, 34] and Aronszajn [4], and later by Bazley and Fox [6, 7] and Beattie [8, 9]. These intermediate problems are constructed in a way which is more specific but usually also more restrictive than the requirements of the general homotopy method mentioned above. For instance, the intermediate problems approach assumes explicit knowledge of eigenvalues *and eigenelements* of the base problem, which causes difficulties, e.g., for many partial differential equation problems. Furthermore, this approach does not provide the optimality properties of Lehmann's method.

In Section 4.6, we will briefly comment on the numerical tools needed to implement the D. Weinstein, Kato, Rayleigh–Ritz, and Lehmann bounds. Section 4.7 will contain some numerical examples with differential equation problems.

4.1 D. Weinstein's bounds

Possibly the simplest result on eigenvalue bounds for problem (4.1) is the following Theorem by D. Weinstein (see [11, Corollary 6.20]):

Theorem 4.1 *Let $\tilde{u} \in D(A) \setminus \{0\}$ and $\tilde{\lambda} \in \mathbb{R}$. Moreover, let $\delta := \|A\tilde{u} - \tilde{\lambda}\tilde{u}\|/\|\tilde{u}\|$. Then, the interval $[\tilde{\lambda} - \delta, \tilde{\lambda} + \delta]$ contains at least one eigenvalue of problem (4.1).*

Proof. $\|A\tilde{u} - \tilde{\lambda}\tilde{u}\|^2 = \sum_{n=1}^{\infty} \langle A\tilde{u} - \tilde{\lambda}\tilde{u}, u_n\rangle^2 = \sum_{n=1}^{\infty} (\lambda_n - \tilde{\lambda})^2 \langle \tilde{u}, u_n\rangle^2 \ge \min_{n\in\mathbb{N}} |\lambda_n - \tilde{\lambda}|^2 \|\tilde{u}\|^2$. ∎

Usually, one will apply this theorem with $(\tilde{u}, \tilde{\lambda})$ denoting some *approximate* eigenpair obtained, e.g., by numerical means. Depending on the accuracy of this approximation, the defect δ will then be "small", so that the enclosing interval has a small diameter.

This simple result has some drawbacks. The most serious (but – as we will see – natural) one is the lack of information about *which* eigenvalue is enclosed in $[\tilde{\lambda} - \delta, \tilde{\lambda} + \delta]$, i.e., which index it has. We postpone the discussion of this problem. Next, the quality of the bounds $(O(\delta))$ is not too good. Thus, there is need for other methods which we will report on in the coming sections. Finally, Theorem 4.1 does not provide enough information if two (or more) enclosing intervals overlap. It is in general then *not* true that their union contains at least two eigenvalues, as can be seen by simple matrix examples (see [28, p. 218]). The following result is a generalization of Theorem 4.1 covering such situations (occurring in the case of multiple or clustered eigenvalues).

Theorem 4.2 *Let $\tilde{u}_1, \ldots, \tilde{u}_l \in D(A)$ be linearly independent and suppose that the maximal eigenvalue ρ of the $l \times l$-matrix $I - U$, where $U_{ij} := \langle \tilde{u}_i, \tilde{u}_j \rangle / (\|\tilde{u}_i\| \cdot \|\tilde{u}_j\|)$, satisfies $\rho < 1$. Moreover, let $\tilde{\lambda}_1, \ldots, \tilde{\lambda}_l \in \mathbb{R}$ be ordered by magnitude, and let $\delta_i := \|A\tilde{u}_i - \tilde{\lambda}_i \tilde{u}_i\| / \|\tilde{u}_i\|$ $(i = 1, \ldots, l)$, $\delta := (1-\rho)^{-1/2} \left[\sum_{i=1}^{l} \delta_i^2\right]^{1/2}$. Then, the interval $[\tilde{\lambda}_1 - \delta, \tilde{\lambda}_l + \delta]$ contains at least l eigenvalues of problem (4.1). More precisely,*

$$\# \{n \in \mathbb{N} : \lambda_n \in [\tilde{\lambda}_1 - \delta, \tilde{\lambda}_l + \delta]\} \geq l$$

The proof of Theorem 4.2 is elementary but a bit lengthy. It can be found in [29, Theorem 4].

Theorem 4.2 will usually be applied with $(\tilde{\lambda}_i, \tilde{u}_i)$ $(i = 1, \ldots, l)$ denoting approximate eigenpairs, with $\tilde{\lambda}_1, \ldots, \tilde{\lambda}_l$ being clustered (i.e., closely neighbouring) so that the intervals $[\tilde{\lambda}_i - \delta_i, \tilde{\lambda}_i + \delta_i]$ (compare Theorem 4.1) overlap, and with $\tilde{u}_1, \ldots, \tilde{u}_l$ being approximately orthogonal (as provided by all "good" numerical approximation methods) so that the matrix $I - U$ in Theorem 4.2 is close to the zero matrix, and ρ is therefore "small". Thus, depending on the accuracy of the approximations, the enclosing interval $[\tilde{\lambda}_1 - \delta, \tilde{\lambda}_l + \delta]$ again has a small diameter (at least if l is not too large and the cluster is not too widely spread).

Concerning the question of the *indices* of the enclosed eigenvalues, or the problem of enclosing the *first* N eigenvalues, we now suppose that some numerical method has provided approximate eigenpairs $(\tilde{\lambda}_1, \tilde{u}_1), \ldots, (\tilde{\lambda}_N, \tilde{u}_N) \in \mathbb{R} \times (D(A) \setminus \{0\})$, with $\tilde{u}_1, \ldots, \tilde{u}_N$ approximately orthogonal. Using Theorems 4.1 and 4.2 we are then able to compute enclosures for – altogether – N eigenvalues of problem (4.1): after computing (upper bounds for) the defects $\delta_1, \ldots, \delta_N$, we first check if the intervals $[\tilde{\lambda}_i - \delta_i, \tilde{\lambda}_i + \delta_i]$ $(i = 1, \ldots, N)$ are pairwise disjoint. If so, they enclose N eigenvalues according to Theorem 4.1. Otherwise, we collect clusters, indicated by overlapping intervals, and apply Theorem 4.2 to them. If the new intervals enclosing the clusters are disjoint from the others and from each other, we have enclosed N eigenvalues altogether. If not, we have to enlarge the clusters, compute new enclosing intervals for them, and check disjointness again. Finally, this process must be successful, at the latest when only one big cluster is left which then contains N eigenvalues.

Let $\bar{\lambda}_N$ denote the maximal point of the intervals obtained in this way, enclosing altogether N eigenvalues of problem (4.1). If we have the a priori information that the eigenvalue λ_{N+1} of problem (4.1) satisfies

$$\lambda_{N+1} > \bar{\lambda}_N, \tag{4.2}$$

then it is easy to conclude that the N enclosed eigenvalues are in fact $\lambda_1, \ldots, \lambda_N$, so that our task is fulfilled.

The fundamental problem of obtaining an a priori lower bound for λ_{N+1}, needed for (4.2), is postponed to Section 4.5.

4.2 Kato's bounds

The quality of D. Weinstein's bounds (measured by the diameter of the enclosing interval) is $O(\delta)$, with δ denoting the defect of some given approximate eigenpair. The following theorem of Kato [23, pp. 336], provides an $O(\delta^2)$ bound which – in computational practice – is not very much different from the quality of Lehmann's optimal bounds (see Section 4.4).

Theorem 4.3 *Let* $\tilde{u} \in D(A) \setminus \{0\}$ *and* $\tilde{\lambda} := \langle A\tilde{u}, \tilde{u}\rangle / \langle \tilde{u}, \tilde{u}\rangle$. *Moreover, let* $\delta := \|A\tilde{u} - \tilde{\lambda}\tilde{u}\| / \|\tilde{u}\| = [\|A\tilde{u}\|^2\|\tilde{u}\|^2 - \langle A\tilde{u}, \tilde{u}\rangle^2]^{1/2} / \|\tilde{u}\|^2$. *Finally, suppose that* $\mu, \nu \in \mathbb{R}$ *are known such that, for some* $n \in \mathbb{N}$,

$$\lambda_{n-1} \leq \mu < \tilde{\lambda} < \nu \leq \lambda_{n+1} \tag{4.3}$$

(with $\lambda_0 := \mu := -\infty$ *if* $n = 1$*). Then*

$$\tilde{\lambda} - \frac{\delta^2}{\nu - \tilde{\lambda}} \leq \lambda_n \leq \tilde{\lambda} + \frac{\delta^2}{\tilde{\lambda} - \mu}. \tag{4.4}$$

Proof. For $\rho \in \{\mu, \nu\}$ and all $i \in \mathbb{N}$, we have $(\lambda_i - \lambda_n)(\lambda_i - \rho)\langle \tilde{u}, u_i\rangle^2 \geq 0$. Summation over i yields $\|A\tilde{u}\|^2 - (\lambda_n + \rho)\langle A\tilde{u}, \tilde{u}\rangle + \lambda_n \rho \|\tilde{u}\|^2 \geq 0$, i.e., $\delta^2 + \tilde{\lambda}^2 - (\lambda_n + \rho)\tilde{\lambda} + \lambda_n \rho \geq 0$. Solving for λ_n we obtain the assertion. ∎

In the case of well separated (not clustered) eigenvalues, the a priori bounds μ and ν satisfying (4.3) can be obtained, for instance, from D. Weinstein's bounds for λ_{n-1} and λ_{n+1}. Of course, the index information on the Weinstein bounds is then required. Thus, according to the discussion in the previous section, a lower bound for λ_{N+1} is needed as a priori information. (For $n = N$, this lower bound can serve directly as ν in (4.3).)

In the case of clustered eigenvalues, one needs a generalized version of Theorem 4.3. Kato's original paper [23] contains such a result which, however, has to be modified for computational practice since it assumes approximate eigenfunctions $\tilde{u}_1, \ldots, \tilde{u}_l$ satisfying $\langle \tilde{u}_i, \tilde{u}_j\rangle = \langle A\tilde{u}_i, \tilde{u}_j\rangle = 0$ $(i \neq j)$ *exactly*. Corresponding modifications can be found in [29], Theorem 5, and [30], Theorem 2. The latter version is moreover formulated in a way which avoids the auxiliary use of D. Weinstein's bounds. Nevertheless, it too needs a lower bound for λ_{N+1} as a priori information.

Finally, we wish to remark that, in particular, Theorem 4.3 states that, for any $\tilde{u} \in D(A) \setminus \{0\}$ and any $\nu \in \mathbb{R}$ satisfying $\langle A\tilde{u}, \tilde{u}\rangle / \langle \tilde{u}, \tilde{u}\rangle < \nu \leq \lambda_2$,

$$\frac{\nu\langle A\tilde{u}, \tilde{u}\rangle - \langle A\tilde{u}, A\tilde{u}\rangle}{\nu\langle \tilde{u}, \tilde{u}\rangle - \langle A\tilde{u}, \tilde{u}\rangle} \leq \lambda_1 \leq \frac{\langle A\tilde{u}, \tilde{u}\rangle}{\langle \tilde{u}, \tilde{u}\rangle}. \tag{4.5}$$

This is the well known fact that λ_1 is enclosed between the Temple quotient and the Rayleigh quotient. (Observe again that a lower bound for λ_2 is needed as a priori information!) Furthermore, (4.5) is a special case – and in some sense the basis – of the Rayleigh–Ritz method (right inequality in (4.5)) and the Lehmann method (left inequality in (4.5)) which we will discuss in the next two sections.

4.3 Rayleigh–Ritz bounds

We have already introduced the Rayleigh–Ritz method in Section 3.1 as a method for *approximating* $(u_1, \lambda_1), \ldots, (u_N, \lambda_N)$. However, the Rayleigh–Ritz method not only provides approximations but *upper bounds* to the eigenvalues, without needing any a priori information: with $\Lambda_1, \ldots, \Lambda_M$ denoting the eigenvalues of problem (3.2), ordered by magnitude, and with A_1, A_2 formed in (3.1) using linearly independent trial functions $\tilde{u}_1, \ldots, \tilde{u}_M$, we have

Theorem 4.4 $\lambda_n \leq \Lambda_n$ *for* $n = 1, \ldots, M$. (4.6)

Of course the practical application of Theorem 4.4 requires the computation of upper bounds for the eigenvalues $\Lambda_1, \ldots, \Lambda_N$ of the *matrix* eigenvalue problem (3.2). To facilitate this nontrivial task, it is often advantageous first to compute approximate eigenpairs (U_n, Λ_n) $(n = 1, \ldots, N)$ as described in Chapter 3, with general trial functions $\tilde{u}_1, \ldots, \tilde{u}_M$, and then set up (3.1) and (3.2) anew, now with $M := N$ and trial functions $\tilde{u}_n := U_n$ $(n = 1, \ldots, N)$. Then, $A_1 \approx \operatorname{diag}(\Lambda_1, \ldots, \Lambda_N)$ and $A_2 \approx I$ (these relations do not hold exactly since the numerical solution of the "old" problem (3.2) involves numerical errors) and the computation of bounds to the eigenvalues of this "new" problem (3.2) can be carried out rather easily (see Section 4.6).

The proof of Theorem 4.4 is based on the following variational principle, which we will use again in Section 4.5.

Theorem 4.5 *For all* $n \in \mathbb{N}$,

$$\lambda_n = \min_{\substack{U \subset D(A) \text{ subspace} \\ \dim U = n}} \max_{u \in U \setminus \{0\}} \frac{\langle Au, u\rangle}{\langle u, u\rangle}. \tag{4.7}$$

Proof. To prove "$\leq$", let $U \subset D(A)$ denote some arbitrary n-dimensional subspace. We choose some $u \in U \setminus \{0\}$ satisfying $\langle u, u_i\rangle = 0$ for $i = 1, \ldots,$

$n-1$. Then,

$$\frac{\langle Au,u\rangle}{\langle u,u\rangle} = \left(\sum_{i=n}^{\infty} \lambda_i \langle u,u_i\rangle^2\right) \Big/ \left(\sum_{i=n}^{\infty} \langle u,u_i\rangle^2\right) \geq \lambda_n .$$

To prove "$\geq$", let $U := \operatorname{span}\{u_1,\ldots,u_n\}$. Then, for all $u \in U \setminus \{0\}$,

$$\frac{\langle Au,u\rangle}{\langle u,u\rangle} = \left(\sum_{i=1}^{n} \lambda_i \langle u,u_i\rangle^2\right) \Big/ \left(\sum_{i=1}^{n} \langle u,u_i\rangle^2\right) \leq \lambda_n .$$

■

Proof of Theorem 4.4. Let $U_1,\ldots,U_M$ be given as indicated after (3.2). Then,

$$\langle U_i,U_j\rangle = \delta_{ij}, \quad \langle AU_i,U_j\rangle = \Lambda_i\delta_{ij} \quad (i,j=1,\ldots,M). \tag{4.8}$$

Let $n \in \{1,\ldots,M\}$ be given, and define $U := \operatorname{span}\{U_1,\ldots,U_n\}$. Using Theorem 4.5 and (4.8) we obtain

$$\begin{aligned}
\lambda_n &\leq \max_{u\in U\setminus\{0\}} \frac{\langle Au,u\rangle}{\langle u,u\rangle} \\
&= \max_{c\in\mathbb{R}^n\setminus\{0\}} \frac{\sum_{i,j=1}^{n} c_i c_j \langle AU_i,U_j\rangle}{\sum_{i,j=1}^{n} c_i c_j \langle U_i,U_j\rangle} = \max_{c\in\mathbb{R}^n\setminus\{0\}} \frac{\sum_{i=1}^{n} \Lambda_i c_i^2}{\sum_{i=1}^{n} c_i^2} \leq \Lambda_n .
\end{aligned}$$

■

4.4 Lehmann's method

As the Rayleigh–Ritz method generalizes the right-hand inequality in (4.5), the Lehmann method generalizes the left-hand one. It also uses linearly independent trial functions $\tilde{u}_1,\ldots,\tilde{u}_N \in D(A)$ (which now may be regarded as approximate eigenfunctions from the beginning) and yields lower bounds to $\lambda_1,\ldots,\lambda_N$, provided that some $\nu \in \mathbb{R}$ satisfying

$$\Lambda_N < \nu \leq \lambda_{N+1}, \tag{4.9}$$

with Λ_N denoting the largest eigenvalue of problem (3.2), is known a priori. (Compare the assumption $\langle A\tilde{u},\tilde{u}\rangle/\langle\tilde{u},\tilde{u}\rangle < \nu \leq \lambda_2$ needed for (4.5).) Besides the matrices A_1 and A_2 in (3.1) (now with $M=N$), we form further matrices

$$A_3 := (\langle A\tilde{u}_i, A\tilde{u}_j\rangle)_{i,j=1,\ldots,N}, \tag{4.10}$$

$$B_1 := A_1 - \nu A_2 = (\langle (A-\nu)\tilde{u}_i, \tilde{u}_j \rangle)_{i,j=1,\dots,N},$$
$$B_2 := A_3 - 2\nu A_1 + \nu^2 A_2 = (\langle (A-\nu)\tilde{u}_i, (A-\nu)\tilde{u}_j \rangle)_{i,j=1,\dots,N}, \tag{4.11}$$

and consider the matrix eigenvalue problem

$$B_1 x = \mu B_2 x. \tag{4.12}$$

Due to the left-hand inequality in (4.9), B_1 is negative definite. Moreover, B_2 is positive definite, since otherwise some nontrivial $\tilde{u} \in \operatorname{span}\{\tilde{u}_1, \dots, \tilde{u}_N\}$ satisfying $(A-\nu)\tilde{u} = 0$ would exist, which contradicts the negative definiteness of B_1. Thus, problem (4.12) has N negative eigenvalues $\mu_1 \le \mu_2 \le \dots \le \mu_N < 0$. Lehmann's theorem now reads:

Theorem 4.6 $\lambda_{N+1-n} \ge \nu + \dfrac{1}{\mu_n}$ *$(n = 1, \dots, N)$.*

The remarks made after Theorem 4.4 concerning the computation of the required matrix eigenvalue bounds (here, for problem (4.12)) hold here as well, with the restriction that A_3 usually deviates from a diagonal matrix not only by the errors made in the numerical solution of the "old" problem (3.2), but also by some defect terms of the approximate eigenpairs $(\tilde{u}_i, \Lambda_i)$.

The *optimality* of Lehmann's method (with respect to fixed trial functions $\tilde{u}_1, \dots, \tilde{u}_N$) has already been mentioned. This has to be understood in the sense that no better lower bounds can be produced by knowledge of $\tilde{u}_1, \dots, \tilde{u}_N$ and $A\tilde{u}_1, \dots, A\tilde{u}_N$ alone; see [27, Section 7], for details.

Proof of Theorem 4.6. Let $x^{(1)}, \dots, x^{(N)} \in \mathbb{R}^N$ denote eigenvectors of problem (4.12) satisfying $(x^{(i)})^\top B_2 x^{(j)} = \delta_{ij}$, $(x^{(i)})^\top B_1 x^{(j)} = \mu_i \delta_{ij}$, and define $U_i := \sum_{k=1}^{N} x_k^{(i)} \tilde{u}_k$ $(i = 1, \dots, N)$. Then,

$$\langle (A-\nu)U_i, (A-\nu)U_j \rangle = \delta_{ij}, \qquad \langle (A-\nu)U_i, U_j \rangle = \mu_i \delta_{ij}. \tag{4.13}$$

Let $n \in \{1, \dots, N\}$ be given, and choose $(c_1, \dots, c_n) \in \mathbb{R}^n \setminus \{0\}$ such that $u := \sum_{i=1}^{n} c_i U_i$ satisfies the $n-1$ homogeneous equations $\langle u, u_k \rangle = 0$ $(k = N+2-n, \dots, N)$. Then, since (4.9) provides $\lambda_i - \nu \ge 0$ for $i \ge N+1$ and, together with Theorem 4.4, $\lambda_i - \nu < 0$ for $i \le N$,

$$\begin{aligned}
\langle (A-\nu)u, u \rangle &\ge \sum_{i=1}^{N+1-n} (\lambda_i - \nu) \langle u, u_i \rangle^2 \\
&\ge \frac{1}{\lambda_{N+1-n} - \nu} \sum_{i=1}^{N+1-n} (\lambda_i - \nu)^2 \langle u, u_i \rangle^2
\end{aligned}$$

$$\geq \frac{1}{\lambda_{N+1-n}-\nu}\langle (A-\nu)u,(A-\nu)u\rangle. \tag{4.14}$$

On the other hand, (4.13) yields

$$\begin{aligned}\langle (A-\nu)u,u\rangle &= \sum_{i=1}^{n} c_i c_j \langle (A-\nu)U_i,U_j\rangle \\ &= \sum_{i=1}^{n}\mu_i c_i^2 \leq \mu_n \sum_{i=1}^{n} c_i^2 = \mu_n\langle (A-\nu)u,(A-\nu)u\rangle\end{aligned} \tag{4.15}$$

and $(A-\nu)u \neq 0$. (4.14) and (4.15) therefore provide $(\lambda_{N+1-n}-\nu)^{-1} \leq \mu_n$ and thus, since both sides of this inequality are negative, the assertion is proved. ∎

4.5 The homotopy method

In this section, we will discuss possibilities for the solution of the fundamental problem of calculating an a priori lower bound for λ_{N+1}, as needed for Weinstein's (see (4.2)), for Kato's (see the remarks after Theorem 4.3), and for Lehmann's (see (4.9)) methods, as well as for any other method providing lower eigenvalue bounds.

Since the required lower bound for λ_{N+1} need not be *very* accurate, it can sometimes be computed with the aid of a *comparison problem*.

Let $(H_0, \langle\cdot,\cdot\rangle_0)$ denote a second pre-Hilbert space, $D(A_0) \subset H_0$ a dense subspace, $A_0 : D(A_0) \to H_0$ a linear, symmetric operator, with the same general properties as in the original problem (4.1): $\dim H_0 = \infty$, H_0 has an orthonormal basis of eigenelements of A_0, the eigenvalue sequence $(\lambda_n^0)_{n\in\mathbb{N}}$ converges to $+\infty$ and is ordered by magnitude. The connection to the original problem is reflected in the following assumption:

$$H_0 \supseteq H,\ D(A_0) \supseteq D(A),\ \frac{\langle A_0 u,u\rangle_0}{\langle u,u\rangle_0} \leq \frac{\langle Au,u\rangle}{\langle u,u\rangle} \quad \text{for all } u \in D(A)\setminus\{0\} \tag{4.16}$$

implying the following

Theorem 4.7 $\lambda_n^0 \leq \lambda_n$ *for all* $n \in \mathbb{N}$.

Proof. In the variational characterization of λ_n given in Theorem 4.5, we diminish the Rayleigh quotient by the inequality in (4.16), and then diminish the *min* replacing $D(A)$ by the bigger space $D(A_0)$. The result is the variational characterization of λ_n^0. ∎

In particular, Theorem 4.7 provides $\lambda_{N+1}^0 \leq \lambda_{N+1}$. Our problem is therefore solved if we can compute a lower bound ν for λ_{N+1}^0 (e.g., by choosing A_0 such that its eigenvalues are known in closed form) which in turn has to satisfy the other required condition, e.g., $\nu > \Lambda_N$ for Lehmann's method (see (4.9)).

Since $\Lambda_N \geq \lambda_N$ due to Theorem 4.4, the *existence* of such a ν (regardless of its *computation*) requires

$$\lambda_N < \lambda^0_{N+1}, \tag{4.17}$$

so that A_0 must not be "too far away" from A. To choose A_0 having this property on the one hand, and allowing the computation of a lower bound ν for λ^0_{N+1} (by closed form solution or other means) on the other hand, is of course not always possible, even if many eigenvalue problems have been treated in this way (see [16–19] and Example 4 in Section 4.7).

The severe restriction (4.17) can be overcome by the *homotopy method* developed independently in [16] and in [29, 30]. Again we suppose that $(H_0, \langle\cdot,\cdot\rangle_0)$, $D(A_0)$, A_0 can be constructed such that (4.16) holds and lower bounds $\underline{\lambda}^0_n \leq \lambda^0_n$ are available, but we release (4.17) so that A_0 and A may be "far apart". We suppose that the two eigenvalue problems may be connected by a homotopy, i.e., that families $((H_s, \langle\cdot,\cdot\rangle_s))_{s\in[0,1]}$ of Hilbert spaces, $(D(A_s))_{s\in[0,1]}$ of dense subspaces $D(A_s) \subset H_s$, and $(A_s)_{s\in[0,1]}$ of linear, symmetric operators $A_s : D(A_s) \to H_s$ exist which "start" at $(H_0, \langle\cdot,\cdot\rangle_0)$, $D(A_0)$, A_0 and "end" at $(H_1, \langle\cdot,\cdot\rangle_1) = (H, \langle\cdot,\cdot\rangle)$, $D(A_1) = D(A)$, $A_1 = A$, and which satisfy

$$H_s \supseteq H_t,\ D(A_s) \supseteq D(A_t),\ \frac{\langle A_s u, u\rangle_s}{\langle u, u\rangle_s} \leq \frac{\langle A_t u, u\rangle_t}{\langle u, u\rangle_t} \quad \text{for all } u \in D(A_t)\setminus\{0\},\ 0 \leq s \leq t \leq 1. \tag{4.18}$$

By the arguments used in the proof of Theorem 4.7, we obtain

$$\lambda^0_n \leq \lambda^s_n \leq \lambda^t_n \leq \lambda_n \quad \text{for all } n \in \mathbb{N},\ 0 \leq s \leq t \leq 1. \tag{4.19}$$

We assume moreover that for each fixed $n \in \mathbb{N}$, λ^s_n depends continuously on $s \in [0,1]$, or at least does not jump by "too large" amounts – recall that (4.19) forbids discontinuities other than jumps – which allows "fine" step functions, for example. The following description of the homotopy method will clarify what is really needed.

In practice, it is often possible to choose $(H_0, \langle\cdot,\cdot\rangle_0) := (H, \langle\cdot,\cdot\rangle)$, $D(A_0) := D(A)$ in the base problem. In such cases, a homotopy can easily be constructed by choosing $(H_s, \langle\cdot,\cdot\rangle_s) := (H, \langle\cdot,\cdot\rangle)$, $D(A_s) := D(A)$, and

$$A_s := (1-s)A_0 + sA \tag{4.20}$$

for all $s \in [0,1]$. Obviously, this choice satisfies (4.18) – provided, of course, that the inequality required in (4.16) holds.

We start the homotopy algorithm choosing some $M \geq N$ such that λ^0_M and λ^0_{M+1} are "well separated" and moreover,

$$\lambda_N < \lambda^0_{M+1} \tag{4.21}$$

(with a "not too small" difference $\lambda^0_{M+1} - \lambda_N$). One can check (4.21) (i.e. find an appropriate M) using the lower bound $\underline{\lambda}^0_{M+1} \leq \lambda^0_{M+1}$ which we assumed to be known, and e.g. the Rayleigh–Ritz upper bound $\Lambda_N \geq \lambda_N$. Eqn. (4.21) replaces the restrictive condition (4.17); obviously, (4.21) is *always* satisfied for some M since $\lambda^0_n \to \infty$ for $n \to \infty$.

Next, we look for some $s_1 \in (0,1]$ such that

$$\lambda^{s_1}_M < \lambda^0_{M+1} \tag{4.22}$$

which is close to $\sup\{s \in (0,1] : \lambda^s_M < \lambda^0_{M+1}\}$. Such an s_1 exists since λ^0_M and λ^0_{M+1} are "well separated", and λ^s_M is "not too discontinuous", as a function of s. (4.22) can be checked by using the known lower bound $\underline{\lambda}^0_{M+1} \leq \lambda^0_{M+1}$ and e.g. the Rayleigh–Ritz upper bound $\Lambda^{s_1}_M \geq \lambda^{s_1}_M$, i.e., by testing the inequality $\Lambda^{s_1}_M < \underline{\lambda}^0_{M+1}$. Since (4.19) provides $\lambda^0_{M+1} \leq \lambda^{s_1}_{M+1}$, we can then satisfy (4.9) (with A_{s_1} in place of A and M in place of N) with $\nu := \underline{\lambda}^0_{M+1}$, so that we can apply Lehmann's method to obtain lower bounds $\underline{\lambda}^{s_1}_1, \ldots, \underline{\lambda}^{s_1}_M$ for $\lambda^{s_1}_1, \ldots, \lambda^{s_1}_M$. The situation is similar if e.g. Weinstein's or Kato's bounds are used.

Of course, everything is finished if $s_1 = 1$. Otherwise, we are in an analogous situation as at the start, now with s_1 in place of 0, and M in place of $M+1$. Thus, if $\lambda^{s_1}_{M-1}$ and $\lambda^{s_1}_M$ are "well separated", we can go on looking for some $s_2 \in (s_1, 1]$ such that

$$\lambda^{s_2}_{M-1} < \lambda^{s_1}_M \tag{4.23}$$

which is close to $\sup\{s \in (0,1] : \lambda^s_{M-1} < \lambda^{s_1}_M\}$. To check (4.23) we use the lower bound $\underline{\lambda}^{s_1}_M \leq \lambda^{s_1}_M$ and e.g. the Rayleigh–Ritz upper bound $\Lambda^{s_2}_{M-1} \geq \lambda^{s_2}_{M-1}$. Since (4.19) provides $\lambda^{s_1}_M \leq \lambda^{s_2}_M$, (4.9) is then satisfied (with A_{s_2} in place of A and $M-1$ in place of N) for $\nu := \underline{\lambda}^{s_1}_M$, so that Lehmann's method provides lower bounds $\underline{\lambda}^{s_2}_1, \ldots, \underline{\lambda}^{s_2}_{M-1}$ for $\lambda^{s_2}_1, \ldots, \lambda^{s_2}_{M-1}$ (similarly for Weinstein's or Kato's method).

If $\lambda^{s_1}_{M-1}$ and $\lambda^{s_1}_M$ are *not* well separated, i.e., if they belong to some cluster $\lambda^{s_1}_{M-K_1}, \ldots, \lambda^{s_1}_M$ of eigenvalues, we look for s_2 satisfying, instead of (4.23),

$$\lambda^{s_2}_{M-K_1-1} < \lambda^{s_1}_{M-K_1}$$

(regard $\lambda^{s_1}_{M-K_1-1}$ and $\lambda^{s_1}_{M-K_1}$ as "well separated") and obtain, as before, lower bounds $\underline{\lambda}^{s_2}_1, \ldots, \underline{\lambda}^{s_2}_{M-K_1-1}$ for $\lambda^{s_2}_1, \ldots, \lambda^{s_2}_{M-K_1-1}$.

If $s_2 < 1$, we go on with this algorithm until $s_R = 1$ for some $R \in \mathbb{N}$ (or until the algorithm breaks down since no eigenvalue is left to continue). With $\bar{K}$ denoting the total number of eigenvalues which have been "dropped" during the algorithm according to the above description, we are then done if $M - \bar{K} \geq N$. Indeed, condition (4.21) ensures that this goal is reached if the gaps between right-hand and left-hand sides in (4.22), (4.23), and so on, are not too large, and if all computed bounds are sufficiently accurate. If $M - \bar{K} < N$, the algorithm has to be restarted, with some larger M.

4.6 Numerical tools

In this short section, we will briefly comment on some of the numerical tools needed for the practical implementation of the methods described in Sections 4.1–4.5, in particular on those concerned with safe bounds.

As already discussed in Section 4.3 the *approximate* eigenpairs needed for the enclosure methods discussed here can be computed by the Rayleigh–Ritz (approximation) method with suitable basis functions $\tilde{u}_1, \ldots, \tilde{u}_M$, e.g. finite element, spline, polynomial or trigonometric polynomial basis functions in the case of a function space problem. For "large" problems (e.g., partial differential equations), one will combine it with *inverse* (subspace) *iteration.*

The matrix elements $\langle A\tilde{u}_i, \tilde{u}_j \rangle$ and $\langle \tilde{u}_i, \tilde{u}_j \rangle$ are *integrals* if $(H, \langle \cdot, \cdot \rangle)$ is some L_2-space, as usual for differential equation problems. They can therefore be approximated by quadrature formulas or, in more specific cases, be calculated in closed form by computer algebra packages.

For the *enclosure* methods discussed in this chapter we need, of course, enclosures, i.e. bounds, for the numbers $\langle A\tilde{u}_i, \tilde{u}_j \rangle$, $\langle \tilde{u}_i, \tilde{u}_j \rangle$, $\langle A\tilde{u}_i, A\tilde{u}_j \rangle$, $||A\tilde{u}_i - \tilde{\lambda}_i \tilde{u}_i||$, $||\tilde{u}_i||$ (depending on which method is to be applied), with $(\tilde{u}_i, \tilde{\lambda}_i)$ now denoting approximate eigenpairs. In the L_2-space case, this can again be carried out by quadrature formulas. Now, however, the remainder terms also have to be controlled, i.e., rough uniform bounds are needed for some higher derivative of the integrands. This can be achieved by means of *interval arithmetic* (see [1, 3] for general descriptions, [22, 24, 25] for practical implementations) which also controls all *rounding errors* made in the calculations (e.g., in the evaluation of the quadrature formulas). Of course, the alternative of calculating the integrals – if possible – in closed form by computer algebra, may be applied here, too. Then, interval arithmetic is needed only for rounding error control.

The computation of bounds for the eigenvalues of the *matrix* eigenvalue problems (3.2) (Rayleigh–Ritz) and (4.12) (Lehmann) is a nontrivial task of its own which we will not discuss in too much detail here. The most efficient and reliable method for treating symmetric matrix eigenvalue problems has been proposed by Behnke [10]; it is based on $LDL^{\top}$-decomposition and on Lehmann's method again. (Note that the fundamental difficulty of knowing a lower bound for some higher eigenvalue a priori can be solved much more easily for matrix problems.)

Since, due to our approach via approximate eigenpairs, the matrices in (3.2) and (4.12) are almost diagonal (with the mentioned restriction for the matrix B_2), we can also obtain eigenvalue enclosures by the following simple method based on Gerschgorin's theorem: e.g. for problem (3.2), where $A_2 \approx I$, let $\Delta := I - A_2$. Then, if $||\Delta||_\infty < 1$, $A_2^{-1} = \sum_{k=0}^{\infty} \Delta^k = \sum_{k=0}^{m-1} \Delta^k + \Delta^m (I-\Delta)^{-1}$ and thus,

$$||A_2^{-1} - \sum_{k=0}^{m-1} \Delta^k||_\infty \le \frac{||\Delta^m||_\infty}{1 - ||\Delta||_\infty} \quad \text{for arbitrary} \quad m \in \mathbb{N},$$

providing an accurate enclosure for A_2^{-1}. Carrying out the product $A_2^{-1}A_1$ in interval arithmetic, we obtain an enclosure for $A_2^{-1}A_1$, which is still almost diagonal; enclosures for its eigenvalues (and thus, for the eigenvalues of problem (3.2)) can now be obtained from Gerschgorin's theorem.

4.7 Numerical examples

We close this article with some concrete examples of eigenvalue problems for differential equations. First we consider *Sturm–Liouville* problems

$$-(pu')' + qu = \lambda wu \quad \text{on} \quad (a,b) \tag{4.24}$$

which are "very" regular in the sense that $p \in C_1[a,b]$ and $q, w \in C[a,b]$ are real-valued, and $p(x) > 0$, $w(x) > 0$ on $[a,b]$ (see Section 2.2), or more generally, *elliptic* problems

$$-\sum_{\mu,\nu=1}^{m} \frac{\partial}{\partial x_\mu}\left(p_{\mu\nu}\frac{\partial u}{\partial x_\nu}\right) + qu \equiv -\text{div}\,(P\nabla u) + qu = \lambda wu \quad \text{on} \quad \Omega, \tag{4.25}$$

with Ω denoting some domain in $\mathbb{R}^m$ with sufficient boundary regularity (e.g., C_2-domains or convex polygonal domains in $\mathbb{R}^2$ will do), $P = (p_{\mu\nu}) \in C_1(\overline{\Omega})^{m,m}$, $q, w \in C(\overline{\Omega})$ are real-valued, $P(x)$ Hermitian and positive definite, $w(x) > 0$ on $\overline{\Omega}$ (see Section 2.3).

In addition, we require boundary conditions, which for technical simplicity we assume to be of Dirichlet type. Then, for

$$H := L_2(\Omega), \quad \langle u, v \rangle := \int_\Omega wu\bar{v}\,dx,$$

$$D(A) := H_2(\Omega) \cap \mathring{H}_1(\Omega), \quad Au := \frac{1}{w}\left[-\text{div}\,(P\nabla u) + qu\right],$$

our general assumptions on the spectral properties of A are satisfied, according to regularity theory for elliptic problems (see [15]). (Since problem (4.24) is contained in (4.25) as a special case, we do not refer to (4.24) specifically.)

Let $p_0 > 0$, $w_0 > 0$, $q_0 \in \mathbb{R}$ denote numbers satisfying

$$P(x) - p_0 I \text{ positive semidefinite}, \quad w(x) \le w_0, \quad \frac{q(x)}{w(x)} \ge \frac{q_0}{w_0} \quad \text{for } x \in \overline{\Omega} \tag{4.26}$$

and define, for $s \in [0,1]$,

$$P_s(x) := (1-s)p_0 I + sP(x), \quad q_s(x) := (1-s)q_0 + sq(x),$$

$$w_s(x) := (1-s)w_0 + sw(x) \quad (x \in \overline{\Omega}),$$

$$H_s := L_2(\Omega), \quad \langle u, v\rangle_s := \int_\Omega w_s u \bar{v}\, dx,$$

$$D(A_s) := D(A), \quad A_s u := \frac{1}{w_s}\left[-\text{div}\,(P_s \nabla u) + q_s u\right],$$

Then, for all $u \in D(A) \setminus \{0\}$ and $s \in [0,1]$,

$$\frac{\langle A_s u, u\rangle_s}{\langle u, u\rangle_s}$$
$$= \frac{q_0}{w_0} + \int_\Omega \left[(\nabla u)^\top P_s(\nabla \bar{u}) + s\left(q - \frac{q_0}{w_0} w\right)|u|^2\right] dx \Big/ \int_\Omega w_s |u|^2\, dx$$

which is monotonically nondecreasing with respect to s according to (4.26). Thus, (4.18) holds true, so that the homotopy method is applicable. Furthermore,

$$A_0 u = \frac{1}{w_0}(-p_0 \Delta u + q_0 u) \quad (u \in D(A)) \tag{4.27}$$

so that its eigenvalues are known in closed form if $m = 1$ (i.e., problem (4.24) is considered) or if Ω is "simple" (e.g., a rectangle or a ball). If Ω is not "simple", one may try to compute bounds for them using a second homotopy: one chooses some "simple" domain Ω_0 containing Ω, and some family $(\Omega_s)_{0 \le s \le 1}$ of (regular) domains connecting Ω_0 to $\Omega_1 = \Omega$ by continuous deformation, and satisfying $\Omega_s \supset \Omega_t$ for $s \le t$. Since the Dirichlet eigenvalues of $-\Delta$ increase if the domain shrinks, we obtain (4.19) also for this domain homotopy. Thus, starting with the known eigenvalues of A_0 on Ω_0 in place of Ω, we may apply the homotopy method to compute the required (lower) bounds for the eigenvalues of A_0 (on Ω).

Example 1:

$$-u'' = \lambda(1 + \alpha \sin x)u \quad (0 < x < \pi), \quad u(0) = u(\pi) = 0.$$

This eigenvalue problem describes the bending of a beam with variable bending strength [12, p. 169]; compare also Section 1.3. We computed eigenvalue enclosures for the two cases $\alpha = 1$ and $\alpha = 5$. Starting with $M + 1 = 10$ eigenvalues $\lambda_n^0 = n^2/(1+\alpha)$ of the "base problem" $-u'' = \lambda(1+\alpha)u$

on $(0, \pi)$, $u(0) = u(\pi) = 0$, and using the mentioned combination of Weinstein's and Kato's bounds (with 30 polynomial basis functions in the approximation procedure), we computed enclosures for the first 9 (in the case $\alpha = 1$) and 8 (in the case $\alpha = 5$) eigenvalues, respectively. For $\alpha = 1$, condition (4.17) is satisfied for $N = 9$ since $\lambda_{10}^0 = 50$, so that the base problem is actually a direct comparison problem, and the homotopy is not really needed. For $\alpha = 5$, however, the homotopy algorithm is "active" in the sense that, for some s_1, λ_9^s "hits" $\lambda_{10}^0 = 16.\overline{6}$ and has to be dropped according to the description in Section 4.5.

n	λ_n for $\alpha = 1$	λ_n for $\alpha = 5$
1	0.54031885955_5^9	0.19007990_{18}^{21}
2	2.3717868832_4^8	0.8959316_5^7
3	5.4486361_{47}^{51}	2.1112791_0^7
4	9.7629721_4^7	3.831816_2^5
5	15.312608_{27}^{36}	6.055505_0^8
6	22.096731_5^9	8.78115_1^4
7	30.114984_1^9	12.0079_{76}^{81}
8	39.3671_{89}^{92}	15.7354_2^5
9	49.853_1^3	—

The next two examples are elliptic problems on $\Omega := (0, \pi)^2$. In both cases, we can choose the base problem $-\Delta u = \lambda u$ on Ω, $u = 0$ on $\partial\Omega$, with eigenvalues $n_1^2 + n_2^2$ $(n_1, n_2 \in \mathbb{N})$. Starting with $M+1 = 9$ of its eigenvalues, and using again (cluster versions of) Weinstein's and Kato's bounds (with 8×8 bi-quintic rectangular finite elements in the approximation procedure), we computed enclosures for the first 6 (Example 2) and 5 (Example 3) eigenvalues. In Example 2, the value s_1 (see (4.22)) equals 0.7; since $\lambda_7^{0.7}$ and $\lambda_8^{0.7}$ form a cluster, both have to be dropped, and the homotopy algorithm continues with 6 eigenvalues up to $s = 1$. In Example 3, three values $s_1 = 0.5$, $s_2 = 0.54$, $s_3 = 0.94$ occur where eigenvalues have to be dropped, so that the algorithm ends with 5 eigenvalues at $s = 1$.

Example 2:

$$-\frac{\partial}{\partial x_1}\left(p\,\frac{\partial u}{\partial x_1}\right) - \frac{\partial}{\partial x_2}\left(p\,\frac{\partial u}{\partial x_2}\right) = \lambda u \text{ on } \Omega,\ u = 0 \text{ on } \partial\Omega,$$

$$p(x_1, x_2) := 1 + \left(\frac{2}{\pi}\right)^4 x_1(\pi - x_1)x_2(\pi - x_2).$$

n	λ_n^0	$\lambda_n^{0.7}$	λ_n
1	2	2.50936^{59}_{46}	2.70383^{23}_{09}
2	5	6.5959^{640}_{596}	7.22084^{68}_{09}
3	5	6.59596^{40}_{16}	7.22084^{68}_{37}
4	8	10.26782^{61}_{24}	11.1500^{435}_{372}
5	10	13.349^{5092}_{4995}	14.6809^{39}_{05}
6	10	13.57481^{85}_{33}	14.9570^{39}_{28}
7	13	16.755^{84}_{61}	—
8	13	16.755^{84}_{72}	—
9	17	—	—

Example 3: $-\Delta u = \dfrac{\lambda u}{1+\frac{1}{\pi^2}(x_1^2+2x_2^2)}$ on Ω, $u = 0$ on $\partial\Omega$.

n	λ_n^0	$\lambda_n^{0.5}$	$\lambda_n^{0.54}$	$\lambda_n^{0.94}$	λ_n
1	2	2.54041^{36}_{21}	2.59529^{62}_{47}	3.28359^{71}_{52}	3.41364^{19}_{00}
2	5	6.28387^{58}_{17}	6.41333^{41}_{00}	8.02608^{55}_{03}	8.32907^{66}_{13}
3	5	6.4522^{512}_{489}	6.6034^{605}_{581}	8.58491^{72}_{42}	8.98114^{51}_{19}
4	8	10.277^{2050}_{1997}	10.52081^{96}_{41}	13.8133^{93}_{79}	14.4838^{78}_{60}
5	10	12.6403^{27}_{06}	12.8958^{71}_{49}	15.942^{180}_{094}	16.499^{62}_{35}
6	10	12.7508^{40}_{28}	13.0370^{63}_{50}	16.82^{37}_{15}	—
7	13	16.514^{46}_{34}	16.88^{933}_{871}	—	—
8	13	16.97^{97}_{71}	—	—	—
9	17	—	—	—	—

Example 4: Our final example is the *singular* Sturm–Liouville problem

$$-((1-x^7)u')' = \lambda x^7 u \quad \text{on} \quad (0,1) \tag{4.28}$$

known as *Latzko*'s equation (see [26]). It can be shown that together with the boundary conditions

$$u(0) = 0, \quad \lim_{x\to 1}[(1-x^7)u'(x)] = 0, \tag{4.29}$$

problem (4.28) meets our general hypotheses. More precisely, with

$$L_2(0,1;x^7) := \left\{ u : (0,1) \to \mathbb{C} : \ u \ \text{ measurable}, \ \int_0^1 x^7 |u|^2 \, dx < \infty \right\}$$

and with $AC_{\rm loc}(0,1)$ denoting the space of locally absolutely continuous functions on $(0,1)$, the operator $\hat{A}$ defined by

$$\hat{A}u := -x^{-7}((1-x^7)u')',$$

$$D(\hat{A}) := \{u \in L_2(0,1;x^7): \quad u,(1-x^7)u' \in AC_{\rm loc}(0,1), \\ x^{-7}((1-x^7)u')' \in L_2(0,1;x^7), \\ u \text{ satisfies } (4.29)\}$$

is selfadjoint in $L_2(0,1;x^7)$ and has an orthonormal basis of eigenfunctions, with eigenvalue sequence converging to $+\infty$.

As a comparison problem, we choose the problem

$$-((1-x^7)u')' = \lambda x^5 u \text{ on } (0,1), \quad u(0)=0, \quad \lim_{x\to 1}[(1-x^7)u'(x)] = 0 \tag{4.30}$$

with the operator

$$\hat{A}_0 u := -x^{-5}((1-x^7)u')',$$

$$D(\hat{A}_0) := \{u \in L_2(0,1;x^5): \quad u,(1-x^7)u' \in AC_{\rm loc}(0,1), \\ x^{-5}((1-x^7)u')' \in L_2(0,1;x^5), \\ u \text{ satisfies } (4.29)\}$$

which is selfadjoint in $L_2(0,1;x^5)$ and has an orthonormal basis of eigenfunctions, with eigenvalue sequence converging to $+\infty$.

Fortunately, problem (4.30) can be solved in closed form: the eigenvalues are $\lambda_n^0 = 7n(7n-6)$, and the eigenfunctions are certain Jacobi polynomials (see [14]).

However, $L_2(0,1;x^5) \subsetneq L_2(0,1;x^7)$ which is the wrong order for satisfying assumption (4.16) with these (pre-) Hilbert spaces H_0 and H. This difficulty can be solved by choosing, instead,

$$H := H_0 := L_2(0,1;x^7) \cap C[0,1) = L_2(0,1;x^5) \cap C[0,1),$$

$$\langle u,v\rangle := \int_0^1 x^7 u\bar{v}\,dx, \quad \langle u,v\rangle_0 := \int_0^1 x^5 u\bar{v}\,dx,$$

$$D(A) := D(\hat{A}) \cap C[0,1), \quad A := \hat{A}\big|_{D(A)},$$

$$D(A_0) := D(\hat{A}_0) \cap C[0,1), \quad A_0 := \hat{A}_0\big|_{D(A_0)}.$$

It is easy to check that (4.16) is now satisfied. Furthermore, the differential equations (4.28) and (4.30) show that all eigenfunctions of both problems are continuous on $[0,1)$, so that A and A_0 still have the same eigenpairs as $\hat{A}$ and $\hat{A}_0$, respectively, and therefore satisfy our general assumptions.

Fortunately, problem (4.30) is "close enough" to problem (4.28), (4.29), so that (4.17) is satisfied for $N=5$. Thus, the homotopy algorithm is not needed here. The following results were computed by G. Goldstein in her diploma thesis [20] supervised by Friedrich Goerisch. She used the Rayleigh–Ritz method for upper bounds and Lehmann's method for lower bounds. The basis functions for the numerical approximations were polynomials, so that all integrals could be computed in closed form. Rounding errors were neglected (so that final rigour is not given here) which has the effect that, for the first three eigenvalues, the presented upper and lower "bounds" coincide within the available number of decimals.

For comparison, we list the approximations obtained by Bailey, Everitt, and Zettl [5] with their package SLEIGN2, in the last column of the table.

n	λ_n^0	λ_n	SLEIGN2
1	7	8.72747035399	8.7274702
2	112	152.42307087863	152.423014
3	315	435.06333217590	435.060768
4	616	855.68572526^{82}_{20}	855.6817
5	1015	1414.1428^{24}_{08}	1414.12619
6	1512	—	—

Bibliography

1. Adams, E., and Kulisch U. (eds.): *Scientific Computing with Automatic Result Verification. 1. Language and Programming Support for Verified Scientific Computation*, 2. *Enclosure Methods and Algorithms with Automatic Result Verification*, 3. *Applications in the Engineering Sciences.* Academic Press, San Diego, 1993.
2. Adams, R.A. *Sobolev spaces.* Academic Press, New York, 1975.
3. Alefeld, G., and Herzberger, J. *Einführung in die Intervallrechnung.* Bibliographisches Institut (Reihe Informatik, Nr. 12), Mannheim Wien Zürich, 1974.
4. Aronszajn, N. *Approximation eigenvalues of completely continuous symmetric operators.* In: Proc. of the Spectral Theory and Differential Problems, pp. 179–202. Stillwater, Oklahoma, 1951.
5. Bailey, P.B., Everitt, W.N., and Zettl, A. *Computing Eigenvalues of Singular Sturm–Liouville Problems.* Results in Math. **20** (1991), 391–423.
6. Bazley, N.W., and Fox, D.W. *A procedure for estimating eigenvalues.* J. Math. Phys. **3** (1962), 469–471.
7. Bazley, N.W., and Fox, D.W. *Comparison operators for lower bounds to eigenvalues.* J. reine angew. Math. **223** (1966), 142–149.
8. Beattie, C. *An extension of Aronszajn's Rule: Slicing the spectrum for intermediate problems.* SIAM J. Numer. Anal. **24** (4) (1987), 828–843.

9. Beattie, C., and Goerisch, F. *Methods for computing lower bounds to eigenvalues of self-adjoint operators.* Numer. Math. **72** (1995), 143–172.

10. Behnke, H. *Inclusion of Eigenvalues of General Eigenvalue Problems for Matrices.* In: Kulisch, U., Stetter, H.J. (eds.): Scientific Computation with Automatic Result Verification, Computing, Suppl. **6** (1988), 69–78.

11. Chatelin, F. *Spectral Approximations of Linear Operators.* Academic Press, New York, 1983.

12. Collatz, L. *Eigenwertaufgaben mit technischen Anwendungen.* Akademische Verlagsgesellschaft Geest & Portig, K.G., Leipzig, 1949.

13. Courant, R., and Hilbert, D. *Methods of Mathematical Physics*, Vol. 1. Interscience Publ., New York, 1953.

14. Fettis, H.E. *On the eigenvalue Latzko's differential equation.* ZAMM **37** (1957), 398–399.

15. Gilbarg, D., and Trudinger, N.S. *Elliptic partial differential equations of second order.* 2nd edn., Springer, Berlin, Heidelberg, 1983.

16. Goerisch, F. *Ein Stufenverfahren zur Berechnung von Eigenwertschranken.* In: Albrecht, J., Collatz, L., Velte, W., Wunderlich, W. (eds.): Numerical Treatment of Eigenvalue Problems, Vol. 4, pp. 104–114, ISNM 83, Birkhäuser, Basel, 1987.

17. Goerisch, F., and Albrecht, J. *Eine einheitliche Herleitung von Einschließungssätzen für Eigenwerte.* In: Albrecht, J., Collatz, L., Velte, W. (eds.): Numerical Treatment of Eigenvalue Problems, Vol. 3, pp. 55–88, ISNM 69, Birkhäuser, Basel, 1984.

18. Goerisch, F., and Haunhorst, H. *Eigenwertschranken für Eigenwertaufgaben mit partiellen Differentialgleichungen.* ZAMM **65** (1985), 129–135.

19. Goerisch, F., and Albrecht, J. *The convergence of a new method for calculating lower bounds to eigenvalues.* Equadiff 6 (Brno, 1985), pp. 303–308, Lect. Notes in Math. 1192, Springer, Berlin–New York, 1986.

20. Goldstein, G. *Eigenwertschranken für singuläre Sturm–Liouville-Eigenwertaufgaben.* Diploma thesis, TU Braunschweig, 1993.

21. Greenberg, L., and Marletta, M. *Oscillation Theory and Numerical Solution of Fourth Order Sturm–Liouville Problems.* IMA J. Numer. Anal. **15** (1995), 319–356.

22. IBM: ACRITH-XSC: *IBM High Accuracy Arithmetic – Extended Scientific Computation.* Version 1, Release 1. IBM Deutschland GmbH (Schönaicher Straße 220, 71032 Böblingen), 1990. 1. General Information. GC 33-6461-01. 2. Reference, SC 33-6462-00. 3. Sample Pro-

grams, SC 33-6463-00. 4. How To Use, SC 33-6464-00. 5. Syntax Diagrams, SC 33-6466-00.

23. Kato, T. *On the upper and lower bounds of eigenvalues.* J. Phys. Soc. Japan **4** (1949), 334–339.
24. Klatte, R., Kulisch, U., Neaga, M., Ratz, D., and Ullrich, Ch. *PASCAL-XSC – Language Reference with Examples.* Springer, Berlin Heidelberg New York, 1992.
25. Klatte, R., Kulisch, U., Lawo, C., Rauch, M., and Wiethoff, A. *C-XSC, A C++ Class Library for Extended Scientific Computing.* Springer, Berlin Heidelberg New York, 1993.
26. Latzko, H. *Der Wärmeübergang an einem turbulenten Flüssigkeits- oder Gasstrom.* ZAMM **1** (1921), 268–290.
27. Lehmann, N.J. *Optimale Eigenwerteinschließungen.* Numer. Math. **5** (1963), 246–272.
28. Parlett, B.N. *The symmetric eigenvalue problem.* Prentice-Hall, Englewood Cliffs, N.J. 07632, 1980.
29. Plum, M. *Eigenvalue inclusions for second-order ordinary differential operators by a numerical homotopy method.* ZAMP **41** (1990), 205–226.
30. Plum, M. *Bounds for eigenvalues of second-order elliptic differential operators.* ZAMP **42** (1991), 848–863.
31. Rektorys, K. *Variational methods in Mathematics, Science and Engineering.* Second Edition, D. Reidel Publ. Co., Dordrecht, 1980.
32. Riesz, F., and Nagy, B. *Leçons d'analyse fonctionnelle.* 5me. ed., Paris, 1968.
33. Weinstein, A. *On the Sturm–Liouville theory and the eigenvalues of intermediate problems.* Numer. Math. **5** (1963), 238–245.
34. Weinstein, A., and Stenger, W. *Methods of Intermediate Problems for Eigenvalues.* Academic Press, New York, 1972.
35. Weyl, H. *Über gewöhnliche Differentialgleichungen mit Singularitäten und die zugehörigen Entwicklungen willkürlicher Funktionen.* Math. Ann. **68** (1910), 220–269.

Hierarchic modelling in mechanics

Christoph Schwab

SAM, ETH Zürich, Rämistrasse 101, CH-8092 Zürich, Switzerland

1 Introduction. What is hierarchic modelling?

Hierarchic modelling is a circle of ideas in the context of numerical simulation of physical systems and not simply a specific mathematical algorithm for a specific problem class. The circle of ideas may be explained starting with the observation that the numerical modelling of physical systems commonly proceeds in several steps.

Firstly, the system is mapped to a *mathematical model* which should represent all known physical aspects of the system. This model is usually quite complex and possibly includes many details that are irrelevant for the quantitative simulation of those aspects of the physical system which are of primary interest.

Therefore, secondly, one replaces the mathematical model by a simplified *working model*. This is usually an (initial-) boundary value problem of (partial) differential equations, describing only some aspects of interest in the underlying physical system.

Finally, the working model is discretized, or replaced, by a *numerical model* such as a finite element discretization, that may be solved to obtain quantitative information on the relevant aspects of the physical system.

Roughly speaking there are three kinds of errors associated with this methodology.

- **Error A** arises from incomplete knowledge of the system to be modelled. For example, certain physical phenomena are either unknown or inaccurately represented by the mathematical model.
- **Error B**, or modelling error, arises from the reduction of the *mathematical model* to the simplified *working model.*
- **Error C** arises from numerical inaccuracies due to discretization error and, possibly, round-off errors.

The rapid development of *a posteriori* error estimators and adaptive discretization methods for ordinary and partial differential equations in the past 15 years was driven by a desire to control Type C errors. Today, for many currently used linear working models, the control of Type C errors is possible [3, 16, 50]. General principles exist showing how this can be achieved for finite element discretizations of linear as well certain nonlinear initial boundary value problems of mathematical physics. However, the technical details necessary in the application of these principles might very well be nontrivial.

Nowadays, Type A errors are practically negligible for numerous three-dimensional boundary value problems in engineering. For example, few doubt that the Navier–Stokes equations are a correct description of viscous, incompressible fluid flow (at least for reasonable values of the Reynolds number) or that the Lamé equations are a correct description of the linearly elastic deformation of a body.

Therefore, modelling or Type B errors, are mainly responsible for any quantitative discrepancy between numerical results and the actual observed behaviour of the physical system. Examples include the error in the transition from Navier–Stokes to the 'shallow water' equations; the error in the transition from three-dimensional elasticity to certain lower dimensional theories, such as beams, plates and shells; and, the transition from nonlinear to linear elasticity.

In applications, if results obtained with simplified working models are inaccurate due to Type B error, then often the only alternative is to tackle the full mathematical model computationally. This may be prohibitive for a number of reasons such as limited computer resources, stiff behaviour and so on. As a result, one has to settle for a *slightly* more sophisticated working model which, hopefully, represents the phenomena under consideration more faithfully and consequently reduces Type B errors. Obviously, this process could then be reiterated.

A natural need arises for a finely graded hierarchy of working models to map classes of physical problems onto the computer with an increasing degree of resolution. Such model hierarchies can be derived by a variety of mathematical techniques. For example, *physical intuition* often allows relevant effects to be identified along with terms in the governing equations that may be neglected; *Galerkin projection* onto the relevant modes results in a reduction in the system complexity; and, finally, *asymptotic analysis* is often relevant when small or large parameters are present in the mathematical model.

Model hierarchies should satisfy certain minimal requirements:

(a) the (initial) boundary value problems should be well-posed, and

(b) a mechanism should exist to couple differing models from the hierarchy in separate subdomains while preserving the global well-posedness.

The latter requirement involves coupling conditions and often is mathematically nontrivial. For example, the Euler equations provide a mathematical working model for the Navier–Stokes equations but are of entirely different type, rendering coupling between subdomains difficult; or, in shell theory, the limiting, zero thickness models obtained by asymptotic analysis are often posed on different spaces than the original, three-dimensional problem.

Once a model hierarchy is available, the natural question arises of how to obtain a *quantitative estimate for the error* B to help decide whether to accept the simulation or to adopt the next model in the hierarchy. Once flexible model hierarchies and *a posteriori* modelling error estimators become available, this decision should, and increasingly will, be made by the computer rather than the so-called expert.

In the present note, we illustrate these ideas for problems in mechanics posed on thin domains. We start with the simple model problem of heat conduction in laminated, or *sandwich*, plates. The hierarchic models are two dimensional approximations of the original three dimensional problems and will be obtained by a *dimensional reduction* consisting of semi-discretization in the transverse direction in conjunction with Galerkin projection. In order to ensure optimality of the models, it is crucial that proper subspaces are used in the semi-discretization. To this end, we perform an *asymptotic analysis of the three dimensional problem* [49]. For homogeneous material, we find that polynomials in the transverse direction give asymptotically (as the thickness t tends to zero) optimal dimensionally reduced models, whereas for laminated material, the asymptotic approach gives uniquely defined, non-polynomial functions.

The class of problems considered is singularly perturbed due to the presence of the small thickness parameter t. Consequently, the asymptotic expansion also contains a *boundary layer* term which describes, in a sense, the edge effects in the sandwich. This edge effect has a complicated, three dimensional structure which is only poorly resolved by two dimensional hierarchic models. Therefore, it is mandatory to allow for *variable model order* in various subregions in order to resolve these effects. Ideally, the model orders should be selected adaptively requiring suitable *modelling error indicators* and *estimators* which will be derived later. We establish asymptotic and spectral exactness of these estimators in strong, exponentially weighted energy norms. This means that an increase in the model order on subregions where a large error is indicated will indeed result in improved accuracy.

We perform the modelling error analysis under the *standing assumption that the two dimensional models are solved exactly.* That is to say, it is assumed that Type C errors are negligible. Of course, this is not possible in practice and an additional finite element discretization in the midsurface must be performed to obtain a fully discrete problem. However, the finite element discretization is delicate since the hierarchic models of thin solids are *coupled elliptic systems* which are *singularly perturbed.* Therefore, their solutions exhibit *boundary layers* and *corner layers.* Once more making use of asymptotic analysis for the solution of the reduced problem [17, 52] we show how to design *hp*-finite element discretizations which converge at an *arbitrarily high, algebraic rate.* These results were obtained in the recent

Ph.D. thesis [52]. We show how our modelling error estimators can be combined with robust *a posteriori* error estimators for the finite element discretization of the hierarchic models to give robust *a posteriori* error estimation procedures for the fully three-dimensional problem.

In addition, for elastic plates, the problem of *shear locking* arises in the finite element discretization of the working models. For shells, *membrane locking* also arises. In both cases, high order finite element discretization removes, or at least alleviates, the locking problems.

It is important to realize that control of Type B error can only succeed if the Type C discretization error is controlled in a quantitative way, such as by adaptive finite element discretizations. While, in principle, this is possible (see, for example, [3, 16] and the references therein), we point out that many of the currently used *a posteriori* discretization error estimators exhibit a striking nonrobustness when applied to singularly perturbed elliptic problems, so that much remains to be done here too.

Finally, the question arises of how to solve the linear system resulting from a fully discrete hierarchic model. It turns out that thanks to the way the models have been derived (through Galerkin projection and semi-discretization), the linear system has a special *block structure* with each block corresponding to a subspace used in the derivation of the hierarchic model. Parallel and sequential subspace correction schemes (using the terminology introduced by J. Xu [53]) can be analyzed and sharp bounds on their convergence rate may be derived using Fourier analysis [36]. As a consequence, it is possible to exhibit a particular basis for which the subspace correction schemes have a *contraction rate which decreases with the thickness.*

Acknowledgement: The research of the author leading to the results presented in this work was supported by the AFOSR under grant No. F49620-J-0100.

2 Poisson problem in a thin plate

2.1 Notation and problem formulation

The *midsurface* $\omega \subset \mathbb{R}^n$ is a bounded domain with Lipschitz boundary $\gamma = \partial\omega$. Given a parameter $d > 0$, the plate corresponding to ω is the $(n+1)$-dimensional domain

$$\Omega = \{(x, y) \mid x \in \omega, y \in (-d, d)\} = \omega \times (-d, d) \tag{2.1}$$

of thickness $t = 2d$, with lateral boundary, or *edge*,

$$\Gamma = \gamma \times (-d, d) \tag{2.2}$$

and *faces*

$$R_{\pm} = \{(x, y) \mid x \in \omega, y = \pm d\}\,. \tag{2.3}$$

The boundary γ is partitioned into a Dirichlet and a Neumann part, denoted by γ_D and γ_N respectively, such that

$$\gamma = \overline{\gamma}_D \cup \overline{\gamma}_N, \quad \text{meas}\,(\gamma_D) > 0 \tag{2.4}$$

and we define $\Gamma_D = \gamma_D \times (-d,d)$ and $\Gamma_N = \gamma_N \times (-d,d)$. Consider the following model problem, describing, for example, stationary heat conduction in the plate Ω

$$\begin{aligned} Lu &= 0 && \text{in } \Omega, \\ \gamma_0 u &= 0 && \text{on } \Gamma_D, \\ \gamma_1 u &= 0 && \text{on } \Gamma_N, \\ \gamma_1 u = a(\pm 1)\frac{\partial u}{\partial y} &= f^{\pm} && \text{on } R_{\pm}. \end{aligned} \tag{2.5}$$

We assume the material constituting Ω to be transversely orthotropic and laminated, so that

$$Lu = -\frac{\partial}{\partial y}\left(a\left(\frac{y}{d}\right)\frac{\partial u}{\partial y}\right) + b\left(\frac{y}{d}\right)Au, \quad Au = -\nabla_x \cdot \mathbf{C}\nabla_x u, \tag{2.6}$$

where γ_0 denotes the trace operator and γ_1 the distributional conormal derivative corresponding to L, defined on

$$H_L(\Omega) = H^1(\Omega) \cap \{u \mid Lu \in L^2(\Omega)\}.$$

The $n \times n$ conductivity matrix $\mathbf{C}$ is assumed to be symmetric and positive definite. The coefficient functions a and b belong to $L^\infty(-1,1)$ and are strictly positive,

$$0 < \underline{A} \le a(z) \le \overline{A}, \qquad 0 < \underline{B} \le b(z) \le \overline{B} \quad \forall z \in (-1,1). \tag{2.7}$$

The variational form of eqn. (2.5) reads: *find* $u \in H^1(\Omega, \Gamma_D)$ *such that*

$$B(u,v) = F(v) \quad \forall v \in H^1(\Omega, \Gamma_D), \tag{2.8}$$

where

$$H^1(\Omega, \Gamma_D) = H^1(\Omega) \cap \{u \mid \gamma_0 u = 0 \text{ on } \Gamma_D\},$$

$$B(u,v) = \int_\Omega \left\{a\left(\frac{y}{d}\right)\frac{\partial u}{\partial y}\frac{\partial v}{\partial y} + b\left(\frac{y}{d}\right)\nabla_x u \cdot C(x)\nabla_x v\right\} dydx,$$

$$F(v) = \int_\omega \left(f^+(x)v(x,d) + f^-(x)v(x,-d)\right) dx.$$

From eqn. (2.7) together with the continuity of the trace operator and the Lax–Milgram Lemma we get

Theorem 2.1 *For every f^+, $f^- \in L^2(\omega)$ the boundary value problem eqn. (2.5) admits a unique weak solution $u \in H^1(\Omega, \Gamma_D)$.*

We assume in what follows, that the coefficients $a(z)$ and $b(z)$ are even,

$$a(z) = a(-z), \quad b(z) = b(-z), \qquad z \in (-1, 1). \tag{2.9}$$

This implies that any solution $u(x, y)$ of eqn. (2.5) can be split into a symmetric and an antisymmetric part which are mutually orthogonal with respect to the energy inner product induced by the bilinear form $B(\cdot, \cdot)$.

Proposition 2.2 *Let $u = u^I + u^{II}$ with*

$$\begin{aligned} u^I(x,y) &= \frac{1}{2}\left(u(x,y) - u(x,-y)\right), \\ u^{II}(x,y) &= \frac{1}{2}\left(u(x,y) + u(x,-y)\right) \end{aligned} \tag{2.10}$$

and let

$$\begin{aligned} \mathcal{H}^I(\Omega) &= H^1(\Omega, \Gamma_D) \cap \{u \mid u(x,y) = -u(x,-y)\}, \\ \mathcal{H}^{II}(\Omega) &= H^1(\Omega, \Gamma_D) \cap \{u \mid u(x,y) = u(x,-y)\}. \end{aligned} \tag{2.11}$$

Then there holds

$$B(u, v) = 0 \quad \forall u \in \mathcal{H}^I(\Omega), v \in \mathcal{H}^{II}(\Omega) \tag{2.12}$$

and u^I, u^{II} solve: find $u^j \in \mathcal{H}^j(\Omega)$ such that

$$B(u^j, v) = F^j(v) \qquad \forall v \in \mathcal{H}^j(\Omega), \quad j \in \{I, II\} \tag{2.13}$$

with

$$\begin{aligned} F^I(v) &= \int_\omega f^I(x)(v(x,d) - v(x,-d))dx\ , \\ F^{II}(v) &= \int_\omega f^{II}(x)(v(x,d) + v(x,-d))dx \end{aligned} \tag{2.14}$$

and

$$f^I(x) = \frac{1}{2}(f^+(x) + f^-(x)), \quad f^{II}(x) = \frac{1}{2}(f^+(x) - f^-(x)).$$

2.2 Asymptotic structure of the solution

The boundary value problem eqn. (2.5) is singularly perturbed due to the parameter d being small. In this section we will present a complete asymptotic expansion of the solution u with respect to d which is valid for general Lipschitz domains ω and sufficiently smooth data f^+, f^-. This insight into the asymptotic structure of the solution will be useful later on in the proper design of lower dimensional, hierarchic models with certain optimality properties.

As is well known (see, e.g. [18]) high order asymptotic expansions are only justified if the boundary data f^+, f^- satisfy certain compatibility conditions which, however, are rarely satisfied in practice. Therefore, the asymptotic expansions obtained below contain *boundary layers* defined as solutions of so-called *Saint–Venant* problems with homogeneous data on the faces $R_\pm$ with inhomogeneous boundary data on Γ.

2.2.1 *A sequence of director functions*

The asymptotic expansion of u^j, where $j \in \{I, II\}$, in powers of d will depend on a sequence $\{\psi_k^j\}$ of functions of the scaled transverse coordinate y/d depending only on the material parameters a and b appearing in eqn. (2.6). In particular, the sequence is independent of Ω and the boundary data f^+, f^-. The director functions are recursively defined to be solutions of a sequence of Neumann problems on the *unit-fiber* $(-1, 1)$.

$$\psi_{-1}^j = 0, \quad j \in \{I, II\}, \qquad \psi_0^I(z) = 0, \tag{2.15}$$

$$\mathbf{a}(\psi_k^j, v) + \mathbf{b}(\psi_{k-1}^j, v) = \delta_{1k} \begin{cases} v(1) - v(-1) & \text{for } j = I \\ v(1) + v(-1) & \text{for } j = II \end{cases} \tag{2.16}$$

for all $v \in H^1(-1, 1)$ and $k = 0, 1, 2, ..., j \in \{I, II\}$ where the bilinear forms $\mathbf{a}(\cdot, \cdot)$ and $\mathbf{b}(\cdot, \cdot)$ are given by

$$\mathbf{a}(u, v) = \int_{-1}^{1} a(z) u'(z) v'(z) dz, \quad \mathbf{b}(u, v) = \int_{-1}^{1} b(z) u(z) v(z) dz.$$

Some properties of the sequences $\{\psi_k^j\}_{k=0}^\infty$ are collected in the following

Theorem 2.3 *The Neumann problems eqn. (2.16) with eqn. (2.15) define the sequences $\{\psi_k^j\}_{k=0}^\infty$, $j \in \{I, II\}$, uniquely. In particular*

$$\psi_0^{II}(z) = \textit{constant}, \tag{2.17}$$

$$\psi_k^I(z) = -\psi_k^I(-z), \quad \psi_k^{II}(z) = \psi_k^{II}(-z) \tag{2.18}$$

and

$$\int_{-1}^{1} b(z) \psi_k^j(z) dz = 0 \qquad \begin{array}{l} j = I, k = 0, 1, 2, ... \\ j = II, k = 1, 2, ... \end{array} \tag{2.19}$$

Moreover, the system $\{\psi_k^j\}$ is dense in $L^2(-1, 1)$.

The density property of the sequence is fundamental and was proved in [49].

Remark 1 If the coefficient functions $a(\cdot)$, $b(\cdot)$ appearing in the definition of $\mathbf{a}(\cdot, \cdot)$ and $\mathbf{b}(\cdot, \cdot)$ are (piecewise) constant, the director functions ψ_k^j are (piecewise) polynomials of degree at most $2k$. The *exact solutions* of the recursion eqn. (2.15), eqn. (2.16) can be efficiently computed by a p-version finite element method in one dimension provided that the layer interfaces are taken as nodes in the mesh.

2.2.2 *Asymptotic expansion of u*

We are now in a position to present the asymptotic expansion of u.

Theorem 2.4 *Let $N \in \mathbb{N}_0$ and $f^+, f^- \in H^{2N}(\omega)$. Then u admits the decomposition*

$$u^j = u_N^j + U_N^j + e_N^j, \qquad j \in \{I, II\} \tag{2.20}$$

with the asymptotic components

$$\begin{aligned} u_N^I(x,y) &= \sum_{k=1}^{N} d^{2k-2}\psi_k^I\left(\frac{y}{d}\right) A^{k-1} f^I(x), \\ u_N^{II}(x,y) &= \sum_{k=0}^{N} d^{2k-1}\psi_k^{II}\left(\frac{y}{d}\right) A^{k-1} f^{II}(x) \end{aligned} \tag{2.21}$$

and the boundary layers

$$U_N^I(x,y) = \sum_{k=1}^{N} d^{2k-2}\mathbf{U}_k^I(x,y), \quad U_N^{II}(x,y) = \sum_{k=1}^{N} d^{2k-1}\mathbf{U}_k^{II}(x,y) \tag{2.22}$$

where the functions $\mathbf{U}_k^j$ are solutions of the Saint–Venant problems

$$\begin{aligned} L\mathbf{U}_k^j &= 0 && \text{in } \Omega, \\ \gamma_1\mathbf{U}_k^j &= 0 && \text{on } R_\pm, \\ \gamma_0\mathbf{U}_k^j &= -\psi_k^j\left(\tfrac{y}{d}\right)\gamma_0\left(A^{k-1}f^j\right)(s) && \text{on } \Gamma_D, \\ \gamma_1\mathbf{U}_k^j &= -\psi_k^j\left(\tfrac{y}{d}\right)\gamma_1\left(A^{k-1}f^j\right)(s) && \text{on } \Gamma_N. \end{aligned} \tag{2.23}$$

For the remainder terms e_N^j in eqn. (2.20) the following estimates hold

$$\left\| e_N^j \right\|_{E(\Omega)} \leq C(N) \begin{cases} d^{2N-1/2}\left\|A^N f^I\right\|_{L^2(\omega)} & \text{for } j = I, N \geq 1 \\ d^{2N+1/2}\left\|A^N f^{II}\right\|_{L^2(\omega)} & \text{for } j = II, N \geq 0 \end{cases} \tag{2.24}$$

where

$$C(N) = \int_{-1}^{1} a(z)\left(\frac{d}{dz}\psi_{N+1}^j(z)\right)^2 dz.$$

The proof uses a classical asymptotic expansion and a so-called *zooming* of the domain Ω to unit thickness. See [33, 32, 35, 49] for details.

Remark 2 If the data f^+, f^- satisfy compatibility conditions then the boundary layer terms U_N^j vanish. More precisely, from eqn. (2.23) we see immediately that

$$\gamma_0\left(A^{k-1}f^j\right)=0 \text{ on } \Gamma_D, \quad \gamma_1\left(A^{k-1}f^j\right)=0 \text{ on } \Gamma_N \tag{2.25}$$

are necessary and sufficient for U_N^j to vanish. However, these conditions are not satisfied in general.

The corrections U_N^j may be seen to arise due to the incompatibility of the data at the edge γ. However, it is not obvious that they are really *boundary* effects. This will be discussed in the next section.

2.3 A priori estimates for the boundary layers

In this section we show that the solutions of the Saint–Venant problems eqn. (2.24) and the boundary layer terms eqn. (2.22) really are boundary effects. More precisely, we show that U_N^j decays exponentially away from the lateral boundary γ.

In classical constructions of boundary correctors (see, e.g. [18]) one assumes a smooth boundary γ and reduces the problem via a partition of unity and local coordinate transforms to a problem on the half-space, where one must solve an initial value problem for a system of ordinary differential equations the solutions of which decay exponentially owing to the ellipticity of the boundary value problem under consideration.

Here we present a different approach. We exhibit the boundary layer behavior of the solutions to the Saint–Venant problems by means of *a priori* estimates in exponentially weighted energy norms. These estimates hold under the general assumptions in Section 2.1, and therefore, in particular also for Lipschitz domains ω, for which the classical boundary corrector construction is not applicable. Another approach to obtain results of this kind uses the Fourier analysis of the problem on certain unbounded domains and the Payley–Wiener Theorem. This approach will be presented in the context of the elasticity problem in Section 4.

2.3.1 *An abstract stability estimate*

We consider the bilinear form $B(\cdot,\cdot)$ from eqn. (2.8) in weighted Sobolev spaces. First we introduce some notions that will be used in the sequel. Let $B(\cdot,\cdot)$ be a continuous bilinear form on a pair of Hilbert spaces $H_1 \times H_2$ equipped with the norms $||\cdot||_1$ respectively $||\cdot||_2$.

We say that $B(\cdot,\cdot)$ is (C,δ)-regular if there exist positive constants C and δ such that

$$|B(u,v)| \le C||u||_1||v||_2 \quad \forall u \in H_1, \forall v \in H_2\,, \tag{2.26}$$

with

$$\inf_{||u||_1=1}\ \sup_{||v||_2=1} |B(u,v)| \ge \delta > 0\,, \tag{2.27}$$

and

$$\sup_{\|u\|_1=1} |B(u,v)| > 0 \quad \forall 0 \neq v \in H_2 \,. \tag{2.28}$$

It is well known (see, e.g. [6]) that for a (C,δ)-regular form $B(\cdot,\cdot)$, corresponding to every bounded linear functional $F(\cdot)$ on H_2 there exists a unique $u \in H_1$ such that

$$B(u,v) = F(v) \qquad \forall v \in H_2.$$

Moreover, if

$$\sup_{\|v\|_2=1} |F(v)| \leq A, \tag{2.29}$$

then there the *a priori* estimate holds

$$\|u\|_1 \leq A/\delta. \tag{2.30}$$

Let $\varphi \in W^{1,\infty}(\omega)$ be a positive weight function. In what follows we will use frequently the space

$$H_\varphi = \left\{ u \in H^1(\Omega,\Gamma) \mid \int_{-d}^{d} b\left(\frac{y}{d}\right) u(x,y)dy = 0 \quad \text{a.e.} x \in \omega \right\} \tag{2.31}$$

which we equip with the weighted energy norm

$$\begin{aligned} \|u\|_\varphi^2 \;:=\; & \int_\omega \varphi^2(x) \int_{-d}^{d} \left\{ a\left(\tfrac{y}{d}\right) \left(\frac{\partial u}{\partial y}\right)^2 \right. \\ & \left. + b\left(\tfrac{y}{d}\right) \nabla_x u \cdot C(x) \nabla_x u \right\} dy dx \,. \end{aligned} \tag{2.32}$$

In particular, we will be interested in *exponentially increasing or decreasing* weight functions φ. We have

Lemma 2.5 *Let $u \in H_\varphi$. Then, for every open subset $\sigma \subseteq \omega$,*

$$\begin{aligned} & \int_{\sigma\times(-d,d)} b\left(\frac{y}{d}\right) \varphi^2(x) |u(x,y)|^2 dy dx \leq \\ & \Lambda^2 d^2 \int_{\sigma\times(-d,d)} a\left(\frac{y}{d}\right) \varphi^2(x) \left(\frac{\partial u}{\partial y}\right)^2 dy dx \end{aligned} \tag{2.33}$$

with

$$\frac{1}{\Lambda^2} = \inf_{\psi} \frac{\int_{-1}^{1} a(z)(\psi')^2 dz}{\int_{-1}^{1} b(z)\psi^2(z) dz}$$

and the infimum is taken over all $\psi \in H^1(-1,1) \cap \{\psi \mid \int_{-1}^{1} b(z)\psi(z)dz = 0\}$.

Proof For smooth functions u and every $x \in \omega$ we have, after a scaling and thanks to the definition of Λ, that

$$\int_{-d}^{d} b\left(\frac{y}{d}\right) u^2(x,y)dy \leq \Lambda^2 d^2 \int_{-d}^{d} a\left(\frac{y}{d}\right)\left(\frac{\partial u}{\partial y}\right)^2 dy.$$

We multiply both sides by φ^2, integrate over $\sigma \subseteq \omega$ and obtain the assertion using a density argument. □

Remark 3 We denote the constant in eqn. (2.33) by Λ^j if the infimum is taken only over $H_\varphi \cap \mathcal{H}^j(\Omega)$. Also note that if $a = b = 1$ we have $\Lambda^I = 1/\pi$, $\Lambda^{II} = 2/\pi$.

Now we can state the stability result.

Theorem 2.6 *Let $0 < \varphi \in W^{1,\infty}(\omega)$ be a weight function and*

$$Q = \max_{1 \leq i \leq n} \left\| \frac{\partial \varphi}{\partial x_i} / \varphi(x) \right\|_{L^\infty(\omega)} \leq \frac{1}{2C_4(\omega,n)d}, \quad C_4(\omega,n) = \sqrt{n\overline{C}}\Lambda \tag{2.34}$$

with Λ as in eqn. (2.33). Then the bilinear form $B(\cdot,\cdot)$ is $(1,\delta)$-regular uniformly with respect to d on $H_\varphi \times H_{\varphi^{-1}}$ with

$$\delta = \left(1 - \sqrt{n\overline{C}}\Lambda Q d\right)\left(1 + 4\sqrt{n\overline{C}}\Lambda Q d(1 + \sqrt{n\overline{C}}\Lambda Q d)\right)^{-1/2}. \tag{2.35}$$

For a proof, we refer to [35].

Remark 4 We give several examples for admissible weight functions φ.
1°. Let

$$\varphi(x) = 1. \tag{2.36}$$

Then $Q = 0$ and Theorem 2.6 implies that $B(\cdot,\cdot)$ is $(1,1)$-regular on $H^1(\Omega,\Gamma_D) \times H^1(\Omega,\Gamma_D)$. In this case we recover the coercivity of the bilinear form $B(\cdot,\cdot)$.
2°. For our boundary layer estimates the following choice will be most useful. Let

$$\varphi(x) = \exp\{\beta \mathrm{dist}(x,\gamma)\}, \quad \beta \in \mathbb{R}. \tag{2.37}$$

For Lipschitz boundaries γ we have $\varphi \in W^{1,\infty}(\omega)$ (see, e.g., [42] Chap. 6.3). Therefore in this case

$$Q = |\beta| R(\omega), \quad R(\omega) = \max_{1 \leq i \leq n} \left\| \frac{\partial}{\partial x_i} \mathrm{dist}(x,\gamma) \right\|_{L^\infty(\omega)}. \tag{2.38}$$

3°. Let $x_0 \in \overline{\omega}$ be any point in the midsurface. Then we write

$$\varphi(x) = \exp\{\beta\, \mathrm{dist}(x,x_0)\}, \quad \beta \in \mathbb{R}. \tag{2.39}$$

Here

$$Q = |\beta|. \tag{2.40}$$

2.3.2 *Estimates for the boundary layers*

We will now utilize Theorem 2.6 to show that U_N^j is indeed a boundary layer effect. To this end we will select the weight function eqn. (2.36). According to Theorem 2.6, the form $B(\cdot,\cdot)$ is $(1,\delta)$ regular with $\delta > 0$ independent of d provided that

$$|\beta| < 1/(2C_4 R(\omega) d). \tag{2.41}$$

The proof of the boundary layer character of U_N^j consists of establishing *a priori* estimates in the norm eqn. (2.32) with the exponential weight eqn. (2.36).

We start with a trace lemma for functions in H_φ.

Lemma 2.7 *Let $U(x,y) \in H^1(\Omega)$ and assume that*

$$\int_{-d}^{d} b\left(\frac{y}{d}\right) U(x,y)dy = 0 \qquad a.e.\ x \in \omega. \tag{2.42}$$

Then we have, with $\varphi(x)$ in eqn. (2.36), β in eqn. (2.41) and for every open subset $\emptyset \neq \tilde{\gamma} \subseteq \gamma$, that

$$\|\gamma_0 U\|_{L^2(\tilde{\Gamma})} \le C \|U\|_\varphi \,. \tag{2.43}$$

Here the constant C depends only on $\tilde{\gamma}$, n and the coefficients of L. In particular, it is independent of d.

The proof is completely analogous to the proof of Lemma 4.2 in [33]. In the next lemma we show that the orthogonality eqn. (2.42) over each fiber follows from a global orthogonality with respect to the bilinear form $B(\cdot,\cdot)$.

Lemma 2.8 *Assume that for $U \in H^1(\Omega, \Gamma_D)$,*

$$B(U,V) = 0 \qquad \forall V = V(x) \in H^1(\omega, \gamma_D)$$

holds. Then eqn. (2.42) holds.

Proof The condition $B(U,V) = 0$ becomes, because of the special form of V,

$$0 = \int_\omega \nabla_x V(x) \cdot C(x) \nabla_x \left(\int_{-d}^{d} b\left(\frac{y}{d}\right) U(x,y)dy \right) dx = 0. \tag{2.44}$$

According to our assumptions $U \in H^1(\Omega, \Gamma_D)$, we also have

$$V(x) = \int_{-d}^{d} b\left(\frac{y}{d}\right) U(x,y)dy \in H^1(\omega, \gamma_D).$$

We make this choice of $V(x)$ in eqn. (2.44) and obtain from Poincaré's inequality that

$$\|V\|_{L^2(\omega)} = 0$$

and the assertion then follows. □

Lemma 2.9 *Let $g \in L^2(\omega)$ with $Ag \in L^2(\omega)$ and $\psi \in H^1(-1,1)$ with*

$$\int_{-1}^{1} b(z)\psi(z)dz = 0. \tag{2.45}$$

Further, let U denote the solution of the Saint–Venant problem

$$\begin{aligned} LU &= 0 \text{ in } \Omega, \\ \gamma_0 U &= \gamma_0 g(s)\psi\left(\frac{y}{d}\right) \text{ on } \Gamma_D, \\ \gamma_1 U &= \gamma_1 g(s)\psi\left(\frac{y}{d}\right) \text{ on } \Gamma_N, \\ \gamma_1 u &= 0 \text{ on } R_+ \cup R_-. \end{aligned} \tag{2.46}$$

Then the a priori estimate eqn. (2.30) holds for U in the form

$$||U||_\varphi \leq C_8(\omega, n)d^{-1/2}||\psi||_{H^1(-1,1)}||Ag||_{L^2(\omega)} \tag{2.47}$$

for every β satisfying eqn. (2.41).

Proof By assumption $g \in H^2(\omega)$ so we have $\gamma_1 g \in L^2(\gamma)$ and

$$||\gamma_1 g||_{L^2(\gamma_N)} \leq C(\omega)\, ||Ag||_{L^2(\omega)} \,. \tag{2.48}$$

We write eqn. (2.46) in variational form: *find $U \in H^1(\Omega)$ such that $U = g(s)\psi(\frac{y}{d})$ on Γ_D and*

$$B(U,V) = G(V) \qquad \forall V \in H^1(\Omega, \Gamma_D) \tag{2.49}$$

where

$$G(V) = \int_{\Gamma_N} (\gamma_1 g)\,(s)\psi\left(\frac{y}{d}\right)V(s,y)dsdy.$$

To obtain the *a priori* estimate, we first construct a particular solution $\tilde{U} \in H_\varphi$ of eqn. (2.49) which satisfies the nonhomogeneous Dirichlet data on Γ_D.

The function $\tilde{U}(x,y) = g(x)\psi(\frac{y}{d})/\varphi(x)$ belongs to H_φ due to eqn. (2.45). Furthermore,

$$\left\|\tilde{U}\right\|_\varphi \leq C(\omega)d^{-1/2}\left\|\sqrt{a}\,\dot{\psi}\right\|_{L^2(-1,1)}||g||_{H^1(\omega)}\,. \tag{2.50}$$

Since the bilinear form $B(\cdot,\cdot)$ is $(1,\delta)$-regular on $H_\varphi \times H_{1/\varphi}$, with eqn. (2.50) we have that

$$\begin{aligned} B(\tilde{U},V) &\leq \left\|\tilde{U}\right\|_\varphi ||V||_{1/\varphi} \\ &\leq C(\omega)d^{-1/2}\left\|\sqrt{a}\,\dot{\psi}\right\|_{L^2(-1,1)}||g||_{H^1(\omega)}\,||V||_{1/\varphi} \end{aligned} \tag{2.51}$$

for all $V \in H_{1/\varphi}$.

Now we decompose the weak solution U of eqn. (2.49) according to $U = W + \tilde{U}$ with $W \in H^1(\Omega, \Gamma)$. Then we have for W: *find* $W \in H^1(\Omega, \Gamma)$ *such that*

$$B(W,V) = H(V) = G(V) - B(\tilde{U},V) \qquad \forall V \in H^1(\Omega,\Gamma). \qquad (2.52)$$

Since $B(W,V) = H(V) = 0$ for all $V = V(x) \in H^1(\omega, \gamma_D)$, the condition eqn. (2.42) holds for W according to Lemma 2.8. The proof will be completed with an *a priori* estimate for W analogous to eqn. (2.50). To show it, we start with

$$|G(V)| \leq \|V\|_{1/\varphi} \quad \forall V \in H_{1/\varphi}.$$

This follows from eqn. (2.48) and from

$$|G(V)| \leq \left(\int_{-d}^{d} \left(\psi(\frac{y}{d})\right)^2 dy \int_{\gamma_N} (\gamma_1 g)^2 \, ds \right)^{1/2} \left(\int_{\Gamma_N} (\gamma_0 V)^2 \, ds dy \right)^{1/2}.$$

Since $V \in H_\varphi$, we have eqn. (2.42) and with the trace estimate Lemma 2.7 we get

$$\begin{aligned} |G(V)| &\leq \sqrt{d}\, \|\psi\|_{L^2(-1,1)} \|\gamma_1 g\|_{L^2(\gamma_N)} \|\gamma_0 V\|_{L^2(\Gamma_N)} \\ &\leq C(\omega, n)\sqrt{d}\, \|\psi\|_{L^2(-1,1)} \|\gamma_1 g\|_{L^2(\gamma_N)} \|V\|_{1/\varphi}. \end{aligned}$$

This implies, with eqn. (2.51), the estimate

$$|H(V)| \leq C d^{-1/2} \|Ag\|_{L^2(\omega)} \|V\|_{1/\varphi} \qquad \forall V \in H_\varphi. \qquad (2.53)$$

Now we again use the $(1, \delta)$-regularity of the form $B(\cdot, \cdot)$ on $H_\varphi \times H_{1/\varphi}$. In particular the solution W of eqn. (2.52) is in H_φ and the *a priori* estimate eqn. (2.30) holds in the form

$$\|W\|_\varphi \leq \frac{1}{\delta} \sup_V \frac{|H(V)|}{\|V\|_{1/\varphi}}$$

where the supremum is taken over all $0 \neq V \in H_\varphi$. Now the assertion follows from eqn. (2.53). □

We are now in a position to show the announced *a priori* estimate for the boundary layers U_N^j.

Theorem 2.10 *Let* $A^N f \in L^2(\omega)$ *for some* $N \geq 1$ *and* φ *be a weight function as in* Theorem 2.6. *Then*

$$\|U_N^j\|_\varphi \leq C_8 \begin{cases} d^{-1/2}\Phi(d, I, N) \text{ for } j = I, \\ d^{1/2}\Phi(d, II, N) \text{ for } j = II \end{cases} \qquad (2.54)$$

where the functionals

$$\Phi(d,j,N) = \sum_{k=1}^{N} d^{2k-2} \|\psi_k^j\|_{H^1(-1,1)} \|A^k f^j\|_{L^2(\omega)},$$

are bounded independently of d *as* $d \to 0$.

Proof It follows from eqn. (2.23) that $\mathbf{U}_k^j$ satisfies the assumptions of Lemma 2.9 with $g = A^{k-1} f^j$. Consequently we have, for $j \in \{I, II\}$ and $k = 1, ..., N$, the estimates

$$\left\| \mathbf{U}_k^j \right\|_{\varphi} \leq C d^{-1/2} \|\psi\|_{H^1(-1,1)} \left\| A^k f^j \right\|_{L^2(\omega)}.$$

The assertion follows with eqn. (2.23). □

Remark 5 Theorem 2.10 holds in particular for $\varphi(x) = \exp\{\beta \text{dist}(x,\gamma)\}$ with $\beta \geq 0$ as in eqn. (2.41).

Remark 6 The structure eqn. (2.21) of the asymptotic parts shows that their regularity depends only on that of the data f^j. The result of Theorem 2.10 implies in particular that perturbations of the three-dimensional solution due to the edge singularities at $\gamma \times \{\pm d\}$ and the edge and vertex singularities at those parts of Γ where γ is unsmooth, are confined to a $O(d)$-neighborhood of Γ, in the sense that these terms decay exponentially into the interior of ω. The regularity of u in the interior of Ω is essentially governed by the regularity of the data f^j.

2.4 Hierarchic models

In the asymptotic analysis of the exact solution u we have seen that in the interior of the domain Ω the solution essentially consists of the asymptotic components u_N^j in eqn. (2.21) that have a *product structure*. That is, they are products of the director functions ψ_k^j, defined in Theorem 2.3, with functions of x, namely higher derivatives of the data $f^{\pm}$. As we showed in Theorem 2.10, the boundary layers $\mathbf{U}_k^j$ are confined to the vicinity of the edge Γ. The product structure of the asymptotic solution components u_N^j in eqn. (2.21) motivates approximating eqn. (2.8) by dimensionally reduced models in the form of elliptic boundary value problems posed only on the midsurface ω. These boundary value problems will be obtained by semi-discretization of eqn. (2.8) in the transverse direction in conjunction with a Galerkin procedure consisting of an energy projection. As is well known, this means the reduced solution has some optimality properties. In particular, *since the director functions* ψ_k^j *are selected as transverse variations of the projected solution*, the reduced solution will exhibit certain optimality properties, as $d \to 0$, in the absence of boundary layers [49].

However, the asymptotics of the three-dimensional solution in Theorem 2.4 show that the structure of the solution in the vicinity of the lateral edge Γ is quite different from that in the interior. Therefore, an increased order of the hierarchic models in the boundary layer is admitted in order to capture equally well both the asymptotic part u_N^j in the interior and the boundary effects. To this end, define a partition of ω

$$\mathcal{P} = \{\omega_i \mid \omega_i \subseteq \omega, i = 1, \ldots, M\} \tag{2.55}$$

into M disjoint Lipschitz domains as, for example, in a finite element triangulation of ω. A local model order, $n_i \in \mathbb{N}_0$, is associated with each subdomain $\omega_i \in \mathcal{P}$. In particular, for uniform model order one has $\mathcal{P} = \{\omega\}$. The model orders of all subdomains are combined into a vector $\mathbf{n} = \{n_1, ..., n_M\}$. Given a dense sequence $\{\varphi_k\}_{k=0}^{\infty} \subseteq H^1(-1,1)$ of linearly independent *director functions*, define

$$\begin{aligned} \mathcal{H}(\mathbf{n}) &= \Big\{ u(x,y) \mid u|_{\omega_i} = \textstyle\sum_{k=0}^{n_i} X_k(x)\varphi_k\left(\frac{y}{d}\right), \\ & \qquad X_k \in H^1(\omega, \gamma_D) \Big\} \cap H^1(\Omega, \Gamma_D) . \end{aligned} \tag{2.56}$$

Proposition 2.11 *$\mathcal{H}(\mathbf{n})$ is a closed, linear subspace of $H^1(\Omega, \Gamma_D)$ and, as $\mathbf{n} \to \infty$, the sequence $\{\mathcal{H}(\mathbf{n})\}$ is dense in $H^1(\Omega, \Gamma_D)$.*

For a proof, we refer to [35] or [34]. The density of the family $\{\mathcal{H}(\mathbf{n})\}$ is a consequence of the density of the sequence ψ_k^j in $L^2(-1,1)$.

The *dimensionally reduced solution* is now obtained as follows: *find $u(\mathcal{P}, \mathbf{n}) \in \mathcal{H}(\mathbf{n})$ such that*

$$B(u(\mathcal{P}, \mathbf{n}), v) = F(v) \qquad \forall v \in \mathcal{H}(\mathbf{n}). \tag{2.57}$$

Proposition 2.11 implies, with Theorem 2.1, that eqn. (2.57) admits a unique solution.

Remark 7 A special case of the concept of hierarchic modelling is *partial* dimensional reduction. Here, the exact solution u is approximated by functions of the form eqn. (2.56) only in certain subdomains, whereas in other parts of the domain the full three-dimensional form is kept. In these subdomains we may set $n_i = \infty$ owing to the density of the director sequence $\{\varphi_k\}_{k=0}^{\infty}$ in $H^1(-1,1)$.

If the underlying partition is evident from the context then we write $u(\mathbf{n})$ for $u(\mathcal{P}, \mathbf{n})$. Under the assumptions of Theorem 2.1 there exists a unique solution $u(\mathcal{P}, \mathbf{n}) \in \mathcal{H}(\mathbf{n})$ of eqn. (2.57). Furthermore, since $\mathcal{H}(\mathbf{n})$ is a closed, linear subspace of $H^1(\Omega, \Gamma_D)$ the *modelling error* $e(\mathcal{P}, \mathbf{n}) = u - u(\mathcal{P}, \mathbf{n})$ satisfies the usual relation

$$B(e(\mathcal{P}, \mathbf{n}), v) = 0 \qquad \forall v \in \mathcal{H}(\mathbf{n}). \tag{2.58}$$

The reduced solution $u(\mathcal{P}, \mathbf{n})$ is the best approximation of u with respect to the energy norm:

$$\|u - u(\mathcal{P}, \mathbf{n})\|_{E(\Omega)} \leq \|u - v\|_{E(\Omega)} \qquad \forall v \in \mathcal{H}(\mathbf{n}). \tag{2.59}$$

As in eqn. (2.10), the approximation $u(\mathbf{n})$ can be decomposed into two parts. These parts are, due to eqn. (2.12), mutually independent energy projections of u^j onto $\mathcal{H}^j(\mathbf{n}) = \mathcal{H}(\mathbf{n}) \cap \mathcal{H}^j(\Omega)$.

The span of the director functions φ_k in eqn. (2.56) defines $u(\mathcal{P}, \mathbf{n})$ uniquely. The selection of the director functions is made on the basis of the asymptotics of the three-dimensional solution in Theorem 2.1:

$$\{\varphi_k\}_{k=0}^{\infty} = \{\psi_0^{II}, \psi_0^{I}, \psi_1^{II}, \psi_1^{I}, \ldots\}. \tag{2.60}$$

Then, in particular, we have

$$\mathcal{H}^I(\mathbf{n}) = \left\{ u \mid u\big|_{\omega_i} = \sum_{k=0}^{\lfloor n_i/2 \rfloor} X_k(x)\psi_k^I\left(\frac{y}{d}\right), \quad X_k(x) \in H^1(\omega, \gamma_D) \right\}$$

and

$$\mathcal{H}^{II}(\mathbf{n}) = \left\{ u \mid u\big|_{\omega_i} = \sum_{k=0}^{\lfloor n_i/2 \rfloor - 1} X_k(x)\psi_k^{II}\left(\frac{y}{d}\right), \quad X_k(x) \in H^1(\omega, \gamma_D) \right\}.$$

Obviously $\mathcal{H}(\mathbf{n})$ is not a finite dimensional subspace of $H^1(\Omega, \Gamma_D)$: the determination of the unknown coefficients X_k in $u(\mathcal{P}, \mathbf{n})$ still requires the solution of an elliptic boundary value problem on ω. We will describe the structure of the boundary value problem in the case when $\mathcal{P} = \{\omega\}$ and the model order is uniform $\mathbf{n} = \{2N\}$. Let φ denote the vector of director functions in eqn. (2.60) and observe that the matrices

$$\mathbf{A} = \int_{-1}^{1} b(z)\varphi(z)\varphi(z)^{\top} dz, \quad \mathbf{B} = \int_{-1}^{1} a(z)\varphi'(z)\varphi'(z)^{\top} dz$$

are independent of d. The Euler–Lagrange equations determine the vector X, of unknown coefficient functions, as the solution of an elliptic system with constant coefficients

$$\begin{aligned} &-d\mathbf{A}\nabla \cdot \mathbf{C}\nabla X(x) + \frac{1}{d}\,\mathbf{B}X(x) = f^{+}(x)\varphi(1) - f^{-}(x)\varphi(-1) \\ &\gamma_0 X = 0 \text{ on } \gamma_D\,, \qquad \gamma_1 X = 0 \text{ on } \gamma_N\,. \end{aligned} \tag{2.61}$$

Remark 8 The elliptic system eqn. (2.61) is singularly perturbed rendering accurate numerical solution, for example by finite elements, nontrivial for small values of the parameter d. Nevertheless, the robust and accurate finite element approximation of eqn. (2.61) is possible with the use of high order elements and nonuniform grids, the so-called *hp* finite element method (see [38, 43] and Section 2.9 below).

When $j = II$, $\mathcal{P} = \{\omega\}$ and $N = 0$, the hierarchic models exactly reproduce the first term u_0^{II} in the asymptotic expansion eqn. (2.21).

Proposition 2.12 *There holds*

$$u^{II}(\{\omega\}, 0) = u_0^{II}.$$

Proof From the definition eqn. (2.57) of $u^{II}(\{\omega\}, 0)$ we obtain

$$u^{II}(\{\omega\}, 0)(x, y) = X_0(x)\psi^{II}(y/d).$$

From the solvability of eqn. (2.16) for $k = 1$ and $j = II$ it follows that

$$\frac{1}{2}\,\psi_0^{II} = \int_{-1}^{1} b(z)dz.$$

From eqn. (2.61) we find the Euler–Lagrange equation for X_0:

$$d\,\psi_0^{II} \int_{-1}^{1} b(z)dz AX_0 = 2f^{II}(x).$$

Hence $X_0(x) = \frac{1}{d}\,A^{-1}f^{II}(x)$ and the assertion follows. □

Another important property of the modelling error $e(\mathcal{P}, \mathbf{n})$ is given in

Proposition 2.13

$$\int_{-d}^{d} e(\mathbf{n})(x, y)dy = 0 \ a.e.\ x \in \omega.$$

Proof The assertion follows from eqn. (2.58) and the fact that ψ_0^{II} is constant along with Lemma 2.8. □

2.5 *A priori* estimates of the modelling error

In the present section we prove various *a priori* error estimates for the modelling error $e(\mathcal{P}, \mathbf{n})$. First, for three-dimensional solutions that are free of boundary layers, we show that hierarchic models converge to the three-dimensional solution as $t \to 0$ at an optimal rate. Then we consider the general case with incompatible data and show that hierarchic models with uniform model order converge to optimal order on interior subdomains of the midsurface ω. In particular, inaccurate modelling at the edge of the plate does not pollute the quality in the interior. Finally we show that by increasing the model order in a boundary strip of width $O(t)$, the optimal asymptotic convergence rates given in eqn. (2.24) can be recovered by the hierarchic models.

2.5.1 *Compatible data*

Selecting the director functions as in eqn. (2.60) means that the asymptotic parts u_N^j of the exact solution u can be approximated to optimal order in d.

Proposition 2.14 *Let $\mathcal{P} = \{\omega\}$ and assume the data f^j are smooth, so that $A^N f^j \in L^2(\omega)$, and compatible so that eqn. (2.25) holds for $k = 0, ..., N-1$ and $j = I, II$. Then, with model order $\mathbf{n} = \{2N\}$, the modelling errors $e^j(\mathbf{n}) = u^j - u^j(\mathbf{n})$ may be estimated by the asymptotic, a priori bound*

$$\left\| e^j(\mathbf{n}) \right\|_{E(\Omega)} \leq C(N) \begin{cases} d^{2N-1/2} \left\| A^N f^I \right\|_{L^2(\omega)} & \text{for } j = I, N \geq 1 \\ d^{2N+1/2} \left\| A^N f^{II} \right\|_{L^2(\omega)} & \text{for } j = II, N \geq 0 \end{cases} \tag{2.62}$$

Proof The assertion follows immediately from the optimality condition eqn. (2.59) and the fact that, with the choice eqn. (2.60) for the director functions, the asymptotic parts u_N^j in the decomposition eqn. (2.20) belong to the subspaces $\mathcal{H}^j(\mathbf{n})$ and, owing to the compatibility conditions eqn. (2.25), the boundary layers U_N^j vanish. Therefore we select $v = u_N^j$ in eqn. (2.59) and obtain the assertion with the aid of eqn. (2.24). □

Remark 9 If $N = 0$ then eqn. (2.62) holds for all $f^{II} \in L^2(\omega)$.

2.5.2 *Boundary layer resolution*

If the data f^j are incompatible but smooth then, as we showed in Theorem 2.4, the three-dimensional solution contains the boundary layer components U_N^j. We now show that if a three-dimensional approximation, or an increased model order, is used in the boundary layer then the models converge with the optimal asymptotic rate given in eqn. (2.24) in the whole of the domain Ω. This means that the boundary layers have been resolved.

Partition the midsurface ω as follows

$$\mathcal{P} = \{\omega_t, \sigma_t\}, \quad \mathbf{n} = \{N, M\}, \tag{2.63}$$

where $t > 0$ is a parameter to be selected, and set $\omega_t = \{x \in \omega \mid \text{dist}(x, \gamma) > t\}$ along with $\sigma_t = \omega \backslash \overline{\omega}_t$. Consequently, N denotes the model order in the interior and M denotes model order in the boundary layer.

Theorem 2.15 *Let the data be regular, $A^N f^j \in L^2(\omega)$, yet incompatible, so that eqn. (2.25) fails at least for one value of j with $0 \leq j \leq N-1$. Further, let $\mathcal{P}, \mathbf{n}$ be as in eqn. (2.63) with $M = \infty$. That is, a three-dimensional approximation is used in the boundary strip σ_t. Then there holds*

$$\|u - u(\mathcal{P}, \mathbf{n})\|_{E(\Omega)} \leq C_N \begin{cases} t^{2N-1/2} & \text{for } j = I, \\ t^{2N+1/2} & \text{for } j = II \end{cases}$$

provided that

$$t \geq 8NC_4(\omega, n)R(\omega)d\,|\ln d| \tag{2.64}$$

where $R(\omega)$ is as in eqn. (2.38). That is, R essentially only depends on the Lipschitz constant of γ.

Proof The index j is omitted since the argument is identical for $j = I$ and $j = II$. The quasi-optimality eqn. (2.59) means

$$\begin{aligned} \|e(\mathcal{P},\mathbf{n})\|_{E(\Omega)} &\leq \|u-w\|_{E(\Omega)} \\ &\leq \|u-u_N-U_N\|_{E(\Omega)} + \|u_N+U_N-w\|_{E(\Omega)} \,. \end{aligned} \tag{2.65}$$

According to Theorem 2.4 the first term exhibits the desired behavior and it remains to estimate the second one.

Therefore, let $\tilde{\chi} \geq 0$ be a C^∞ cut-off function with

$$\tilde{\chi}(\xi) = \begin{cases} 1 & \text{for} \quad 0 \leq \xi \leq 1/2, \\ 0 & \text{for} \qquad \xi \geq 1 \end{cases}$$

and

$$\chi(x) = \tilde{\chi}(\text{dist}(x,\gamma)/t), \qquad t > 0.$$

For Lipschitz boundaries γ the function dist$(\cdot,\gamma)$ belongs to $W^{1,\infty}(\omega)$, and therefore, we also have $\chi \in W^{1,\infty}(\omega)$. Furthermore

$$\text{supp}(1-\chi(x)) \subseteq \overline{\omega}_{t/2}.$$

Select the function w in eqn. (2.65) to be $w = u_N + \chi U_N \in \mathcal{H}(\mathbf{n})$ to obtain

$$\|u_N+U_N-w\|^2_{E(\Omega)} = \|(1-\chi)U_N\|^2_{E(\Omega_{t/2}\setminus\Omega_t)} + \|U_N\|^2_{E(\Omega_t)} \tag{2.66}$$

where $\Omega_t = \omega_t \times (-d,d)$. We estimate these terms as follows. For the first term,

$$\begin{aligned} \|(1-\chi)U_N\|^2_{E(\Omega_{t/2}\setminus\Omega_t)} &\leq \int_{\Omega_{t/2}\setminus\Omega_t} \Big\{ a\Big(\frac{y}{d}\Big)\Big(\frac{\partial U_N}{\partial y}\Big)^2 \\ &\qquad + \overline{C}b\Big(\frac{y}{d}\Big)|\nabla_x((1-\chi)U_N)|^2 \Big\} dx dy \\ &\leq \int_{\Omega_{t/2}\setminus\Omega_t} \Big\{ a\Big(\frac{y}{d}\Big)\Big(\frac{\partial U_N}{\partial y}\Big)^2 + 2\overline{C}b\Big(\frac{y}{d}\Big)|\nabla_x U_N|^2 \\ &\qquad + \frac{C_9^2(\omega)}{t^2} b\Big(\frac{y}{d}\Big)(U_N)^2 \Big\} dx dy \end{aligned}$$

with $C_9^2 = 2n\overline{C}\|\tilde{\chi}'\|_{L^\infty} R^2(\omega)$ and $R(\omega)$ in eqn. (2.38). With Lemma 2.5 we then obtain the estimate

$$\|(1-\chi)U_N\|^2_{E(\Omega_{t/2}\setminus\Omega_t)} \leq C_{10}^2 \|U_N\|^2_{E(\Omega_{t/2}\setminus\Omega_t)}$$

where

$$C_{10}^2(\omega) = \max\{1 + C_9^2\Lambda^2 d^2 t^{-2}, 2\overline{C}/\underline{C}\}.$$

In particular, due to eqn. (2.64), the constant C_{10} is uniformly bounded as $d \to 0$.

For the second term in eqn. (2.66) we note that the inequality $\exp(\beta t/2) \leq \exp(\beta \mathrm{dist}(x,\gamma))$ is obviously valid on $\omega_{t/2}$. For $\beta \geq 0$ as in eqn. (2.38) we therefore have, with Theorem 2.10,

$$e^{\beta t}\|U_N\|^2_{E(\Omega_{t/2}\setminus\Omega_t)} \leq \|U_N\|^2_{\varphi} \leq C_8^2 \begin{cases} d^{-1}(\Phi(d,I,N))^2 \text{ for } j = I, \\ \\ d(\Phi(d,II,N))^2 \text{ for } j = II \end{cases}$$

provided that β satisfies the condition eqn. (2.41). In an analogous fashion we obtain also the estimate

$$\|U_N\|^2_{E(\Omega_t)} \leq e^{-2\beta t} C_8^2 \begin{cases} d^{-1}(\Phi(d,I,N))^2 \text{ for } j = I, \\ \\ d(\Phi(d,II,N))^2 \text{ for } j = II \end{cases} \tag{2.67}$$

Altogether,

$$\|u_N + U_N - w\|_{E(\Omega)} \leq d^{1/2} e^{-\beta t/2} C_8 (1 + C_{10}^2)^{1/2} \Phi(d,j,N)\,, \quad j \in \{I, II\}\,. \tag{2.68}$$

Now we choose the parameter t in eqn. (2.64) such that eqn. (2.68) and eqn. (2.24) are of the same order of magnitude in d:

$$C_N d^{2N} \|A^N f\|_{L^2(\omega)} \sim e^{-\beta t/2} C_8 (1 + C_{10}^2)^{1/2} \Phi(d,\cdot,N). \tag{2.69}$$

This leads to $\beta = 1/(2C_4 R(\omega) d)$ as in eqn. (2.41) and, with t as in eqn. (2.64) and $0 < d \leq 1$, we obtain

$$\begin{aligned} e^{-\beta t/2} C_8 (1 + C_{10}^2)^{1/2} \Phi(N,f,d) &\leq \exp\left(-\frac{C_{11}|\ln d|}{4C_4 R}\right) C_8 (1 + C_{10}^2)^{1/2} \Phi \\ &= d^{\frac{C_{11}}{4C_4 R}} C_8 (1 + C_{10}^2)^{1/2} \Phi. \end{aligned}$$

Making the selection $C_{11} = 8NC_4 R(\omega)$ gives eqn. (2.69) for sufficiently small d. Finally, we add eqn. (2.68) and eqn. (2.24) in Theorem 2.4 and obtain the assertion with $C_{12} = C_N \|A^N f\|_{L^2(\omega)} + C_8 (1 + C_{10}^2)^{1/2} \Phi(d,\cdot,N)$ and the functional Φ in eqn. (2.54). Since Φ, and thus also C_{12}, is uniformly bounded as $d \to 0$, the proof is complete. □

The proof also shows that for uniform model order N the modelling error is of optimal asymptotic order outside of a boundary layer.

Corollary 2.16 *Let $\mathcal{P} = \{\omega\}$. If t satisfies eqn. (2.64) then the following error estimates hold:*

$$\|u - u^j(N)\|_{E(\Omega_t)} \leq C_N \begin{cases} d^{2N-1/2} & j = I, \\ d^{2N+1/2} & j = II\,. \end{cases}$$

2.6 *A posteriori* estimates of the modelling error

The *a priori* error estimates obtained in the previous section were concerned with the asymptotic behavior of the modelling error as $d \to 0$ under various assumptions on the data f^j. There are several disadvantages to a purely d-asymptotic analysis of the modelling error. For instance, the thickness $t = 2d$ of the structure is fixed *a priori* and cannot be reduced if the error in the asymptotic expansion is found to be too large. Further, numerical values of the constants in the error estimates are in general not known explicitly and any estimate will be overly conservative. Therefore, in the present section, we derive computable *a posteriori* estimators for the modelling error $\|e(\mathcal{P}, \mathbf{n})\|$. In particular, we again focus on the exponentially weighted energy norm eqn. (2.32). Due to Proposition 2.2 the modelling error admits the decomposition

$$\|e(\mathcal{P},\mathbf{n})\|_\varphi^2 = \left\|e^I(\mathcal{P},\mathbf{n})\right\|_\varphi^2 + \left\|e^{II}(\mathcal{P},\mathbf{n})\right\|_\varphi^2 \tag{2.70}$$

and each part will have a separate estimator.

The estimators for the modelling error we propose here are based on the residual when $u(\mathcal{P}, \mathbf{n})$ is inserted into the governing equations eqn. (2.5). The derivation begins with a variational characterization of the modelling error.

2.6.1 *Representation of the residual*

Assume for now that the material is homogeneous and isotropic, so that

$$a = b = 1, \quad C = I \tag{2.71}$$

so that the operator L in eqn. (2.6) is the Laplace operator $-\Delta$. Furthermore, as was pointed out in Remark 1, the directors in Theorem 2.3 are polynomials. In particular, the director functions φ_k are chosen to be Legendre polynomials on $(-1, 1)$,

$$\varphi_k = L_k, \qquad k = 0, 1, 2, \dots \ . \tag{2.72}$$

The space $\mathcal{H}^I(\mathbf{n})$ is formed using only Legendre polynomials of odd order and $\mathcal{H}^{II}(\mathbf{n})$ is formed using only even order polynomials. As already mentioned in Section 2.4, the approximation $u(\mathcal{P}, \mathbf{n})$ depends only on the space spanned by the director functions. However, in the following lemma we show that choosing Legendre polynomials is advantageous for obtaining a representation of the residual.

Lemma 2.17 *Let the model order be uniform in ω, i.e. $\mathcal{P} = \{\omega\}$ and $\mathbf{n} = \{N\}$ with odd N for $j = I$ and even N otherwise. Then the modelling errors $e^j(\mathcal{P}, \mathbf{n})$, $j \in \{I, II\}$, are solutions of the variational problems*

$$e^j \in \mathcal{H}^j(\mathbf{n}) \qquad B(e^j, v) = R^j(v) \qquad \forall v \in \mathcal{H}^j(\mathbf{n}) \tag{2.73}$$

where the residual $R^j(v)$ *is given by*

$$R^j_{\mathbf{n}}(v) = \int_\omega r^j_{\mathbf{n}}(x)\Phi^j_{\mathbf{n}}[v](x)dx \tag{2.74}$$

and the residual fluxes are given by

$$r^j_{\mathbf{n}}(x) = f^j(x) - \gamma_1 u(\mathcal{P}, \mathbf{n})(x) \text{ on } R_+.$$

with

$$\Phi^j_{\mathbf{n}}[v](x) = v(x,d) \pm v(x,-d) - \int_{-d}^{d} v(x,y)\frac{\partial}{\partial y}\left(L_{N+1}\left(\frac{y}{d}\right)\right)dy$$

(with − for $j = I$ *and + for* $j = II$*).*

Proof Using eqn. (2.57) it follows immediately that $e^j(\mathbf{n})$ satisfies eqn. (2.73) with

$$\begin{aligned} R^j_{\mathbf{n}}(v) &= \int_\omega r^j_{\mathbf{n}}(x)\left(v(x,d) \pm v(x,-d)\right)dx \\ &+ \int_\Omega v(x,y)\Delta u^j(\mathbf{n})(x,y)dxdy \\ &- \int_{\Gamma_N} \gamma_1 u^j(\mathbf{n})\gamma_0 v dyds. \end{aligned} \tag{2.75}$$

Since $\left(\gamma_1 u^j(\mathbf{n})\right)(s,y) = \left(\gamma_1 X(s)\right)^\top \varphi(\frac{y}{d})$ and $\gamma_1 X(s) = 0$ (see eqn. (2.62)), the residual vanishes on the Neumann boundary.

Now let $j = I$ and $N = 2m+1$ with $m \in \mathbb{N}_0$. Then

$$\Delta u^I(\mathbf{n}) = \sum_{k=0}^{N} A_{jk}(x)L_k\left(\frac{y}{d}\right)$$

with appropriate coefficients $A_{jk} \in H^{-1}(\omega)$. To determine the coefficients we use eqn. (2.57),

$$R^I_{\mathbf{n}}(v) = 0 \qquad \forall v \in \mathcal{H}^I(\mathbf{n}) \cup \mathcal{H}^{II}(\Omega).$$

Choose the test function $v(x,y) = V(x)L_{2k}\left(\frac{y}{d}\right) \in \mathcal{H}^{II}(\Omega)$ with arbitrary $V \in H^1(\omega,\gamma)$ to obtain $A_{Ik} = 0$ for odd k. For $k = 2\ell+1$ and $0 \le \ell \le m$ we find

$$0 = \int_\omega V(x)\left\{2r^I_{\mathbf{n}}(x) + A_{I,2\ell+1}(x)d\int_{-1}^{1}\left(L_{2\ell+1}(z)\right)^2 dz\right\}dx.$$

Since $V \in H^1(\omega,\gamma)$ was arbitrary, it follows that

$$A_{I,2\ell+1} = -\frac{1}{d}\left(2(2\ell+1)+1\right)r^I_{\mathbf{n}}(x).$$

Hence

$$\Delta u^I(\mathbf{n})(x,y) = -\frac{1}{d} r^I_{\mathbf{n}}(x) \sum_{\ell=0}^{m} (4\ell+3) L_{2\ell+1}\left(\frac{y}{d}\right) = -r^I_{\mathbf{n}}(x) \frac{d}{dy}\left(L_{2m+2}\left(\frac{y}{d}\right)\right).$$

Inserting into eqn. (2.75) yields eqn. (2.74). For $j = II$ and even N we proceed analogously. □

Remark 10 Integration by parts with respect to y in the functional $\Phi^j_{\mathbf{n}}[v]$ in eqn. (2.74) yields an alternative representation of the residual which is sometimes more suitable:

$$\Phi^j_{\mathbf{n}}[v](x) = \int_{-d}^{d} \frac{\partial v}{\partial y}(x,y) L_{N+1}(\frac{y}{d}) dy. \tag{2.76}$$

Remark 11 Lemma 2.17 only considered the case when homogeneous Neumann data are prescribed on Γ_N. We now sketch the necessary modifications of Lemma 2.17 when the non-homogeneous boundary condition

$$\gamma_1 u = g \quad \text{on } \Gamma_N, \tag{2.77}$$

with $g \in L^2(\Gamma_N)$, is prescribed in eqn. (2.5). Then from eqn. (2.75) and eqn. (2.76) we have

$$B(e^j(\mathbf{n}), v) = R^j_{\mathbf{n}}(v) = \int_\omega r^j_{\mathbf{n}}(x) \Phi^j_{\mathbf{n}}[v](x) dx + \int_{\Gamma_N} (g - \gamma_1 u^j(\mathbf{n}))\, \gamma_0 v dy ds.$$

It is easily seen from eqn. (2.58) and eqn. (2.76) that $\Phi^j_{\mathbf{n}}[v] = 0$ for $v \in \mathcal{H}(\mathbf{n})$, and it follows that

$$\int_{\Gamma_N} (g - \gamma_1 u^j(\mathbf{n}))\, \gamma_0 v dy ds = 0 \qquad \forall v \in \mathcal{H}(\mathbf{n}).$$

Remark 12 The residual fluxes $R^j(v)$ in eqn. (2.73) contain no contributions from the boundary Γ. This is a consequence of the assumption that the hierarchic models are solved exactly. For a complete discretization, where the reduced models are discretized by finite elements, further residual terms on Γ appear which are entirely due to the finite element discretization of the reduced model.

Therefore, in particular,

$$\int_{-d}^{d} (g - \gamma_1 u^j(\mathbf{n}))\, (s,y) \varphi_k(\frac{y}{d}) dy = 0 \qquad \text{a.e. } s \in \gamma_N, k = 0, \ldots, N$$

so that $\gamma_1 u^j(\mathbf{n})$ is the 'fiberwise' $L^2(-d,d)$-projection of the data g onto the span $\{\varphi_0, \ldots, \varphi_N\}$ of the director functions.

Furthermore, it is possible to derive the contributions of the volume forces to the residual (see eqn. (2.5)). For the details we refer to Section 3.

2.6.2 *Derivation of the modelling error estimators*

Based on Lemma 2.17 we now derive an *a posteriori* estimator for the modelling error $\|e(\mathcal{P},\mathbf{n})\|_\varphi$ in the exponentially weighted energy norm. We again assume uniform model order, $\mathcal{P}=\{\omega\}$, and homogeneous, isotropic material as in eqn. (2.71).

Observe that due to $e^j \in H_\varphi$, and the stability of the bilinear form $B(\cdot,\cdot)$ on $H_\varphi \times H_{1/\varphi}$ in Theorem 2.6, we have for every admissible weight function φ,

$$\delta \left\|e^j(\mathbf{n})\right\|_\varphi \le \sup_{0\neq\|v\|_{1/\varphi}} \frac{\left|B(e^j(\mathbf{n}),v)\right|}{\|v\|_{1/\varphi}} \qquad j\in\{I,II\} \tag{2.78}$$

with δ as in eqn. (2.35). It follows from eqn. (2.73) that

$$\begin{aligned}\delta^2 \left\|e^j(\mathbf{n})\right\|_\varphi^2 &\le \sup_{0\neq\|v\|_{1/\varphi}} \frac{\left(R^j_{\mathbf{n}}(v)\right)^2}{\|v\|^2_{1/\varphi}} \\ &\le \sup_{0\neq\|v\|_{1/\varphi}} \frac{\left(R^j_{\mathbf{n}}(v)\right)^2}{\displaystyle\int_\Omega \varphi^{-2}\left(\frac{\partial v}{\partial y}\right)^2(x,y)\,dydx}\end{aligned} \tag{2.79}$$

where the supremum is taken over the set

$$M = L^2(\omega, H^1(-d,d)) \cap \left\{v \mid \int_{-d}^{d} v(x,y)dy = 0 \text{ a.e. } x\in\omega\right\}.$$

By eqn. (2.76) and the Schwarz inequality we have

$$\left|\Phi^j_{\mathbf{n}}[v](x)\right|^2 \le \int_{-d}^{d}\left(\frac{\partial v}{\partial y}\right)^2 dy \int_{-d}^{d}(L_{N+1}(y/d))^2\,dy = \frac{2d}{2N+3}\int_{-d}^{d}\left(\frac{\partial v}{\partial y}\right)^2 dy.$$

Hence

$$\begin{aligned}\left|R^j_{\mathbf{n}}(v)\right|^2 &\le \left(\int_\omega \left|r^j_{\mathbf{n}}(x)\right|\left|\Phi^j_{\mathbf{n}}[v](x)\right|dx\right)^2 \\ &\le \left\|\varphi r^j_{\mathbf{n}}\right\|^2_{L^2(\omega)} \int_\omega \varphi^{-2}(x)\left|\Phi^j_{\mathbf{n}}[v]\right|^2 dx \\ &\le \frac{2d}{2N+3}\left\|\varphi r^j_{\mathbf{n}}\right\|^2_{L^2(\omega)} \int_\omega \varphi^{-2}(x)\int_{-d}^{d}\left(\frac{\partial v}{\partial y}\right)^2 dydx.\end{aligned}$$

Inserting this in eqn. (2.79) gives:

Theorem 2.18 *Let $f^j \in L^2(\omega)$ and let φ be any weight function which is admissible in Theorem 2.6. Let δ be defined as in eqn. (2.35). Then the modelling error $\left\|e^j(\mathbf{n})\right\|_\varphi$ of the hierarchic model of uniform order $\mathbf{n}=\{N\}$*

may be bounded a posteriori by

$$\delta^2 \left\| e^j(\mathbf{n}) \right\|_\varphi^2 \leq dC_N \int_\omega \varphi^2(x)(r_\mathbf{n}^j(x))^2 dx, \qquad j \in \{I, II\} \tag{2.80}$$

with

$$C_N = \frac{2}{2N+3}. \tag{2.81}$$

2.6.3 *Laminated materials*

So far, the *a posteriori* estimator has been analyzed only for homogeneous and isotropic materials. We now show how the results can be transferred with minor changes to the general case eqn. (2.6).

Assume that a basis $\{\psi_j\}$ of director functions is given such that the linear hull coincides with the span of the asymptotically optimal director functions constructed in Theorem 2.3 and normalized as follows: for $j = 0, 1, 2, ...$

$$\psi_{2j}(z) = \psi_{2j}(-z), \quad \psi_{2j+1}(z) = \psi_{2j+1}(-z) \tag{2.82}$$

and

$$\int_{-1}^{1} b(z)\psi_j(z)\psi_k(z)dz = \frac{2}{2j+1}\delta_{jk}. \tag{2.83}$$

This is the same normalization as the Legendre polynomials. Due to the recurrence relation eqs. (2.15)–(2.16) for the asymptotically optimal basis functions we obtain, after a calculation identical to the homogeneous case, the following representation for the residual:

$$\begin{aligned} R_\mathbf{n}^j(v) &= \int_\omega r_\mathbf{n}^j(x) \{v(x,d) \pm v(x,-d)\}\, dx \\ &+ \sum_{j=0}^{q} a_{ij} \int_\omega r_\mathbf{n}^j(x) \int_{-d}^{d} b\left(\frac{y}{d}\right) \psi_j\left(\frac{y}{d}\right) v(x,y) dy dx \end{aligned} \tag{2.84}$$

with the minus sign corresponding to $j = I$ and the plus sign to $j = II$. Using the Galerkin orthogonality $R_\mathbf{n}^j(v) = 0$ for all $v \in \mathcal{H}(\mathbf{n})$, we find from eqn. (2.83) that

$$a_{ik} = -(2k+1)\frac{1}{d}\psi_k(1) \qquad 0 \leq k \leq q,$$

with $i = 1$ for even k and $i = 2$ for odd k. The residual fluxes $r_\mathbf{n}^j$ are now given by

$$r_\mathbf{n}^j(x) = f^j(x) - a(1)\frac{\partial u^j(\mathbf{n})}{\partial \vec{n}}(x,d). \tag{2.85}$$

If the model order is uniform, $\mathcal{P} = \{\omega\}$, we argue as in eqs. (2.78)–(2.79) to obtain

$$\gamma_0 \left\| e^j(\mathbf{n}) \right\|_\varphi^2 \leq \sup_{0 \neq v \in M} \left(\Pi_{\mathbf{n}}^j [v] \right)^2 \int_\omega \varphi^2(x) \left(r_{\mathbf{n}}^j(x) \right)^2 dx \tag{2.86}$$

with

$$\Pi_{\mathbf{n}}^j [v] = \frac{v(d) \pm v(-d) - \sum_{j=0}^{n} a_{ij} \int_{-d}^{d} b\left(\frac{y}{d}\right) \psi_j \left(\frac{y}{d}\right) v(y) dy}{\left\{ \int_{-d}^{d} a\left(\frac{y}{d}\right) \left(\frac{\partial v}{\partial y}(y)\right)^2 dy \right\}^{1/2}}$$

where the supremum is now taken over the set

$$M = H^1(-d, d) \cap \left\{ v \mid \int_{-d}^{d} b\left(\frac{y}{d}\right) v(y) dy = 0 \right\}.$$

The problem $\sup_M \Pi_{\mathbf{n}}^j [v]$ again admits a unique maximizing element v^{j*} which is independent of x and satisfies the Euler–Lagrange equations

$$\begin{aligned} \frac{\partial}{\partial y} \left(a\left(\frac{y}{d}\right) \frac{\partial v^{j*}}{\partial y} \right) &= \sum_{j=0}^{n} a_{ij} b\left(\frac{y}{d}\right) \psi_j \left(\frac{y}{d}\right), \\ a(\pm 1) \frac{\partial v^{j*}}{\partial y} &= \begin{cases} 1 & x_3 = d, \quad j = I, \\ \pm 1 & x_3 = -d \quad j = II. \end{cases} \end{aligned} \tag{2.87}$$

The constant in the error estimators is given by

$$\left(\Pi_{\mathbf{n}}^j \left[v^{j*} \right] \right)^2 =: dC^j(\mathbf{n}) \tag{2.88}$$

where $C^j(\mathbf{n})$ is independent of d and must be numerically computed for each choice of material properties a, b. Summarizing,

Theorem 2.19 *Under the assumptions of Theorem* 2.18 *there holds, for* $j \in \{I, II\}$,

$$\gamma_0 \left\| e^j(\mathbf{n}) \right\|_\varphi^2 \leq dC^j(\mathbf{n}) \int_\omega \varphi^2(x) \left(r_{\mathbf{n}}^j(x) \right)^2 dx \tag{2.89}$$

where φ *is as in Theorem* 2.6 *and the residual fluxes* $r_{\mathbf{n}}^j$ *are as in eqn. (2.85).*

2.7 Asymptotic exactness of the error estimator

In the present section we investigate the asymptotic exactness of the modelling error estimator E defined in eqn. (2.80),

$$E^2(j,\mathbf{n},\mathcal{P},f^j) := d\frac{2}{2N+3}\int_\omega \varphi^2(x)\left(r^j_{\mathbf{n}}(x)\right)^2 dx. \tag{2.90}$$

We assume once more that the material is homogeneous and isotropic, so that $L = -\Delta$, and that the model order is uniform, so that $\mathcal{P} = \{\omega\}$ and $\mathbf{n} = \{N\}$. First, we will introduce some notions in connection with *a posteriori* error estimation.

Let $\|\cdot\|$ denote an arbitrary norm on $H^1(\Omega, \Gamma_D)$ and let E be an *a posteriori* estimator for the modelling error $\|e(\mathcal{P},\mathbf{n})\|$. Then we define the *effectivity index* Θ of E with respect to the norm $\|\cdot\|$ to be

$$\Theta = \frac{E}{\|e(\mathcal{P},\mathbf{n})\|}. \tag{2.91}$$

The estimator is a guaranteed upper estimator if $\Theta \geq 1$. It is (κ_-, κ_+)-proper with respect to a class T of data if

$$0 < \kappa_- \leq \Theta \leq \kappa_+ < \infty \qquad \forall f \in T. \tag{2.92}$$

Further, E is *asymptotically exact* if

$$\Theta \longrightarrow 1 \text{ as } d \longrightarrow 0 \quad \forall f \in T. \tag{2.93}$$

The estimator is *locally asymptotically exact* with respect to the class T if eqn. (2.93) holds for the exponentially weighted norms $\|\cdot\|_\varphi$ with φ given in eqn. (2.37).

With these definitions we have immediately

Proposition 2.20 *Let $\varphi = 1$. Then the estimator E, defined by eqn. (2.90), is a guaranteed upper estimator for $\|e(\mathcal{P},\mathbf{n})\|_{E(\Omega)}$.*

For the proof of asymptotic exactness we introduce the set

$$T_\beta = \left\{ f \mid \text{ either } r^j_{\mathbf{n}}(f) = 0 \text{ or } \left\|r^j_{\mathbf{n}}\right\|_{1,\varphi} / \left\|r^j_{\mathbf{n}}\right\|_{0,\varphi} \leq \beta < \infty \right\} \tag{2.94}$$

where

$$\left\|r^j_{\mathbf{n}}\right\|^2_{k,\varphi} = \int_\omega (\varphi(x))^2 \left|\nabla^k_x r^j_{\mathbf{n}}(x)\right|^2 dx, \quad k \in \{0,1\} \text{ and } j \in \{I, II\}.$$

Thus we have

Theorem 2.21 *Let Θ^j, $j \in \{I, II\}$, denote the effectivity indices in eqn. (2.91) with respect to the weighted energy norm eqn. (2.32).*

1°. *If* $f^j \in L^2(\omega)$, *then, for* $j \in \{I, II\}$

$$\Theta^j \geq \kappa_-^j := \left(1 - \frac{4\Lambda^j}{\sqrt{2}} Qd\right)\left(1 + 4\sqrt{2}\Lambda^j Qd\left(1 + \sqrt{2}\Lambda^j Qd\right)\right)^{-1/2}. \quad (2.95)$$

2°. *If* $f^j \in T_\beta$ *then*

$$\Theta^j \leq \kappa_+^j := \left(1 + 6d^2 D_N(\beta^2 + 4Q^2)\right)^{1/2}, \quad j \in \{I, II\} \quad (2.96)$$

with

$$\Lambda^I = \frac{2}{\pi}, \quad \Lambda^{II} = \frac{1}{\pi}, \quad D_N = \frac{1}{(2N+3)^3 - 4}.$$

Moreover, if $\varphi = 1$ *then* $Q = 0$ *and the factor* 6 *in eqn. (2.96) may be replaced by* 2.

Proof 1°. The bound eqn. (2.95) follows from Theorems 2.18 and 2.6 using Remark 3.
2°. To show eqn. (2.96), we select in the error equation (2.73)

$$v(x,y) = \bar{v}(x,y)\varphi^2(x) = v^*\left(\frac{y}{d}\right) r_{\mathbf{n}}^j(x)\varphi^2(x)$$

with $v^*(y/d)$ in eqn. (2.87). Then, using $\left(\Theta_{\mathbf{n}}^j[v^*]\right)^2 = dC_N$ and the $(1,\delta)$-regularity of the bilinear form $B(\cdot,\cdot)$, it follows that

$$\begin{aligned} R_{\mathbf{n}}^j(v) &= dC_N \int_\omega (r_{\mathbf{n}}^j(x))^2(\varphi(x))^2 dx \\ &= B(e^j(\mathcal{P},\mathbf{n}), v) \leq \left\|e^j(\mathcal{P},\mathbf{n})\right\|_\varphi \|v\|_{1/\varphi}. \end{aligned} \quad (2.97)$$

Since

$$|\nabla_x v|^2 \leq \varphi^2\left(3\varphi^2 |\nabla_x \bar{v}|^2 + 6|\bar{v}|^2 |\nabla_x \varphi|^2\right)$$

we can estimate

$$\begin{aligned} \|v\|_{1/\varphi}^2 &= \int_\Omega \varphi^{-2}(x)\Big\{|\nabla_x v|^2 + \left(\frac{\partial v}{\partial y}\right)^2\Big\} dydx \\ &\leq \int_\Omega \Big\{3\varphi^2(x)|\nabla_x \bar{v}|^2 + 6|\bar{v}|^2|\nabla_x\varphi|^2 + \varphi^2\left(\frac{\partial \bar{v}}{\partial y}\right)^2\Big\} dydx \\ &= \int_\Omega \Big\{|v^*|^2\left(3\varphi^2\left|\nabla_x r_{\mathbf{n}}^j\right|^2 + 6|\nabla_x\varphi|^2\left|r_{\mathbf{n}}^j\right|^2\right) + \varphi^2\left(\frac{dv^*}{dy}\right)^2\left|r_{\mathbf{n}}^j\right|^2\Big\} dydx. \end{aligned}$$

Since

$$\int_{-d}^{d}\left(\frac{dv^*}{dy}\right)^2 dy = dC_N$$

and

$$\int_{-d}^{d}(v^*)^2\, dy = 4d^3 C_N D_N,$$

we obtain with D_N in eqn. (2.96)

$$|\nabla_x \varphi(x)|^2 \leq 2Q^2 \varphi^2(x) \qquad x \in \omega$$

and thus

$$\|v\|_{1/\varphi}^2 \leq dC_N \left\| r_{\mathbf{n}}^j \right\|_{0,\varphi}^2 + 6d^3 D_N \left\{ \left\| r_{\mathbf{n}}^j \right\|_{1,\varphi}^2 + 4Q^2 \left\| r_{\mathbf{n}}^j \right\|_{0,\varphi}^2 \right\}. \tag{2.98}$$

Therefore, from the inequality (2.97) we obtain for every $\varepsilon > 0$

$$2dC_N \left\| r_{\mathbf{n}}^j \right\|_{0,\varphi}^2 \leq \varepsilon \left\| e^j(\mathbf{n}) \right\|_{\varphi}^2 + \frac{1}{\varepsilon} \|v\|_{1/\varphi}^2 .$$

Now we select $\varepsilon_0 > 0$ such that

$$\frac{1}{\varepsilon_0} \|v\|_{1/\varphi}^2 \leq dC_N \left\| r_{\mathbf{n}}^j \right\|_{0,\varphi}^2 = E^2 \tag{2.99}$$

and obtain the lower bound

$$E^2 \leq \varepsilon_0 \left\| e^j(\mathbf{n}) \right\|_{\varphi}^2, \qquad j \in \{I, II\}.$$

We may estimate ε_0 using eqn. (2.98) and eqn. (2.99)

$$\varepsilon_0 = \frac{\|v\|_{1/\varphi}^2}{dC_N \left\| r_{\mathbf{n}}^j \right\|_{0,\varphi}^2} \leq 1 + 6d^2 D_N \left\{ 4Q^2 + \frac{\left\| r_{\mathbf{n}}^j \right\|_{1,\varphi}^2}{\left\| r_{\mathbf{n}}^j \right\|_{0,\varphi}^2} \right\}$$

and then, thanks to our assumption $f^j \in T_\beta$, the assertion eqn. (2.96) follows. □

Remark 13 The condition $f^j \in T_\beta$ is only given in terms of the residual fluxes on the faces and therefore cannot be verified *a priori*, except in the case $N = 0$ when the residual flux coincides with the data. Therefore, the main result of Theorem 2.21 is in the possibility of computing a numerical value for β *a posteriori*. This yields a computable estimate for the effectivity index κ_+^j from eqn. (2.96).

Remark 14 If $f^j \in T_\beta$ with $\beta = \bar{\beta}/d^\rho$ for some $\rho < 1$, the estimator E is an asymptotically exact estimator for $\left\| e^j(\mathbf{n}) \right\|_{E(\Omega)}$ with the weight function φ in eqn. (2.39).

If $\beta = \bar{\beta}/d$, the estimator is asymptotically proper. That is, it bounds the modelling error from above and below up to a factor independent of d so that eqn. (2.92) holds with κ_- and κ_+ independent of d.

In the present section the asymptotic exactness for constant model orders in ω has been investigated. In practice, however, the case of variable model order is of substantial interest in the context of adaptive hierarchic modelling. Suitable guaranteed upper error estimators have not, as yet, been developed but will be a generalization of those given here.

2.8 Spectral exactness of the estimator

In the previous section we investigated conditions under which the constants $\kappa_\pm$ tend to 1 when the thickness t tends to zero. In practice, however, d is a fixed parameter over which one has no control. Now we investigate conditions under which the constants $\kappa_\pm$, corresponding to the estimator E in eqn. (2.90), tend to one as the model order tends to infinity, for a fixed, positive thickness d. In such circumstances, we call the estimator E *spectrally exact.* One sees readily from what has already been shown that $\kappa_+ \to 1$ for $N \to \infty$ if the data belong to the class T_β for some fixed $\beta > 0$. To prove that $\kappa_- \to 1$ for $N \to \infty$, however, the dependence of the constant Λ in Lemma 2.5 on N has to be analyzed more carefully. This is the principal purpose of the present section.

In what follows, let the model order $\mathbf{n} = \{N\}$ be uniform and assume the coefficients are as given in eqn. (2.71), so that the material is homogeneous and isotropic. Further, suppose Dirichlet conditions are prescribed on the edges,

$$\gamma_N = \emptyset. \tag{2.100}$$

The eigensolutions of the problem

$$-\Delta\varphi = \lambda\varphi \quad \text{in } \omega, \quad \varphi = 0 \text{ on } \gamma, \tag{2.101}$$

are denoted by φ_k, λ_k for $k = 1, 2, 3, \ldots$ with the eigenvalues enumerated according to increasing magnitude and the eigenfunctions orthonormalized in $L^2(\omega)$. It is well known that the first eigenfunction φ_1 is positive in ω,

$$\varphi_1(x) > 0 \qquad x \in \omega. \tag{2.102}$$

The following result will be used to determine how the constant C_3 appearing in Lemma 2.5 depends on the model order N.

Lemma 2.22 *For all $k \in \mathbb{N}$ and all $x \in \overline{\omega}$ there holds*

$$\left\| \frac{\varphi_k}{\varphi_1} \right\|_{C^0(\overline{\omega})} \leq C(\omega)\theta_k, \qquad \theta_k = \begin{cases} \lambda_k^{1/2} & \text{if } n = 1, \\ \lambda_k^{1+\delta} & \text{if } n = 2 \end{cases}. \tag{2.103}$$

The proof can be found, for example, in [10].

With the eigenfunctions φ_k of $-\Delta$ in ω we can explicitly obtain the eigenfunctions Φ_{jkm} for the boundary value problem eqn. (2.5) in Ω with homogeneous boundary data. We introduce

$$\psi_{Im}\left(\frac{y}{d}\right) = \sin\left(\mu_{Im}\frac{y}{d}\right), \quad \mu_{Im} = \frac{2m-1}{2}\pi \tag{2.104}$$

and

$$\psi_{IIm}\left(\frac{y}{d}\right) = \cos\left(\mu_{IIm}\frac{y}{d}\right), \quad \mu_{IIm} = m\pi \tag{2.105}$$

$m = 1, 2, 3, \ldots$. Then the eigenfunctions of $-\Delta$ in Ω are given by

$$\Phi_{jkm}(x, \frac{y}{d}) = \varphi_k(x)\psi_{jm}\left(\frac{y}{d}\right), \qquad j \in \{I, II\} \tag{2.106}$$

and the corresponding eigenvalues Λ_{jkm} are

$$\Lambda_{jkm} = \lambda_k + (\mu_{jm})^2/d^2, \qquad j \in \{I, II\}, \quad k, m \in \mathbb{N}. \tag{2.107}$$

The sequences $\{\Phi_{jkm}\}_{k,m\in\mathbb{N}}$ are dense in $\mathcal{H}^j(\Omega)$ and we can therefore expand the modelling error with respect to eigenfunctions:

$$e^j(x, y) = \sum_{k,m\in\mathbb{N}} E_{jkm}\Phi_{jkm}(x, \frac{y}{d}), \qquad j \in \{I, II\}. \tag{2.108}$$

The orthogonality relations satisfied by the eigenfunctions give the following result.

Lemma 2.23 *Let*

$$\rho_{jk} = \int_\omega r^j(x)\varphi_k(x)dx \tag{2.109}$$

be the Fourier expansion of the residual with respect to the eigenfunctions φ_j *and*

$$\beta^N_{jm} = \int_{-1}^{1} \psi'_{jm}(z)L_{N+1}(z)dz. \tag{2.110}$$

Then

$$E_{jkm} = \left(d\lambda_k + (\mu_{jm})^2/d\right)^{-1} \beta^N_{jm}\rho_{jk}. \tag{2.111}$$

Proof With eqn. (2.73) and Remark 10 the assertion follows from

$$B(\Phi_{jkm}, \Phi_{j'ln}) = \left(d\lambda_k + (\mu_{jm})^2/d\right)\delta_{jj'}\delta_{kl}\delta_{mn}.$$

□

We now turn to the proof of the spectral exactness of the modelling error estimator eqn. (2.80). As indicated above, we must show that the constant κ_- tends to 1 as $N \to \infty$. From eqn. (2.94) we see that κ_- depends on Λ in eqn. (2.33). The proof of the estimate (2.33) used only that $e^j \in H_\varphi$. However, the stronger condition eqn. (2.58) holds, that is, for the estimation of the modelling error, the infimum in eqn. (2.33) need only be taken over functions e satisfying eqn. (2.58). All of these functions can be represented in the form (2.108). Thus we need only estimate

$$\inf_{x\in\omega}\inf_{e^j} \frac{\int_{-d}^{d}\left(\frac{\partial e^j}{\partial y}(x, y)\right) dy}{\int_{-d}^{d}\left(e^j(x, y)\right)^2 dy} \geq \frac{1}{d^2\left(\Lambda^j_N\right)^2}, \qquad j \in \{I, II\} \tag{2.112}$$

where the second infimum is taken over all e^j of the form (2.108). A direct calculation reveals

$$\int_{-d}^{d} (e^j(x,y))^2 dy = d^3 \sum_{k,l,m} \rho_{jk}\rho_{jl} \left(1 + d^2 \frac{\lambda_k}{(\mu_{jm})^2}\right)^{-1} \left(1 + d^2 \frac{\lambda_l}{(\mu_{jm})^2}\right)^{-1} \frac{(\beta_{jm}^N)^2}{(\mu_{jm})^4} \varphi_k(x)\varphi_l(x) \tag{2.113}$$

and

$$\int_{-d}^{d} \left(\frac{\partial e^j}{\partial y}\right)^2 dy = d \sum_{k,l,m} \rho_{jk}\rho_{jl} \left(1 + d^2 \frac{\lambda_k}{(\mu_{jm})^2}\right)^{-1} \left(1 + d^2 \frac{\lambda_l}{(\mu_{jm})^2}\right)^{-1} \frac{(\beta_{jm}^N)^2}{(\mu_{jm})^2} \varphi_k(x)\varphi_l(x). \tag{2.114}$$

Since we must bound the infimum eqn. (2.112), we estimate eqn. (2.113) from above and eqn. (2.114) from below. Let

$$\Psi = C(\omega) \sum_{k \geq 2} \left(1 + d^2 \frac{\lambda_k}{(\mu_{j1})^2}\right)^{-1} \theta_k \frac{|\rho_{jk}|}{|\rho_{j1}|}, \quad a = 1 + d^2\lambda_1/(\mu_{j1})^2 \tag{2.115}$$

with $C(\omega)$ and θ_k as in eqn. (2.103). Then

$$\int_{-d}^{d} \left(\frac{\partial e^j}{\partial y}\right)^2 dy \geq \frac{d}{a^2} |\rho_{j1}|^2 |\varphi_1(x)|^2 \{1 - 2\Psi - \Psi^2\} \sum_m \frac{(\beta_{jm}^N)^2}{(\mu_{jm})^2} \tag{2.116}$$

provided that $\rho_{j1} \neq 0$. Moreover

$$\int_{-d}^{d} (e^j(x,y))^2 dy \leq \frac{d^3}{a^2} |\rho_{j1}|^2 |\varphi_1(x)|^2 (1 + \Psi)^2 \sum_m \frac{(\beta_{jm}^N)^2}{(\mu_{jm})^4} \tag{2.117}$$

with odd N for $j = I$ and even N for $j = II$. The infimum is therefore essentially determined by the ratio of the sums in eqn. (2.116) and eqn. (2.117). To estimate it, we proceed as follows.

$$\begin{aligned}
\beta_{jm}^N &= \int_{-1}^{1} \psi'_{jm} L_{N+1} dz \\
&= -\int_{-1}^{1} \psi''_{jm} \int^{z} L_{N+1}(\xi) d\xi dz + \psi'_{jm} \int^{z} L_{N+1}(\xi) d\xi \,\Big|_{z=\pm 1} \\
&= (\mu_{jm})^2 \int_{-1}^{1} \psi_{jm}(z) \int^{z} L_{N+1}(\xi) d\xi dz
\end{aligned}$$

where we used $\psi'_{jm}(\pm 1) = 0$ and $-\psi''_{jm} = (\mu_{jm})^2 \psi_{jm}$. Thus the Fourier coefficients of the antiderivative of L_{N+1} are $\left(\beta_{jm}^N\right)^2 / \left(\mu_{jm}\right)^2$, and

$$\int_{-1}^{z} L_{N+1}(\xi) = \frac{L_{N+2}(z) - L_N(z)}{2N+3}.$$

Hence, with D_N as defined in eqn. (2.96), we have

$$\sum_m \frac{(\beta_{jm}^N)^2}{(\mu_{jm})^4} = \frac{K}{(2N+3)^2} \|L_{N+2} - L_N\|_{L^2(-1,1)}^2 = D_N \frac{4K}{2N+3} \tag{2.118}$$

and

$$\sum_m \frac{(\beta_{jm}^N)^2}{(\mu_{jm})^2} = K \, \|L_{N+1}\|_{L^2(-1,1)}^2 = \frac{2K}{2N+3}. \tag{2.119}$$

Here K denotes a constant which depends on the normalization of ψ_{jm}. The precise value of K is immaterial in what follows since we are only interested in the ratio of the sums. The desired estimate results from eqs. (2.116)–(2.119),

$$\left(\Lambda_N^j\right)^2 \leq 2D_N \frac{(1+\Psi)^2}{1 - 2\Psi - \Psi^2}. \tag{2.120}$$

The announced result on the spectral exactness of the modelling error estimator may now be given.

Theorem 2.24
1°. *Let $f^j \in T_\beta$ with T_β as in eqn. (2.94) and let $\beta = \bar{\beta}(N/d)^{1-\varepsilon}$ for some $\varepsilon > 0$ and $\bar{\beta}$ be given independent of d, N. Then $\kappa_+^j \to 1$ as $d \to 0$ and as $N \to \infty$.*
2°. *Moreover, if*

$$0 \neq \rho_{j1} = \int_\omega r_{\mathbf{n}}^j(x) \varphi_1(x) dx$$

and there exists $\varepsilon > 0$ such that for Ψ defined in eqn. (2.115) there holds

$$\Psi \leq \left(\sqrt{2} - 1\right) \left(1 - D_N^{1-\varepsilon}\right),$$

then the estimator is spectrally exact, i.e.

$$\kappa_-^j \to 1 \quad \textit{for } d \to 0, N \to \infty.$$

Proof The first part of the theorem follows immediately from eqn. (2.96). To obtain the second statement, we note that the estimate (2.120) for Λ_N^j in conjunction with our assumption on Ψ implies the bound

$$\left(\Lambda_N^j\right)^2 = \frac{2}{\sqrt{2}-1} D_N^\varepsilon.$$

In particular therefore $\Lambda_N^j \to 0$ for $N \to \infty$, which along with eqn. (2.95) gives the assertion. □

2.9 Numerical solution of hierarchic models by the *hp* version finite element method

So far, we have analyzed the modelling error in the *semidiscrete setting* assuming that the elliptic system eqn. (2.61) was solved exactly. However, in general this system must also be solved numerically using, for example, the *hp* version finite element method. We shall present some results on the regularity, and the *hp* finite element approximation of, the solution of eqn. (2.61). Throughout this subsection, we assume that $C = 1$ in eqn. (2.6), so that $Au = -\Delta u$.

2.9.1 *Model uncoupling*

A change of basis for span$\{\varphi_i : 0 \leq i \leq N\}$ does not affect the approximate solution $u^j(\mathcal{P}, \mathbf{n})$ corresponding to the hierarchic model. A key technical observation is that by using a special basis for the space span$\{\varphi_i\}$ of director functions in eqn. (2.61), the elliptic operator

$$\mathbf{L}X = -\mathbf{A}\Delta X + \frac{1}{d^2}\mathbf{B}X. \tag{2.121}$$

does, in fact, decouple.

Lemma 2.25 *Let $\varphi = \{\varphi_0, ..., \varphi_N\}^\top$ denote any vector of $N+1$ linearly independent functions in $H^1(-1,1)$ with $\varphi_0 = 1$, and define the Gram matrices*

$$\mathbf{A} = \int_{-1}^{1} b(z)\varphi\varphi^\top dz, \qquad \mathbf{B} = \int_{-1}^{1} a(z)\varphi'\varphi'^\top dz.$$

Further, let $\mathbf{Q}$ denote the $(N+1)\times(N+1)$ matrix whose columns q consist of the eigenvectors of

$$\mathbf{B}q = \sigma^2\mathbf{A}q \tag{2.122}$$

with the normalization $q^\top\mathbf{A}q = \mathbf{1}$. The corresponding eigenvalues are denoted by $0 = \sigma_0 < \sigma_1 \leq \sigma_2 \leq ... \leq \sigma_N$. Then the matrix operator $\mathbf{L}$ in eqn. (2.121) is diagonal if the basis

$$\psi = \mathbf{Q}^\top\varphi \tag{2.123}$$

is used for the hierarchic modelling.

Thus, with respect to the basis ψ we have $u(\mathbf{n}) = \sum_{i=0}^{N} X_i(x)\psi_i(y/d)$, with modified X_i, and the reduced system eqn. (2.61) becomes

$$\begin{aligned} -\Delta X_0 &= \frac{1}{d}\left(f^+(x) - f^-(x)\right), \\ -\varepsilon_i^2 \Delta X_i + X_i &= \frac{d}{\sigma_i^2}\left(f^+(x)\psi_i(1) - f^-(x)\psi_i(-1)\right), \\ & \quad i = 1, ..., N \end{aligned} \tag{2.124}$$

where $\varepsilon_i = d/\sigma_i$. Moreover, for later use we record the following consequence of Lemma 2.25.

Corollary 2.26 *With the basis ψ defined in eqn. (2.123), we have for any $X \in \left[H^1(\omega)\right]^{N+1}$ and any $0 < d \leq 1$ with $\sigma_0 = 0$*

$$\left\|X^\top \psi\right\|_{E(\Omega)}^2 = \sum_{i=0}^{N} d\,|X_i|_{H^1(\omega)}^2 + \sigma_i^2 d^{-1}\,\|X_i\|_{L^2(\omega)}^2. \tag{2.125}$$

Remark 15 1°. The ψ_i are approximations to the eigenfunctions of

$$-(a(z)\psi')' + b(z)\psi = \lambda\psi \text{ in } (-1,1) \text{ subject to } a\psi'(\pm 1) = 0 \tag{2.126}$$

from the subspace span$\{\varphi_0, \varphi_1, ..., \varphi_N\}$ spanned by the director functions used in the modelling and σ_i^2 are the corresponding eigenvalue approximations. Evidently, these functions are *not* hierarchic, since they must be computed for each N by an orthogonalization process. Since N is typically small, however, computing the eigenvalues and eigenvectors of eqn. (2.122) is a routine computation with packages like EISPACK.

2°. The basis functions ψ are orthogonal in the $L^2(-1,1)$ inner product as well as in the $H^1(-1,1)$ inner product, weighted with the coefficient functions a and b, respectively. This basis is also most useful for the iterative solution of linear systems resulting from *hp* finite element discretization of hierarchic models, see [36].

2.9.2 *Regularity of the reduced solution*

To discuss the approximate solution of eqn. (2.124) by the *hp* version finite element method, we first address the *regularity of the solution*. Assume for simplicity that

$$\gamma_D = \gamma, \qquad \gamma \text{ is smooth.} \tag{2.127}$$

Unsmooth boundary and mixed boundary conditions will be addressed below. In ω, consider a typical component of eqn. (2.124), for instance

$$-\varepsilon^2 \Delta X + X = f \text{ in } \omega, \quad X = 0 \text{ on } \partial\omega \tag{2.128}$$

where we assume that $f \in C^\infty(\overline{\omega})$ and $\varepsilon = d/\sigma_i$ is small. Then an asymptotic expansion of the solution can be obtained as follows. Introduce *bound-*

ary fitted coordinates ρ, θ in the tubular neighborhood ω_0 of $\partial\omega$,

$$\omega_0 = \{z - \rho \vec{n}_z : z \in \partial\omega, 0 < \rho \leq \rho_0\}, \tag{2.129}$$

where ρ_0 is a constant less than the minimum radius of curvature of $\partial\omega$, by

$$(\rho, \theta) \to z - \rho \vec{n}_z = (x_1(\theta) - \rho x_2'(\theta), x_2(\theta) + \rho x_1'(\theta)) \,.$$

Here $z = z(\theta) = (x_1(\theta), x_2(\theta))$ denotes a point on $\partial\omega$ and $\vec{n}_z$ is the exterior unit normal vector to ω at $z \in \partial\omega$. Now the asymptotic expansion of the solution X of eqn. (2.128) may be stated as follows.

Theorem 2.27 *Assume that in eqn. (2.128) one has* $f \in H^{4M+2}(\omega)$ *for some* $M \in \mathbb{N}$. *Then*

$$X = X_M + \chi X_M^{BL} + r_M \tag{2.130}$$

where $\chi \in C^\infty([0, \infty))$ *is a cut-off function satisfying*

$$\chi(r) = \begin{cases} 1 \text{ for } 0 \leq r \leq \rho_0/3\,, \\ 0 \text{ for } r \geq 2\rho_0/3 \end{cases} \tag{2.131}$$

with $\left|\chi^{(m)}(r)\right| \leq C(\rho_0, m)$, $m = 0, 1, \ldots$ *with the asymptotic part, boundary layer and remainder given by*

$$X_M(x) = \sum_{i=0}^{M} \varepsilon^{2i} \left(\Delta^i f\right)(x),$$

$$X_M^{BL}(x) = \sum_{i=0}^{M} \varepsilon^i \sum_{k=0}^{i} C_{ki}(\theta)(\rho/\varepsilon)^k e^{-\rho/\varepsilon} = \sum_{i=0}^{M} C_i(\theta)\rho^i e^{-\rho/\varepsilon}, \tag{2.132}$$

$$||r_M||_{H^k(\omega)} \leq c\varepsilon^{2M+2-k}, \qquad k = 0, 1, \ldots, 2M+1 \tag{2.133}$$

where the constant c does not depend on ε, *but may depend on* M *and* χ.

The proof uses classical, asymptotic analysis and can be found, for example, in [52].

Remark 16 The regularity result Theorem 2.27 immediately allows one to deduce that for midsurfaces ω with smooth boundary $\partial\omega$, the residual satisfies $r_{\mathbf{n}}^j \in T_\beta$ for some $\beta = \bar{\beta}/d$ where $\bar{\beta}$ depends on N, but not on d. In particular, in the light of Remark 14 this means that the modelling error estimator E in eqn. (2.90) is (κ_-, κ_+)-proper with $\kappa_\pm$ independent of d.

2.9.3 *hp finite element approximation*

The design of the finite element subspace of $\mathcal{H}_{hp}(\mathbf{n}) \subset H_0^1(\omega)$ used to approximate X is dictated by the need for good approximability of the boundary layer X_M^{BL}, since both X_M and r_M are smooth. The subspace for the approximation of X_M^{BL} in eqn. (2.132) will use shape functions which are, in local coordinates, tensor products of polynomials of degree p. To define the elements, assume that $p\varepsilon \le 1$ and partition ω, as in eqn. (2.63), into a boundary strip σ

$$\sigma = \{(\rho, \theta) : 0 \le \theta < L, 0 < \rho < \rho_0 p\varepsilon/2\},$$

and an interior domain $\omega\backslash\sigma$, where ρ_0 is some fixed constant less than the minimal radius of curvature of $\partial\omega$. We triangulate σ by dividing the arclength interval $(0, L]$ quasiuniformly into subintervals (θ_j, θ_{j+1}), $j = 1, ..., m$. Then the elements of the boundary layer mesh are given by $\sigma_j = \{(\rho, \theta) : \theta_j \le \theta \le \theta_{j+1}, 0 \le \rho \le \rho_0\}$. Each σ_j is now subdivided into two elements,

$$\omega_{1j} = \{(\rho, \theta) : \theta_j \le \theta \le \theta_{j+1}, 0 \le \rho \le \rho_0 p\varepsilon/2\},$$

$$\omega_{2j} = \{(\rho, \theta) : \theta_j \le \theta \le \theta_{j+1}, \rho_0 p\varepsilon/2 \le \rho \le \rho_0\}.$$

In this way, we obtain a mesh $\mathcal{M}_\varepsilon^0 = \{\omega_{1j}, \omega_{2j}\}_{j=1}^m$ in the boundary strip σ. We extend this boundary fitted mesh $\mathcal{M}_\varepsilon^0$ by some compatible, quasi-uniform subdivision $\{\omega_k\}_{k=1,...,n}$ consisting of triangles or quadrilaterals, to the whole of ω, resulting in the mesh

$$\mathcal{M}_\varepsilon = \{\omega_{ij}, \omega_k : i = 1, 2, j = 1, ..., m-1, k = 1, ..., n\}.$$

Now, the *linear mapping* in boundary fitted coordinates

$$\xi = \frac{2}{L}\theta - 1, \quad \eta = \frac{2}{\rho_0}\rho - 1 \tag{2.134}$$

will map σ to the reference element $\hat{\sigma} = (-1, 1)^2$ in the (ξ, η) plane. Local element maps F_K for the elements $K = \omega_{ij} \in \mathcal{M}_\varepsilon$ are now readily constructed. In these local (ξ, η) coordinates, X_M^{BL} defined in eqn. (2.132) becomes

$$\tilde{X}_M^{BL}(\xi, \eta) = \exp\left(-\frac{\rho_0(\eta + 1)}{2\varepsilon}\right) \sum_{i=0}^{M} \varepsilon^i \sum_{k=0}^{i} \tilde{C}_{ki}(\xi)\varepsilon^{-k}\pi_k(\eta) \tag{2.135}$$

where π_k denotes a polynomial of degree k and $\tilde{C}_{ki}$ is a smooth function, independent of ε. So, $\tilde{X}_M^{BL}$ still has a product structure so that one-dimensional boundary layer *hp* approximation results recently given in [38] may be brought to bear by use of a tensor product argument. Thus, the *hp* boundary layer approximation result reads as follows.

Lemma 2.28 *Let the finite element space $S_{hp}(\mathcal{M}_\varepsilon)$ on the triangulation $\mathcal{M}_\varepsilon$ be defined as follows:*

$$\begin{aligned} S_{hp}(\mathcal{M}_\varepsilon) = \{u \in H_0^1(\omega) : \\ u|_K \circ F_K \text{ is a polynomial of degree at most } p, \quad \forall K \in \mathcal{M}_\varepsilon\}. \end{aligned} \tag{2.136}$$

Then there holds

$$\inf_{v\in S_{hp}(\mathcal{M}_\varepsilon)} \left\|\chi X_M^{BL} - v\right\|_{\varepsilon,\omega} \leq C\varepsilon^{1/2}p^{-s} \qquad \forall s > 0 \tag{2.137}$$

where $C = C(s, \rho_0)$ is a constant independent of p and ε.

Since the remaining parts of the solution X in eqn. (2.130) are smooth independently of ε, and in particular, the mesh $\mathcal{M}_\varepsilon$ contains a regular, quasiuniform mesh, this implies the following error estimate for a *hp* finite element approximation X^{hp} of X.

Theorem 2.29 *For the mesh $\mathcal{M}_\varepsilon$ above we have*

$$\inf_{v\in S_{hp}(\mathcal{M}_\varepsilon)} \left\|X - X^{hp}\right\|_{\varepsilon,\omega} \leq Cp^{-s} \qquad \forall s > 0 \tag{2.138}$$

where $C = C(s, \rho_0)$ is independent of p and ε.

The full proof can be found in [52]. Theorem 2.29 says that in order to resolve the boundary layer X_{BL}^M in eqn. (2.132) with spectral accuracy, one layer of elements of width $\rho_0 p\varepsilon$ is required in the boundary fitted coordinates (ρ, θ).

Remark 17 We conjecture that we have in fact robust *exponential convergence*; see [19, 38] for one-dimensional results in this direction.

The preceding result for the scalar problem (2.128) has consequences for the design of efficient *hp* finite element methods for the discretization of the elliptic system (2.61). By Lemma 2.25, for *any* selection of the basis for span$\{\varphi_i : 0 \leq i \leq N\}$, the solution vector X of eqn. (2.61) is a linear combination of the X_i in eqn. (2.124). The length scale of X_i is $\varepsilon_i = d/\sigma_i$, with $\varepsilon_i = 0$ if $\sigma_i = 0$. This shows that the boundary layer resolving elements $\sigma_{1\ell} \in \mathcal{M}_\varepsilon$, or the mesh for X_i, must have a width proportional to pd/σ_i. From the variational principle for the Sturm–Liouville problem (2.126), we find that $(\sigma_i^{(N)})^2 \geq \lambda_i \to \infty$ as $i \to \infty$, so that the values of $\sigma_i^{(N)}$ become large with increasing model order. This means that *for the discretization of the full, coupled system (2.61), where each component is a linear combination of the X_i in eqn. (2.124), a boundary fitted mesh $\mathcal{M}_\varepsilon$ with N layers of widths pd/σ_i, $i = 1, ..., N$, should be used* in order to achieve a robust *hp*-FE discretization. Note that since $\sigma_0 = 0$, no mesh refinement is needed for the component X_0.

So far, we have only considered the case when the boundary $\partial\omega$ is smooth. If ω is, say, a polygon, the above considerations are still valid, but in addition a geometric *hp* mesh must be inserted into the boundary strip σ at each vertex. Then Theorem 2.29 remains valid even for polygonal domains. Further details will be found in [39].

2.9.4 *Error estimation for the fully discrete solution*

So far, we have analyzed two kinds of errors described in the introduction: the Type B modelling error of the dimensional reduction from Ω to ω under the assumption that the lower dimensional equations (2.61) are solved exactly, and the Type C discretization error in the $\|\cdot\|_{\varepsilon,\omega}$ norm of an hp finite element approximation of these equations is negligible. We now show how these two error estimates can be combined to give *a priori* and *a posteriori* error estimates of the fully discrete, dimensionally reduced and (hp finite element) discretized problem, that are robust in t, the thickness of the domain. We define the space of three-dimensional fields $X_{FE}^{\top}\psi$ corresponding to a finite element discretization of the hierarchic model by

$$\mathcal{H}_{hp}(\mathbf{n}) = \{u = X^{\top}\psi : X_i \in S_{hp}(\mathcal{M}_\varepsilon) \subseteq \mathcal{H}(\mathbf{n})\}.$$

Lemma 2.30 *Let $\psi = (\psi_0, \psi_1, ..., \psi_N)^{\top}$ denote the basis eqn. (2.123) constructed in Lemma 2.25, let X^{hp} be a finite element approximation of the solution vector X of the reduced system eqn. (2.61) and let $u_{hp}(\mathbf{n}) = {X^{hp}}^{\top}\psi$ be the corresponding fully discrete approximation of the three-dimensional problem. Then there holds:*

$$\begin{aligned} \|u - u_{hp}(\mathbf{n})\|_{E(\Omega)}^2 &= \|u - u(\mathbf{n})\|_{E(\Omega)}^2 + \|u(\mathbf{n}) - u_{hp}(\mathbf{n})\|_{E(\Omega)}^2 \\ &= \|u - u(\mathbf{n})\|_{E(\Omega)}^2 + \sum_{i=0}^{N} d\left|X_i - X_i^{hp}\right|_{1,\omega}^2 \\ &\quad + \frac{\sigma_i^2}{d}\left\|X_i - X_i^{hp}\right\|_{0,\omega}^2 . \end{aligned} \tag{2.139}$$

Proof For the proof, we set $e_{hp}^{\mathbf{n}} = u - u_{hp}(\mathbf{n})$, $e^{\mathbf{n}} = u - u(\mathbf{n})$ and $e_{hp} = u(\mathbf{n}) - u_{hp}(\mathbf{n})$. Then we have, from the definitions of $u(\mathbf{n})$ and $u_{hp}(\mathbf{n})$, the Galerkin orthogonalities

$$B(e^{\mathbf{n}}, v) = 0 \quad \forall v \in \mathcal{H}(\mathbf{n}), \qquad B(e_{hp}, v) = 0 \quad \forall v \in \mathcal{H}_{hp}(\mathbf{n}).$$

From the inclusion $\mathcal{H}_{hp}(\mathbf{n}) \subset \mathcal{H}(\mathbf{n})$ it follows that $B(e^{\mathbf{n}}, e_{hp}) = 0$ and therefore

$$\begin{aligned} B(e_{hp}^{\mathbf{n}}, e_{hp}^{\mathbf{n}}) &= B(e^{\mathbf{n}} + e_{hp}, e^{\mathbf{n}} + e_{hp}) \\ &= B(e^{\mathbf{n}}, e^{\mathbf{n}}) + B(e_{hp}, e_{hp}) + 2B(e^{\mathbf{n}}, e_{hp}) \\ &= \|e^{\mathbf{n}}\|_{E(\Omega)}^2 + \|e_{hp}\|_{E(\Omega)}^2 . \end{aligned}$$

Since $e_{hp} = \left(X - X^{hp}\right)^{\top}\psi$, we use eqn. (2.125) to evaluate $\|e_{hp}\|_{E(\Omega)}^2$. This completes the proof. □

This result immediately allows one to combine the estimates for the modelling and the finite element discretization error and thereby obtain an estimate for the fully discrete scheme consisting of the hierarchic model with finite element discretization of the models. In particular, note that eqn. (2.139) implies no loss in inserting the $u(\mathbf{n})$ into the error $u - u_{hp}(\mathbf{n})$, so that the triangle inequality is an equality here. Moreover, the identity (2.139) allows one to obtain *robust a posteriori error estimators* for the full discretization of the three-dimensional problem. We describe their derivation. Assume that $\omega \subset \mathbb{R}^2$ is a polygon and denote by $\mathcal{M}$ any shape regular triangulation of ω. In particular, no elongated boundary layer resolving elements as in the meshes $\mathcal{M}_\varepsilon$ are allowed. Let Z be the solution of the boundary value problem

$$-\varepsilon^2 \Delta Z + Z = \bar{f} \text{ in } \omega, \quad Z = 0 \text{ on } \partial\omega \tag{2.140}$$

and let $Z^h \in S_h(\mathcal{M})$ be a piecewise linear, continuous finite element approximation on $\mathcal{M}$. We assume that for every element $K \in \mathcal{M}$, an elemental error indicator η_K^Z has been defined and the global error estimator $\eta_\omega^Z(\varepsilon) = \left(\sum_{K\in\mathcal{M}} (\eta_K^Z)^2\right)^{1/2}$ is (c_1, c_2)-proper,

$$0 < c_1 \le \frac{\eta_\omega^Z}{\|Z - Z_h\|_{\varepsilon,\omega}} \le c_2 < \infty \tag{2.141}$$

where the constants c_i do not depend on ε and h. An example of one such estimator will be presented below. The estimator $\eta_\omega(\varepsilon)$ is applied as follows. For $i = 1, ..., N$, denote by Z_i the solutions to

$$-\varepsilon_i^2 \Delta Z_i + Z_i = (f^+(x)\psi_i(1) - f^-(x)\psi_i(-1)) \text{ in } \omega, \quad Z_i = 0 \text{ on } \partial\omega \tag{2.142}$$

where $\varepsilon_i = d/\sigma_i$. Since $\psi_0 = 1$, we have $\sigma_0 = 0$ and obtain

$$-\Delta Z_0 = f^+(x) - f^-(x) \text{ in } \omega, \quad Z_i = 0 \text{ on } \partial\omega. \tag{2.143}$$

Denote by $\eta_\omega^{Z_i}$ the corresponding error estimators which satisfy eqn. (2.141) with constants independent of ϵ_i. Comparing eqn. (2.142) with eqn. (2.124), we see that $Z_0 = dX_0$ and $Z_i = \sigma_i^2 d^{-1} X_i$, $i = 1, ..., N$, hence $\eta_\omega^{Z_0} = d\eta_\omega^{X_0}$ and $\eta_\omega^{Z_i} = \sigma_i^2 d^{-1} \eta_\omega^{X_i}$. Therefore eqn. (2.141) implies

$$\begin{aligned} 0 < c_1 &\le \frac{\eta_\omega^{X_0}}{\left|X_0 - X_0^h\right|_{1,\omega}} \le c_2 < \infty, \\ 0 < c_1 &\le \frac{\eta_\omega^{X_i}}{\left\|X_i - X_i^h\right\|_{d/\sigma_i,\omega}} \le c_2 < \infty, \qquad i = 1, ..., N. \end{aligned} \tag{2.144}$$

Based on eqn. (2.139) and eqn. (2.144), we define the *global error estimator* η_Ω by

$$\eta_\Omega^2 = \frac{2d}{2N+3} \int_\omega (\varphi(x))^2 (r_\mathbf{n}(x))^2 dx + d\eta_\omega^{X_0} + \sum_{i=1}^{N} \frac{\sigma_i^2}{d} \left(\eta_\omega^{X_i}\right)^2. \tag{2.145}$$

Notice that we still omit the superscript j since the estimator eqn. (2.145) applies separately to the cases $j = I$ and $j = II$ with corresponding residuals. From eqn. (2.139), we have

Theorem 2.31 *Under the assumptions of eqn. (2.92) and eqn. (2.141) on the modelling and discretization error estimators, the combined error estimator η_Ω defined in eqn. (2.145) satisfies*

$$0 < \min\{\kappa_-, c_1\} \leq \frac{\eta_\Omega}{\|u - u_{hp}(\mathbf{n})\|_{E(\Omega)}} \leq \max\{\kappa_+, c_2\}. \tag{2.146}$$

In the light of Remark 16, the bounds in eqn. (2.146) do not depend on the thickness of Ω, and therefore η_Ω is a robust, proper error estimator on thin domains in three dimensions.

One drawback of Theorem 2.31 is that the residual fluxes $r^j_{\mathbf{n}}$ are assumed to be computed by eqn. (2.85) with the exact solutions $u(\mathbf{n})$ of the $\mathbf{n}$-model. In practice, however, we have only at our disposal the finite element approximation $u_{hp}(\mathbf{n})$, yielding corresponding approximate residuals $\tilde{r}^j_{\mathbf{n}}$ to be used in eqn. (2.145), thereby giving rise to a perturbed error estimator $\tilde{\eta}^2_\Omega$. Let us therefore assume $\varphi = 1$ in eqn. (2.145) and investigate the impact of the discretization error on the accuracy of the residuals. For convenience, we drop the superscript j. From the definition (2.85) of the residuals, we have using that $\psi_0 = 1$,

$$\begin{aligned}
\|r_{\mathbf{n}}(x) - \tilde{r}_{\mathbf{n}}(x)\|_{0,\omega} &= a(1) \left\| \left(X(x) - X^{hp}(x)\right)^\top \psi'(1) \right\|_{0,\omega} \\
&\leq C_N \sum_{i=1}^{N} \left\| X_i - X_i^{hp} \right\|_{0,\omega} \\
&\leq C_N \sum_{i=1}^{N} \left\| X_i - X_i^{hp} \right\|_{d/\sigma_i,\omega} \\
&\leq \frac{C_N}{c_1} \sum_{i=1}^{N} \eta_\omega^{X_i}
\end{aligned}$$

which implies

$$\left|\eta^2_\Omega - \tilde{\eta}^2_\Omega\right| \leq d\, \frac{C_N}{c_1} \sum_{i=1}^{N} \left(\eta_\omega^{X_i}\right)^2 .$$

Comparing with eqn. (2.145), we see that

$$\frac{\left|\eta^2_\Omega - \tilde{\eta}^2_\Omega\right|}{\eta^2_\Omega} \leq d^2 \frac{C_N}{c_1 \sigma_1^2}$$

We conclude:

a) by eqn. (2.139), *too large a discretization error may cause unreliability of the a posteriori modelling error estimator*; and

b) *the impact of the discretization error on the accuracy of the residuals $r^j_{\mathbf{n}}$ can be monitored using the estimators $\eta^{X_i}_\omega$, $i = 1, .., N$ used to control the discretization errors.*

A fully adaptive loop, with adaptive modelling and adaptive finite element discretization of the models, must always be nested. In the inner loop, the discretization error must be driven below the desired accuracy. Only then can the modelling error be estimated reliably, and even then at best to the accuracy of the finite element discretization. This can also be interpreted from a different point of view. We are effectively treating an anisotropic three-dimensional problem with an anisotropic discretization. For proper anisotropic refinement, we must reliably identify the source of the contributions from both modelling and discretization error.

It remains to present an estimator η^Z_ω for the singularly perturbed problem eqn. (2.140) which satisfies eqn. (2.141). According to [51], we select for $T \in \mathcal{M}$ the elemental error indicator

$$\eta^Z_T(\varepsilon) = a_T\Bigg[\|\bar{f} + \varepsilon^2 \Delta Z^h - Z^h\|^2_{L^2(T)} + \varepsilon^3 \sum_{e \in \partial T} \left\| \left[\frac{\partial Z^h}{\partial n}\right]^2_{L^2(e)} \right\| \Bigg]^{1/2} \tag{2.147}$$

where

$$a_T = \min\{1, h_T/\varepsilon\}.$$

We remark that in addition to eqn. (2.147), other estimators, such as Bank and Weiser's estimator are also robust, provided they are defined with edge and element residuals weighted as in eqn. (2.147).

2.10 Consequences for *hp* FE discretizations of thin structures

The preceding analysis of hierarchic modelling and the subsequent *hp* finite element discretization of the models allows one to draw preliminary conclusions for the direct *hp* finite element discretization of thin, three-dimensional structures. First, a single layer of three-dimensional brick elements should be built up over a mesh in the midsurface ω. If the material is homogeneous, the transverse shape functions should be polynomials, effectively leading to a three-dimensional *p* version finite element method with anisotropic degree distribution. If the material is laminated, the use of the nonpolynomial transverse shape functions obtained in Theorem 2.3 is recommended. These functions are derived by asymptotic analysis and allow the resolution of the transverse variation of the three-dimensional solution with work that is independent of the number of layers. This disregards the one-dimensional, cross-sectional Neumann problems (2.15), (2.16). Overall, the presented approach may be viewed as a three-dimensional *hp* finite element method with special shape functions. Thanks to the anisotropic nature of the problem and the semidiscrete character of the hierarchic modelling, we gave *a posteriori* error estimates for the modelling error, assuming exact solution of the hierarchic models. For the finite element approximation of the models, we showed that proper *hp* discretization leads to robust algebraic convergence of arbitrary order. Moreover, the modelling and the discretization error were shown to be orthogonal with respect to the energy of the three-dimensional problem, allowing the combination of any robust *a posteriori* error estimator for the two-dimensional problem with the modelling error estimators to obtain a robust estimator for the fully discrete solution on the thin domain Ω. A key issue in computational practice is the separation and independent control of these two errors.

3 Elasticity

In this section we show how the concept of hierarchic modelling and *a posteriori* estimation of the modelling error may be applied to linearized, three dimensional elasticity. We present the main ideas and results for clamped plates, and refer the interested reader to [34] for general edge conditions and complete proofs.

3.1 The plate problem

3.1.1 *Governing equations*

The plate problem is a boundary value problem of three dimensional, linearized elasticity in the domain Ω. It consists in finding a displacement field $u : \Omega \to \mathbb{R}^3$ satisfying the governing equations:

(1) Equilibrium conditions:

$$\mathbf{L}u = -\mathrm{div}\sigma[u] = f \quad \text{in } \Omega, \tag{3.1}$$

with the symmetric stress tensor

$$\sigma = \begin{pmatrix} \sigma_{11} & \sigma_{12} & \sigma_{13} \\ \sigma_{12} & \sigma_{22} & \sigma_{23} \\ \sigma_{13} & \sigma_{23} & \sigma_{33} \end{pmatrix}.$$

(2) Constitutive equations (Hooke's Law):

$$\sigma = A\epsilon[u] \tag{3.2}$$

with a fourth order tensor A and the linearized strain tensor

$$\epsilon[u] = \{\epsilon_{ij}[u]\}_{1\le i,j\le 3}, \qquad \epsilon_{ij}[u] = \frac{1}{2}\left(\frac{\partial u_i}{\partial x_j} + \frac{\partial u_j}{\partial x_i}\right).$$

(3) Essential and natural edge boundary conditions:

$$B_0 u = 0 \quad \text{and} \quad B_1 u = 0 \quad \text{on } \Gamma \tag{3.3}$$

where one of the two boundary operators may vanish.

(4) Prescribed normal tractions on the faces:

$$\sigma[u]\vec{n} = g^{+} \quad \text{on } R_{+}, \quad \sigma[u]\vec{n} = g^{-} \quad \text{on } R_{-} \tag{3.4}$$

where $\vec{n}$ denotes the exterior unit normal vector to $R_{+} \cup R_{-}$.

It will be useful to write the six essential components of σ and ϵ as vectors in $\mathbb{R}^6$,

$$\begin{aligned} \sigma[u] &= (\sigma_{11}, \sigma_{22}, \sigma_{33}, \sigma_{12}, \sigma_{13}, \sigma_{23})^{\top}, \\ \epsilon[u] &= (\epsilon_{11}, \epsilon_{22}, \epsilon_{33}, \epsilon_{12}, \epsilon_{13}, \epsilon_{23})^{\top}. \end{aligned} \tag{3.5}$$

We admit materials for which Hooke's law (3.2) can be written in the form

$$\sigma[u] = \begin{pmatrix} a_{11} & a_{12} & a_{13} & a_{14} & 0 & 0 \\ a_{12} & a_{22} & a_{23} & a_{24} & 0 & 0 \\ a_{13} & a_{23} & a_{33} & a_{34} & 0 & 0 \\ a_{14} & a_{24} & a_{34} & a_{44} & 0 & 0 \\ 0 & 0 & 0 & 0 & a_{55} & a_{56} \\ 0 & 0 & 0 & 0 & a_{56} & a_{66} \end{pmatrix} \epsilon[u] \tag{3.6}$$

with a symmetric, positive definite matrix $A \in \mathbb{R}^{6\times 6}_{sym}$. In particular, the thirteen parameters a_{ij} in eqn. (3.6) characterize homogeneous materials with monoclinic symmetry.

To cast the boundary value problem (3.1)–(3.4) into variational form, we use Green's formula

$$\int_{\Omega} v^{\top} \mathbf{L} u \, dx dy = \mathcal{B}(u, v) - \int_{\partial\Omega} v^{\top} \sigma[u] \vec{n} \, do \qquad \forall u \in H_{\mathbf{L}}(\Omega), v \in \left[H^1(\Omega)\right]^3 \tag{3.7}$$

which, in particular, holds on the midsurfaces ω under consideration. In eqn. (3.7), do denotes the surface measure on $\partial\Omega$ and

$$H_{\mathbf{L}}(\Omega) = \left[H^1(\Omega)\right]^3 \cap \left\{u : \mathbf{L}u \in \left[L^2(\Omega)\right]^3\right\}.$$

The bilinear form $\mathcal{B}(\cdot, \cdot)$ is given by

$$\mathcal{B}(u, v) = (\epsilon[v], \sigma[u])$$

with the inner product (ϵ, σ) defined by

$$(\epsilon, \sigma) = \int_{\Omega} \epsilon : \sigma \, dx dy \qquad \sigma, \epsilon \in \left[L^2(\Omega)\right]^{3\times 3} \tag{3.8}$$

where

$$\begin{aligned}\epsilon[u]:\sigma[u] &= \sum_{1\le i,j\le 3} \epsilon_{ij}[u]\sigma_{ij}[u] \\ &= \epsilon_{11}\sigma_{11}+\epsilon_{22}\sigma_{22}+\epsilon_{33}\sigma_{33}+2\left(\epsilon_{12}\sigma_{12}+\epsilon_{13}\sigma_{13}+\epsilon_{23}\sigma_{23}\right).\end{aligned}$$

For prescribed surface tractions $g^+, g^- \in \left[L^2(\omega)\right]^3$ on the faces $R_\pm$ and volume forces $f \in \left[L^2(\Omega)\right]^3$ we define the load functional $\mathcal{F}(u)$ by

$$\begin{aligned}\mathcal{F}(u) &= \int_\Omega u(x,y)^\top f(x,y)dydx + \int_{R_+} g^+(x)^\top u(x,d)ds^+ \\ &\quad + \int_{R_-} g^-(x)^\top u(x,-d)ds^- \\ &= \int_\Omega u(x,y)^\top f(x,y)dydx \\ &\quad + \int_\omega \left\{g^+(x)^\top u(x,d) - g^-(x)^\top u(x,-d)\right\}dx\end{aligned} \tag{3.9}$$

since $ds^+ = -ds^- = dx_1dx_2$. Then Green's formula (3.7) becomes

$$\mathcal{B}(u,v) = \int_\Gamma v^\top \sigma[u]\vec{n}dsdy + \mathcal{F}(v) \quad \forall v \in \left[H^1(\Omega)\right]^3$$

with ds denoting the surface measure on the boundary curve $\partial\omega$. The boundary integral

$$\int_\gamma \int_{-d}^{d} v^\top \sigma[u]\vec{n}dyds \tag{3.10}$$

should vanish according to eqn. (3.3).

3.1.2 *Edge conditions*

In practical applications a variety of *edge conditions* are used since it is well known that the overall solution of the plate problem, and, in particular, its edge behavior, depend strongly on the kind of edge conditions prescribed. For simplicity, we confine ourselves to homogeneous Dirichlet conditions or totally clamped edges. Everything that follows also works in a more general setting [34].

3.1.3 *Existence of weak solutions*

A weak or variational solution of the plate problem is a displacement field $u : \Omega \to \mathbb{R}^3$ minimizing the primal energy

$$\mathcal{G}(u) = \frac{1}{2}\mathcal{B}(u,u) - \mathcal{F}(u) \tag{3.11}$$

over a suitable subset

$$\mathcal{H}(\Omega) = \left\{u \in \left[H^1(\Omega)\right]^3 : \text{trace}(u) = 0 \text{ on } \Gamma\right\} \subset \left[H^1(\Omega)\right]^3$$

of admissible displacement fields. Any minimizer $u \in \mathcal{H}(\Omega)$ of eqn. (3.11) satisfies the variational Euler–Lagrange equations

$$u \in \mathcal{H}(\Omega), \qquad B(u,v) = \mathcal{F}(v) \quad \forall v \in \mathcal{H}(\Omega). \tag{3.12}$$

Theorem 3.1

(1) *$\mathcal{H}(\Omega)$ is a closed, linear subspace of $\left[H^1(\Omega)\right]^3$.*
(2) *There exists a unique weak solution $u \in \mathcal{H}(\Omega)$ of eqn. (3.12).*

Proof

(1) Since $\gamma = \partial\omega$ is Lipschitz then so is $\partial\Omega$ for $0 < d \leq 1$ (see, for example, Appendix C in [35]). Consequently, the trace operator $\gamma_0 : H^1(\Omega) \to L^2(\Gamma)$ is continuous. Therefore the constraint $B_0 u = 0$ on Γ in eqn. (3.3) is well defined. Evidently, $\mathcal{H}(\Omega) \subset \left[H^1(\Omega)\right]^3$ is a linear subspace. It is also closed (see, for example, [35]) by the continuity of the trace operator.
(2) Since $\mathcal{H}(\Omega)$ is a closed, linear subspace of $\left[H^1(\Omega)\right]^3$, we have Korn's inequality

$$\mathcal{B}(u,u) \geq C(t,\omega)\, \|u\|^2_{H^1(\Omega)} \qquad \forall u \in \mathcal{H}(\Omega)$$

(see, for example, Theorem 2.5 in [25]). The assumptions on the data, the continuity of the trace operator on Lipschitz domains and Korn's inequality imply that

$$\begin{aligned} |\mathcal{F}(v)| &\leq C(t) \left(\|f\|^2_{L^2(\Omega)} + \|g^+\|^2_{L^2(\omega)} + \|g^-\|^2_{L^2(\omega)} \right)^{1/2} \|v\|_{H^1(\Omega)} \\ &\leq C(t,f,g^\pm)\, \|v\|_{E(\Omega)} \end{aligned}$$

for every $v \in \mathcal{H}(\Omega)$. As usual, the energy norm $\|v\|_{E(\Omega)}$ is defined as $(\mathcal{B}(v,v))^{1/2}$. Existence and uniqueness of a weak solution to eqn. (3.12) now follows from the Fredholm alternative and the Riesz representation theorem applied to $\mathcal{H}(\Omega)$ equipped with the energy inner product $\mathcal{B}(\cdot,\cdot)$.

□

Remark 18 The assumption that the data $g^\pm$ and f are square integrable can be weakened if necessary.

3.1.4 *Separation of bending and membrane effects*

The variational solution $u \in \mathcal{H}(\Omega)$ of the plate problem can be decomposed into a *membrane part* u^I and a *bending part* u^{II} as follows: let

$$u^I_\alpha(x,y) = u^I_\alpha(x,-y), \quad \alpha = 1,2, \qquad u^I_3(x,y) = -u^I_3(x,-y), \tag{3.13}$$

and

$$u_\alpha^{II}(x,y) = -u_\alpha^{II}(x,-y), \quad \alpha = 1,2, \qquad u_3^{II}(x,y) = u_3^{II}(x,-y). \quad (3.14)$$

We denote the corresponding sets of admissible displacement fields by $\mathcal{H}^I(\Omega)$ and $\mathcal{H}^{II}(\Omega)$, respectively. These subsets of $\mathcal{H}(\Omega)$ are closed and orthogonal with respect to the inner product induced by the bilinear form $\mathcal{B}(\cdot,\cdot)$ on $\mathcal{H}(\Omega)$,

$$\mathcal{H}(\Omega) = \mathcal{H}^I(\Omega) \oplus \mathcal{H}^{II}(\Omega)$$

or

$$\mathcal{B}(u,v) = 0 \quad \forall u \in \mathcal{H}^I(\Omega), v \in \mathcal{H}^{II}(\Omega). \quad (3.15)$$

This is a consequence of the sparsity structure of the constitutive matrix A in eqn. (3.6) and the dependence of the strains $\epsilon_{ij}[u]$ on y implied by eqn. (3.13) and eqn. (3.14). Thus, u^I and u^{II} can be obtained independently of one another provided that the load functional $\mathcal{F}(u)$ is also split into bending and membrane parts as follows:

$$\mathcal{F}(u) = \mathcal{F}^I(u) + \mathcal{F}^{II}(u)$$

where

$$\begin{aligned} \mathcal{F}^I(u) &= \int_\Omega f^I(x,y)^\top v(x,y)\,dx\,dy + \int_\omega \{g_3^I(x)(v_3^+ - v_3^-)(x) \\ &\qquad + g_\alpha^I(v_\alpha^+ + v_\alpha^-)(x)\}\,dx, \\ \mathcal{F}^{II}(u) &= \int_\Omega f^{II}(x,y)^\top v(x,y)\,dx\,dy + \int_\omega \{g_3^{II}(x)(v_3^+ + v_3^-)(x) \\ &\qquad + g_\alpha^{II}(x)(v_\alpha^+ - v_\alpha^-)(x)\}\,dx. \end{aligned} \quad (3.16)$$

in which we set $v^\pm(x) = v(x, \pm d)$ and summation over repeated indices is implied. The membrane and bending loads are given by

$$\begin{aligned} f_\alpha^I(x,y) &= \frac{1}{2}(f_\alpha(x,y) + f_\alpha(x,-y)), \\ f_3^I(x,y) &= \frac{1}{2}(f_3(x,y) - f_3(x,-y)), \\ g_3^I(x) &= \frac{1}{2}(g_3^+(x) + g_3^-(x)), \\ g_\alpha^I(x) &= \frac{1}{2}(g_\alpha^+(x) - g_\alpha^-(x)), \quad \alpha = 1,2 \end{aligned} \quad (3.17)$$

and

$$\begin{aligned} f_\alpha^{II}(x,y) &= \frac{1}{2}(f_\alpha(x,y) - f_\alpha(x,-y)), \\ f_3^{II}(x,y) &= \frac{1}{2}(f_3(x,y) + f_3(x,-y)), \\ g_3^{II}(x) &= \frac{1}{2}(g_3^+(x) - g_3^-(x)), \\ g_\alpha^{II}(x) &= \frac{1}{2}(g_\alpha^+(x) + g_\alpha^-(x)), \quad \alpha = 1,2. \end{aligned} \tag{3.18}$$

3.2 Hierarchic plate models

Hierarchic plate models are obtained by semidiscretization of the plate problem eqn. (3.12) in the transverse direction in conjunction with an energy projection.

Since we are dealing with an elliptic system, different model orders are admitted for the different displacement components. We approximate each component u_i of the displacement field u by a Legendre series in the transverse direction of degree less than or equal to n_i. The maximal degrees n_i, $i = 1, 2, 3$, of the polynomials are collected in the vector $\mathbf{n} \in \mathbb{N}_0^3$. The solution $u^{\mathbf{n}}$ of the dimensionally reduced plate model of order $\mathbf{n}$ is any minimizer of the total energy $\mathcal{G}(u)$ in eqn. (3.11) over the subspace $\mathcal{H}(\mathbf{n}) \subset \mathcal{H}(\Omega)$ of admissible displacement fields of the form

$$\begin{aligned} u_i^{\mathbf{n}}(x,y) &= \sum_{k=0}^{n_i} X_{ik}^{\mathbf{n}}(x)\psi_{ik}\left(\frac{y}{d}\right) = X_i^{\mathbf{n}}(x)^\top L_i\left(\frac{y}{d}\right), \\ & \qquad X_i^{\mathbf{n}} \in H_0^1(\omega), \quad i = 1,2,3\,. \end{aligned} \tag{3.19}$$

Here the coefficient functions $X_{ik}^{\mathbf{n}} \in H_0^1(\omega)$ may be interpreted as generalized rotations and deflections. We have, analogously to Proposition 2.11:

Proposition 3.2 $\mathcal{H}(\mathbf{n}) \subset \mathcal{H}(\Omega)$ *is a closed, linear subspace.*

The solution of the hierarchic plate model is obtained by energy projection onto $\mathcal{H}(\mathbf{n})$ and may be split, as for the full three dimensional solution, into a membrane part $u^I(\mathbf{n})$ and a bending part $u^{II}(\mathbf{n})$. Each part may be obtained independently by solving

$$u^j(\mathbf{n}) \in \mathcal{H}^j(\mathbf{n}) \quad \mathcal{B}(u^j(\mathbf{n}), v) = \mathcal{F}^j(v) \quad \forall v \in \mathcal{H}^j(\mathbf{n}), \; j \in \{I, II\} \tag{3.20}$$

where $\mathcal{H}^j(\mathbf{n}) = \mathcal{H}(\mathbf{n}) \cap \mathcal{H}^j(\Omega)$ is a closed, linear subspace of $\left[H^1(\Omega)\right]^3$ thanks to Proposition 3.2. This implies, as in Theorem 2.1, that for every $\mathbf{n}$ there exists a unique, dimensionally reduced solution $u^j(\mathbf{n}) \in \mathcal{H}^j(\mathbf{n})$.

For homogeneous materials with constitutive law (3.1), as before, we select the Legendre polynomials L_k of degree k as director functions ψ_{ik}. For $\mathbf{n}, \mathbf{m} \in \mathbb{N}_0^3$, we write $\mathbf{n} \succeq \mathbf{m} \Leftrightarrow n_i \geq m_i$, $i = 1, 2, 3$, with the relation $\mathbf{n} \preceq \mathbf{m}$ defined analogously. Similarly, we write $\mathbf{n} \succ \mathbf{m} \Leftrightarrow n_i > m_i$, $i = 1, 2, 3$ etc.

Remark 19 The selection of polynomial director functions for *homogeneous* materials ensures the asymptotic optimality of the hierarchic models. For *laminated* plates, however, this approach is not optimal and the polynomials must be replaced by certain material dependent director functions ψ_{ik}, such as the ψ_k^j obtained in Theorem 2.1, having better approximation properties [1, 32, 44].

Generally we have the following relations for the dependence of the model orders on the maximal transverse polynomial degree q

$$\begin{aligned} \mathbf{n} &= (2\lfloor q/2 \rfloor, 2\lfloor q/2 \rfloor, 2\lfloor (q-1)/2 \rfloor + 1) && \text{for} \quad j = I, \\ \mathbf{n} &= (2\lfloor (q-1)/2 \rfloor + 1, 2\lfloor (q-1)/2 \rfloor + 1, 2\lfloor q/2 \rfloor) && \text{for} \quad j = II \end{aligned} \tag{3.21}$$

where $\lfloor x \rfloor$ denotes the largest integer less than or equal to x. Here, these will be the only model orders considered.

Remark 20 The Reissner–Mindlin (RM) plate model for homogeneous and isotropic plates is, for $\nu > 0$, *not* contained in this hierarchy. This can be readily verified by comparing eqn. (3.22) for $\mathbf{n} = (1, 1, 0)$ with the equations for the RM model. However, the RM model can be derived by suitably modifying the elastic moduli used in the energy minimization. Further details will be found in [34]. However, the estimators derived below will not allow the estimation of the modelling error. Nevertheless, other estimators based on dual variational principles could be used to obtain computable *a posteriori* estimates for the modelling error (see Section 3.4 in [34]).

Equation (3.20) yields, after evaluation of the integrals in the transverse coordinate, a singularly perturbed elliptic system determining the unknown vector functions $X_i^{\mathbf{n}} = \{X_{ik}^{\mathbf{n}}\}_{k=0}^{n_i}$. For isotropic material, defining the Gram-matrices of the director functions to be

$$\mathbf{A}_{ij} = \int_{-1}^{1} \psi_i \psi_j^{\top} \, dz, \quad \mathbf{B}_{ij} = \int_{-1}^{1} \psi_i' \psi_j'^{\top} \, dz, \quad \mathbf{C}_{ij} = \int_{-1}^{1} \psi_i \psi_j'^{\top} \, dz, \quad 1 \le i, j \le 3,$$

we obtain the following strongly elliptic system in ω for the vector functions:

find $X_i^{\mathbf{n}}$, $i = 1,2,3$ *such that*

$$\left\{ -d \begin{pmatrix} \mu\mathbf{A}_{11} & & \\ & \mu\mathbf{A}_{22} & \\ & & \mu\mathbf{A}_{33} \end{pmatrix} \Delta + d^{-1} \begin{pmatrix} \mu\mathbf{B}_{11} & & \\ & \mu\mathbf{B}_{22} & \\ & & (\lambda+2\mu)\mathbf{B}_{33} \end{pmatrix} - \right.$$

$$\left. \begin{pmatrix} d(\lambda+\mu)\mathbf{A}_{11}\partial_{11}^2 & d(\lambda+\mu)\mathbf{A}_{12}\partial_{12}^2 & (\lambda\mathbf{C}_{13}-\mu\mathbf{C}_{31}^{\mathsf{T}})\partial_1 \\ d(\lambda+\mu)\mathbf{A}_{21}\partial_{21}^2 & d(\lambda+\mu)\mathbf{A}_{22}\partial_{22}^2 & (\lambda\mathbf{C}_{23}-\mu\mathbf{C}_{32}^{\mathsf{T}})\partial_2 \\ (\mu\mathbf{C}_{31}-\lambda\mathbf{C}_{13}^{\mathsf{T}})\partial_1 & (\mu\mathbf{C}_{32}-\lambda\mathbf{C}_{23}^{\mathsf{T}})\partial_2 & 0 \end{pmatrix} \right\}$$

$$\begin{pmatrix} X_1^{\mathbf{n}} \\ X_2^{\mathbf{n}} \\ X_3^{\mathbf{n}} \end{pmatrix} = \begin{pmatrix} \int_{-d}^{d} f_1(x,z)\psi_1(\frac{z}{d})dz\psi_1(\frac{y}{d}) \\ \int_{-d}^{d} f_2(x,z)\psi_2(\frac{z}{d})dz\psi_2(\frac{y}{d}) \\ \int_{-d}^{d} f_3(x,z)\psi_3(\frac{z}{d})dz\psi_3(\frac{y}{d}) \end{pmatrix} + \tag{3.22}$$

$$\begin{pmatrix} g_1^+(x)\psi_1(1) & - & g_1^-(x)\psi_1(-1) \\ g_2^+(x)\psi_2(1) & - & g_2^-(x)\psi_2(-1) \\ g_3^+(x)\psi_3(1) & - & g_3^-(x)\psi_3(-1) \end{pmatrix}$$

with the conditions

$$\gamma_0\,(X_{il}^{\mathbf{n}}) = 0 \ \text{on} \ \gamma, \quad l \in \mathcal{I}_i, \quad i \in \{n, t, 3\}$$

where $X_{nl}^{\mathbf{n}} = X_{1l}^{\mathbf{n}} n_1 + X_{2l}^{\mathbf{n}} n_2$, $X_{tl}^{\mathbf{n}} = X_{1l}^{\mathbf{n}} t_1 + X_{2l}^{\mathbf{n}} t_2$.

3.3 *A posteriori* estimation of the modelling error

We turn to the derivation of *a posteriori* estimators for the modelling error

$$e(\mathbf{n}) = u - u(\mathbf{n}) \tag{3.23}$$

of the hierarchic plate models in the energy norm. We utilize the decomposition of the modelling error into a membrane part $e^I(\mathbf{n})$ and a bending part $e^{II}(\mathbf{n})$ which can be estimated independently of one another due to their orthogonality in energy,

$$e^j(\mathbf{n}) = u^j - u^j(\mathbf{n}), \quad j \in \{I, II\}, \qquad \mathcal{B}(e^I, e^{II}) = 0.$$

In particular, this implies

$$\|e(\mathbf{n})\|_{E(\Omega)}^2 = \left\|e^I(\mathbf{n})\right\|_{E(\Omega)}^2 + \left\|e^{II}(\mathbf{n})\right\|_{E(\Omega)}^2. \tag{3.24}$$

Our modelling error estimators will be based on the residual tractions

$$r_{\mathbf{n}}^j = g^j - g_{\mathbf{n}}^j, \qquad j \in \{I, II\}, \tag{3.25}$$

with $g^j_{\mathbf{n}} = \sigma[u^j(\mathbf{n})]e_3 \mid_{y=d}$ denoting the normal tractions corresponding to the $\mathbf{n}$-plate model obtained directly, without reference to the equilibrium equations, from $u(\mathbf{n})$ via Hooke's law (3.6).

The main result on *a posteriori* modelling error estimation is as follows.

Theorem 3.3 *Let the material be homogeneous with a positive definite constitutive matrix (3.6) satisfying $a_{33} > 0$ and let the model order $\mathbf{n}$ be given with dependence on q as in eqn. (3.21). Assume further that the surface and volume forces are square integrable and that the volume forces $f^j(x, y)$ are, for almost every $x \in \omega$, polynomials with respect to y of degree $\mathbf{m} \preceq \mathbf{n}$. Let the midsurface $\omega \subset \mathbb{R}^2$ be an open, bounded Lipschitz domain. Then for any of the variational edge conditions of Section 2 the estimates for the a posteriori modelling error hold:*

$$\begin{aligned}\|e^I(\mathbf{n})\|^2_{E(\Omega)} \leq\ & t\Bigg\{a(q)\int_\omega \begin{pmatrix} r^I_{\mathbf{n}1} \\ r^I_{\mathbf{n}2}\end{pmatrix}^{\mathsf{T}} \begin{pmatrix} a_{55} & a_{56} \\ a_{56} & a_{66}\end{pmatrix}^{-1} \begin{pmatrix} r^I_{\mathbf{n}1} \\ r^I_{\mathbf{n}2}\end{pmatrix} dx \\ & + \frac{b(q)}{2a_{33}}\int_\omega \left(r^I_{\mathbf{n}3}\right)^2 dx\Bigg\}\end{aligned} \tag{3.26}$$

and

$$\begin{aligned}\|e^{II}(\mathbf{n})\|^2_{E(\Omega)} \leq\ & t\Bigg\{b(q)\int_\omega \begin{pmatrix} r^{II}_{\mathbf{n}1} \\ r^{II}_{\mathbf{n}2}\end{pmatrix}^{\mathsf{T}} \begin{pmatrix} a_{55} & a_{56} \\ a_{56} & a_{66}\end{pmatrix}^{-1} \begin{pmatrix} r^{II}_{\mathbf{n}1} \\ r^{II}_{\mathbf{n}2}\end{pmatrix} dx \\ & + \frac{a(q)}{2a_{33}}\int_\omega \left(r^{II}_{\mathbf{n}3}\right)^2 dx\Bigg\}\end{aligned} \tag{3.27}$$

where $a(q)$ and $b(q)$ are defined by

$$a(q) = \left(2\left\lfloor\frac{q}{2}\right\rfloor + \frac{3}{2}\right)^{-1}, \qquad b(q) = \left(2\left\lfloor\frac{q+1}{2}\right\rfloor + \frac{1}{2}\right)^{-1} \tag{3.28}$$

with $\lfloor x \rfloor = \max\{k \in \mathbb{Z} : k \leq x\}$.

The proof of Theorem 3.3 will be sketched in the remainder of this section. First, we show that the residual tractions $r^j_{\mathbf{n}}$ in eqn. (3.25) are always square integrable over ω and we obtain a variational representation for the modelling error in terms of the residual tractions alone. The modelling error estimate then follows directly from the Schwarz inequality. The constant in the estimate will be analyzed by means of a covering $\mathcal{C}$ of ω by a family of small, closed, axiparallel squares Q with disjoint interiors. The contribution to the constant from each box $Q \times (-d, d)$, $Q \in \mathcal{C}$ is then estimated. This is followed by an asymptotic analysis of the constants in these local estimates as the size of Q tends to zero.

3.3.1 *Variational characterization of the modelling error*

We now investigate the regularity of the residual tractions $r^j_{\mathbf{n}}$ upon which our modelling error estimates will be based.

Lemma 3.4 *Assume that the given surface tractions g^+, g^- are square integrable over the faces $R_\pm$. Then, for any admissible midsurface ω and any variational edge condition, the residual tractions $r^j_{\mathbf{n}}$ in eqn. (3.25) are square integrable.*

Proof Thanks to our assumptions, $g^j \in L^2(\omega)$. By eqn. (3.25), it remains to show that

$$g^j_{\mathbf{n}} = \sigma[u^j(\mathbf{n})]e_3 \mid_{y=d} \in L^2(\omega).$$

This follows from the representation (3.19), the smoothness of the director functions $\psi_{ik} = L_k$ and the fact that $X^{\mathbf{n}}_{ik} \in H^1(\omega)$. □

Remark 21 The natural space for the residual tractions is actually larger than $L^2(\omega)$, and is, in fact, a suitable subspace of $H^{-1/2}(\omega)$. However, the corresponding norm is more difficult to evaluate, explaining why our estimators will be based on the $L^2(\omega)$-norm of the residual tractions. Lemma 3.4 ensures that the estimators will be well defined for all $0 < d \leq 1$ and all model orders $\mathbf{n}$. It should be borne in mind, however, that the *convergence* $\|r^j_{\mathbf{n}}\|_{L^2(\omega)} \to 0$ as $d \to 0$ or $\mathbf{n} \to \infty$ may fail due to a lack of regularity of the three dimensional solution caused by the edge and vertex singularities of the plate problem.

Next, we derive a variational characterization of $e(\mathbf{n})$ essential to the development of the *a posteriori* modelling error estimators.

Lemma 3.5 *Let the model order $\mathbf{n}$ be uniform and satisfy eqn. (3.21). Then the membrane and bending part of the modelling error satisfy the residual equation*

$$e^j(\mathbf{n}) \in \mathcal{H}^j(\Omega) \;\; \mathcal{B}(e^j(\mathbf{n}), v) = \mathcal{R}^j_{\mathbf{n}}(v) \;\; \forall v \in \mathcal{H}^j(\omega), \;\; j \in \{I, II\} \tag{3.29}$$

with

$$\mathcal{R}^j_{\mathbf{n}}(v) = \int_\omega r^j_{\mathbf{n}}(x)^\top \Phi^j_{\mathbf{n}}[v](x)dx + \int_\Omega R^j_{\mathbf{n}}[f](x,y)^\top v(x,y)dydx \tag{3.30}$$

and

$$\Phi^I_{\mathbf{n}}[v] = \begin{pmatrix} \displaystyle\int_{-d}^{d} \frac{\partial}{\partial y} v_1(x,y) L_{2\lfloor q/2 \rfloor+1}(\frac{y}{d})dy \\ \displaystyle\int_{-d}^{d} \frac{\partial}{\partial y} v_2(x,y) L_{2\lfloor q/2 \rfloor+1}(\frac{y}{d})dy \\ \displaystyle\int_{-d}^{d} \frac{\partial}{\partial y} v_3(x,y) L_{2\lfloor (q+1)/2 \rfloor}(\frac{y}{d})dy \end{pmatrix}, \tag{3.31}$$

$$\Phi_{\mathbf{n}}^{II}[v] = \begin{pmatrix} \int_{-d}^{d} \frac{\partial}{\partial y} v_1(x,y) L_{2\lfloor (q+1)/2 \rfloor}(\frac{y}{d}) dy \\ \int_{-d}^{d} \frac{\partial}{\partial y} v_2(x,y) L_{2\lfloor (q+1)/2 \rfloor}(\frac{y}{d}) dy \\ \int_{-d}^{d} \frac{\partial}{\partial y} v_3(x,y) L_{2\lfloor q/2 \rfloor +1}(\frac{y}{d}) dy \end{pmatrix} \tag{3.32}$$

and the residual volume forces are given by

$$R_{\mathbf{n}}^{I}[f] = \begin{pmatrix} f_1^I(x,y) - \sum_{k=0}^{\lfloor q/2 \rfloor} \frac{4k+1}{2} L_{2k}(z) dz \int_{-1}^{1} f_1^I(x,zd) L_{2k}\left(\frac{y}{d}\right) \\ f_2^I(x,y) - \sum_{k=0}^{\lfloor q/2 \rfloor} \frac{4k+1}{2} L_{2k}(z) dz \int_{-1}^{1} f_2^I(x,zd) L_{2k}\left(\frac{y}{d}\right) \\ f_3^I(x,y) - \sum_{k=0}^{\lfloor (q-1)/2 \rfloor} \frac{4k+3}{2} L_{2k+1}(z) dz \int_{-1}^{1} f_3^I(x,zd) L_{2k+1}\left(\frac{y}{d}\right) \end{pmatrix}, \tag{3.33}$$

and

$$R_{\mathbf{n}}^{II}[f] = \begin{pmatrix} f_1^{II}(x,y) - \sum_{k=0}^{\lfloor (q-1)/2 \rfloor} \frac{4k+3}{2} L_{2k+1}(z) dz \int_{-1}^{1} f_1^{II}(x,zd) L_{2k+1}\left(\frac{y}{d}\right) \\ f_2^{II}(x,y) - \sum_{k=0}^{\lfloor (q-1)/2 \rfloor} \frac{4k+3}{2} L_{2k+1}(z) dz \int_{-1}^{1} f_2^{II}(x,zd) L_{2k+1}\left(\frac{y}{d}\right) \\ f_3^{II}(x,y) - \sum_{k=0}^{\lfloor q/2 \rfloor} \frac{4k+1}{2} L_{2k}(z) dz \int_{-1}^{1} f_3^{II}(x,zd) L_{2k}\left(\frac{y}{d}\right) \end{pmatrix}. \tag{3.34}$$

The proof is analogous to that of Lemma 2.17 and is therefore omitted. See, however, [34, 35].

Remark 22 We point out that the densities $R_{\mathbf{n}}^{j}[f]$ of the volume residual forces are the remainders of the fiberwise Legendre expansions for the volume forces. In particular, $R_{\mathbf{n}}^{j}[f]$ vanishes for volume forces $f^j(x,y)$ which are, for every $x \in \omega$, polynomials of degree $\mathbf{m} \preceq \mathbf{n}$ in y. In this case

$$\begin{aligned} 0 &= \int_{\omega} r_{\mathbf{n}}^{I}(x)^{\top} \Phi_{\mathbf{n}}^{I}[v](x) dx \\ &= \int_{\Omega} R_{\mathbf{n}}^{I}[f^I](x,y)^{\top} v(x,y) dy dx \quad \forall v \in \mathcal{H}^I(\mathbf{n}), \\ 0 &= \int_{\omega} r_{\mathbf{n}}^{II}(x)^{\top} \Phi_{\mathbf{n}}^{II}[v](x) dx \\ &= \int_{\Omega} R_{\mathbf{n}}^{II}[f^{II}](x,y)^{\top} v(x,y) dy dx \quad \forall v \in \mathcal{H}^{II}(\mathbf{n}). \end{aligned} \tag{3.35}$$

3.3.2 *Basic error estimate*

It is now straightforward to derive the modelling error estimate based on Lemma 3.5.

Lemma 3.6 *Let $M \in \mathbb{R}^{3\times 3}$ be an arbitrary but nonsingular matrix. Assume that the volume forces $f^j(x,y)$ are polynomials of degree $\mathbf{m} \preceq \mathbf{n}$ in the variable y for a.e. $x \in \omega$. Then we have, for every $0 < d \leq 1$, the modelling error estimate*

$$\left\|e^j(\mathbf{n})\right\|_{E(\Omega)} \leq C_j \left\|M r^j_{\mathbf{n}}\right\|_{L^2(\omega)} \qquad j \in \{I, II\}. \tag{3.36}$$

Here the constants C_j are given by

$$(C_j(d,M))^2 = \sup_{0 \neq v \in \mathcal{H}^j_*(\Omega)} \frac{\int_\omega \left|M^{-\top}\Phi^j_{\mathbf{n}}[v](x)\right| dx}{\mathcal{B}(v,v)}$$

and the supremum is taken over the subset $\mathcal{H}^j_*(\Omega) \subset \mathcal{H}^j(\Omega)$ of admissible displacement fields for which

$$\int_{-d}^{d} v_i(x,y) L_j\left(\frac{y}{d}\right) dy = 0 \text{ a.e. } x, \quad j = 0, \ldots, n_i, \quad i = 1,2,3 \tag{3.37}$$

holds.

Proof Since, by assumption, the volume forces are polynomial over each fiber with degree $\mathbf{m} \preceq \mathbf{n}$, the volume residuals $R^j_{\mathbf{n}}[f]$ drop out of the error estimate according to Lemma 3.5 and Remark 22. Thus, from Lemma 3.5 we have for $j = I, II$

$$\|e^j(\mathbf{n})\|_{E(\Omega)} = \sup_{0 \neq v \in \mathcal{H}^j(\Omega)} \frac{\mathcal{B}(e^j(\mathbf{n}), v)}{\|v\|_{E(\Omega)}} = \sup_{0 \neq v \in \mathcal{H}^j(\Omega)} \frac{\mathcal{R}^j_{\mathbf{n}}(v)}{\|v\|_{E(\Omega)}} \tag{3.38}$$

with

$$\mathcal{R}^j_{\mathbf{n}}(v) = \int_\omega r^j_{\mathbf{n}}(x)^\top \Phi^j_{\mathbf{n}}[v](x) dx = \int_\omega \left(M r^j_{\mathbf{n}}(x)\right)^\top \Phi^j_{\mathbf{n}}[M^{-\top} v](x) dx$$

where $M \in \mathbb{R}^{3\times 3}$ is arbitrary but nonsingular. By the Schwarz inequality,

$$\left(\mathcal{R}^j_{\mathbf{n}}(v)\right)^2 \leq \left\|M r^j_{\mathbf{n}}\right\|^2_{L^2(\omega)} \int_\omega \left|\Phi^j_{\mathbf{n}}[M^{-\top} v](x)\right|^2 dx$$

and therefore, from eqn. (3.38), we obtain the *a posteriori* estimate

$$\|e^j(\mathbf{n})\|_{E(\Omega)} \leq C_j \left\|M r^j_{\mathbf{n}}\right\|_{L^2(\omega)}$$

as asserted. □

It remains only to estimate the constants $C_j(d,M)$ in eqn. (3.36). This is achieved by covering the midsurface ω by families of small, axiparallel squares defined as follows. Let $\varepsilon > 0$ and let $\mathcal{M}(\varepsilon)$ denote a covering of $\mathbb{R}^2$ by closed, axiparallel squares $\mathbf{q}$ of edgelength 2ε with origin at a vertex. Then we define

$$\mathcal{C}(\varepsilon) = \{\mathbf{q} \in \mathcal{M}(\varepsilon) : \mathbf{q} \cap \omega \neq \emptyset\}$$

and

$$\omega_\varepsilon = \text{interior}\left(\bigcup_{\mathbf{q}\in\mathcal{C}(\varepsilon)} \mathbf{q}\right) \supseteq \omega.$$

Thanks to the construction of ω_ε, there exists a constant $C(\gamma)$, independent of ε, such that

$$|\omega_\varepsilon \backslash \omega| \leq C(\gamma)\varepsilon. \tag{3.39}$$

Let Ω_ε denote the set

$$\Omega_\varepsilon = \omega_\varepsilon \times (-d, d) \supseteq \Omega.$$

For $\mathbf{q} \in \mathcal{C}(\varepsilon)$ we denote the set $\mathbf{q} \times (-d, d)$ by Q. We will also need prolongations $\tilde{v}$ of $v \in \mathcal{H}^j_*(\Omega)$ to $\mathcal{H}^j_*(\Omega_\varepsilon)$.

Lemma 3.7 *Assume that $\partial\Omega = \partial(\omega \times (-d, d))$ is Lipschitz. Let $v \in \left[H^1(\Omega)\right]^3$ be such that eqn. (3.37) holds. Let $\tilde{\omega} \supseteq \omega$ be open and bounded, and denote $\tilde{\Omega} = \tilde{\omega} \times (-d, d)$. Then the zero extension $\tilde{v}$ of v to $\left[H^1(\tilde{\Omega})\right]^3$, satisfies eqn. (3.37).*

We are now in a position to derive an estimate for the constants $C_j(M, d)$ in eqn. (3.36).

Let $0 < \theta \leq 1$ be a parameter. For a given $v \in \mathcal{H}^j(\Omega)$, let $\tilde{v} \in \mathcal{H}^j_*(\Omega_d)$ denote the zero extension of v to Ω_d. The evaluation of the bilinear form $\mathcal{B}$ over a subset $\Omega' \neq \Omega$ is written as $\mathcal{B}(\Omega'; u, v)$. Then

$$\begin{aligned} \int_\omega \left|M^{-\top}\Phi^j_{\mathbf{n}}[v]\right|^2 dx &\leq \int_{\omega_{\theta d}} \left|M^{-\top}\Phi^j_{\mathbf{n}}[\tilde{v}]\right|^2 dx \\ &= \sum_{\mathbf{q}\in\mathcal{C}(\theta d)} \int_{\mathbf{q}} \left|M^{-\top}\Phi^j_{\mathbf{n}}[\tilde{v}]\right|^2 dx\,. \end{aligned} \tag{3.40}$$

Now, for every $\mathbf{q} \in \mathcal{C}(\theta d)$

$$\int_{\mathbf{q}} \left|M^{-\top}\Phi^j_{\mathbf{n}}[\tilde{v}]\right|^2 dx \leq D\mathcal{B}(Q; v, v)$$

where

$$D = \sup_{0\neq v\in\mathcal{H}^j_*(Q)} \frac{\int_{\mathbf{q}} \left|M^{-\top}\Phi^j_{\mathbf{n}}[\tilde{v}]\right|^2 dx}{\mathcal{B}(Q; v, v)}.$$

Since the elasticity tensor A in the bilinear form does not depend on x, the supremum D is independent of the particular cube $\mathbf{q}$. Moreover, scaling the variables of integration by $1/d$ we find that $D = d/\Lambda^j(M,\theta)$, where the constants $\Lambda^j(M,\theta)$ are defined by

$$\Lambda^j(M,\theta) = \inf_v \frac{\displaystyle\int_{\mathbf{K}(\theta)} \epsilon[v] : \sigma[v] d\bar{x} d\bar{y}}{\displaystyle\int_{\mathbf{k}(\theta)} (\overline{\Phi}^j_{\mathbf{n}}[v])^\top M^{-\top} M^{-1} \overline{\Phi}^j_{\mathbf{n}}[v] d\bar{x}} \tag{3.41}$$

with $\mathbf{k}(\theta) = (-\theta,\theta)^2$ and $\mathbf{K}(\theta) = \mathbf{k}(\theta) \times (-1,1)$. The functionals $\overline{\Phi}^j_{\mathbf{n}}[v]$ are defined in eqs. (3.31), (3.32) with $d = 1$ and the infimum in eqn. (3.41) is taken over all $v \in \mathcal{H}^j_*(\mathbf{K}(\theta))$.

Now, with eqn. (3.40) we have that

$$\begin{aligned} \int_\omega \left|M^{-\top} \Phi^j_{\mathbf{n}}[v]\right|^2 dx &\leq \frac{d}{\Lambda^j(M,\theta)} \sum_{\mathbf{q}\in\mathcal{C}(\theta d)} \mathcal{B}(Q;\tilde{v},\tilde{v}) \\ &= \frac{d}{\Lambda^j(M,\theta)} \mathcal{B}(\Omega_{\theta d};\tilde{v},\tilde{v}) \end{aligned} \tag{3.42}$$

which implies

$$(C_j)^2 \leq \frac{d}{\Lambda^j(M,\theta)} \sup_{0\neq v\in\mathcal{H}^j_*(\Omega)} \frac{\mathcal{B}(\Omega_{\theta d};\tilde{v},\tilde{v})}{\mathcal{B}(v,v)}. \tag{3.43}$$

The idea is now to let θ tend to zero for fixed v in the supremum in eqn. (3.43). For this purpose, we first investigate the infima $\Lambda^j(M,\theta)$ in eqn. (3.41) in more detail.

3.3.3 *Analysis of* $\Lambda^j(M,\theta)$

We now investigate the constants $\Lambda^j(M,\theta)$ for $\theta \in (0,1]$. Our first result shows that for $\theta \in (0,1]$ the infimum in eqn. (3.41) is positive and is indeed attained. It also characterizes the functions v for which it is attained.

Since eqn. (3.41) is formally a Rayleigh quotient, we consider the eigenvalue problem associated with it. Define bilinear forms

$$b(u,v) = \int_{\mathbf{K}(\theta)} \epsilon[v] : \sigma[u] dx dy \tag{3.44}$$

and

$$c^{j\mathbf{n}}(u,v) = \int_{\mathbf{k}(\theta)} (\overline{\Phi}^j_{\mathbf{n}}[v])^\top M^{-\top} M^{-1} (\overline{\Phi}^j_{\mathbf{n}}[u]) dx \tag{3.45}$$

and consider the eigenvalue problem: *find* $0 \neq u \in W$ *and* $\Lambda \in \mathbb{R}^+_0$ *such that*

$$b(u,v) = \Lambda c^{j\mathbf{n}}(u,v) \qquad \forall v \in W \tag{3.46}$$

where the admissible displacements W are given by

$$W = [H^1(\mathbf{K}(\theta))]^3 \cap \{u \mid \overline{\Phi}^j_{\mathbf{n}}[u] \neq 0\}.$$

Lemma 3.8 *Choose $q \geq 1$ in eqn. (3.21) and let $0 < \theta \leq 1$. Then, for every model order $\mathbf{n}$ satisfying eqn. (3.21), we have:*

1°. *the spectrum of the eigenvalue problem (3.46) is discrete and consists of a sequence $\{\Lambda^j_k\}_{k=1}^{\infty}$ of real eigenvalues which accumulate only at infinity,*

2°. *the eigenfunctions u^j_k corresponding to Λ^j_k are of the form*

$$u^I_k(x,y) = \begin{pmatrix} U^I_{1k}(x)Q_{2\lfloor q/2\rfloor+1}(y) \\ U^I_{2k}(x)Q_{2\lfloor q/2\rfloor+1}(y) \\ U^I_{3k}(x)Q_{2\lfloor (q+1)/2\rfloor}(y) \end{pmatrix},$$

$$u^{II}_k(x,y) = \begin{pmatrix} U^{II}_{1k}(x)Q_{2\lfloor (q+1)/2\rfloor}(y) \\ U^{II}_{2k}(x)Q_{2\lfloor (q+1)/2\rfloor}(y) \\ U^{II}_{3k}(x)Q_{2\lfloor q/2\rfloor+1}(y) \end{pmatrix} \tag{3.47}$$

where $Q_k = (L_{k+1} - L_{k-1})/(2k+1)$ is an antiderivative of the Legendre polynomial of degree k and the U^j_{ik} are certain functions in $H^1(\mathbf{k}(\theta))$.

The constants $\Lambda^j(M,\theta)$ could, for fixed $0 < \theta \leq 1$, be approximated numerically in the usual way by discretizing the eigenvalue problem (3.46) via restriction of both forms to a finite dimensional subspace and numerically solving the resulting finite dimensional, generalized eigenvalue problem. In certain cases, this is sufficient to produce a modelling error estimator.

Thanks to Lemma 3.8, $\Lambda^j(M,\theta)$ is a continuous and positive function of $\theta \in (0,1]$. We now ask whether $\lim_{\theta\to 0+} \Lambda^j(M,\theta)$ exists and, if so, whether it is positive. The answer is given in the following lemma.

Lemma 3.9 *Under the assumptions of Theorem 3.3, and with*

$$M = \begin{pmatrix} m_1 \begin{pmatrix} a_{55} & a_{56} \\ a_{56} & a_{66} \end{pmatrix} & 0 \\ 0 & m_2 a_{33} \end{pmatrix}^{-1/2} \tag{3.48}$$

for arbitrary $m_1, m_2 > 0$ the limits

$$\Lambda^j(M,0) = \lim_{\theta\to 0+} \Lambda^j(M,\theta)$$

exist and are equal to

$$\Lambda^I(M,0) = \min\left\{\frac{1}{2m_1 a(q)}, \frac{1}{m_2 b(q)}\right\} \tag{3.49}$$

and

$$\Lambda^{II}(M,0) = \min\left\{\frac{1}{2m_1 b(q)}, \frac{1}{m_2 a(q)}\right\} \tag{3.50}$$

respectively, where $a(q)$ and $b(q)$ are as in (3.28).

Proof Let $j = I$, $\ell = 2\lfloor \frac{q}{2} \rfloor + 1$, $m = 2\lfloor \frac{q+1}{2} \rfloor$ and let

$$\tilde{A} = \begin{pmatrix} a_{55} & a_{56} \\ a_{56} & a_{66} \end{pmatrix}. \tag{3.51}$$

For $0 < \theta \leq 1$, let $u^I(\bar{x}, \bar{y})$ denote the eigenfunction corresponding to the smallest eigenvalue of the spectral problem (3.46). Then, according to Lemma 3.8,

$$u^I(\bar{x}, \bar{y}) = \begin{pmatrix} U_1(\bar{x})Q_\ell(\bar{y}) \\ U_2(\bar{x})Q_\ell(\bar{y}) \\ U_3(\bar{x})Q_m(\bar{y}) \end{pmatrix}$$

and

$$\begin{aligned} \epsilon[u^I] &= \Big(\partial_1 U_1 Q_\ell, \partial_2 U_2 Q_\ell, U_3 Q'_m, \frac{\partial_2 U_1 + \partial_1 U_2}{2} Q_\ell\,, \\ &\quad \frac{U_1 Q'_\ell + \partial_1 U_3 Q_m}{2}, \frac{U_2 Q'_\ell + \partial_2 U_3 Q_m}{2}\Big)^{\mathsf{T}}. \end{aligned}$$

Thus, using

$$\epsilon_{ij}[u^I]\sigma_{ij}[u^I] = \epsilon_{11}\sigma_{11} + \epsilon_{22}\sigma_{22} + \epsilon_{33}\sigma_{33} + 2\left(\epsilon_{12}\sigma_{12} + \epsilon_{13}\sigma_{13} + \epsilon_{23}\sigma_{23}\right),$$

and noting the form of $u^I(\bar{x}, \bar{y})$, after performing the scaling $x = \bar{x}/\theta$, $dx = \theta^{-2} d\bar{x}$ we get, with $U = (U_1, U_2, U_3)^{\mathsf{T}}$, that

$$\mathcal{B}(\mathbf{k}(\theta); u^I, u^I) = \mathcal{B}_0(\mathbf{k}; U, U) + \theta \mathcal{B}_1(\mathbf{k}; U, U) + \theta^2 \mathcal{B}_2(\mathbf{k}; U, U). \tag{3.52}$$

Here the bilinear forms $\mathcal{B}_i(\mathbf{k}; U, U)$ are independent of θ and are explicitly given by

$$\begin{aligned} \mathcal{B}_0(\mathbf{k}; U, U) &= \int_{-1}^{1} (Q_\ell)^2 dz \int_{\mathbf{k}} \Big\{ a_{11}(\partial_1 U_1)^2 + 2a_{12}\partial_1 U_1 \partial_2 U_2 \\ &\quad + a_{22}(\partial_2 U_2)^2 + \frac{3}{2}\left(\partial_1 U_2 + \partial_2 U_1\right)\left(a_{14}\partial_1 U_1 + a_{24}\partial_2 U_2\right) \\ &\quad + \frac{a_{44}}{2}\left(\partial_2 U_1 + \partial_1 U_2\right)^2 \Big\} dx \\ &+ \int_{-1}^{1} (Q_m)^2 dz \int_{\mathbf{k}} \Big\{ \frac{a_{55}}{2}(\partial_1 U_3)^2 + a_{56}\partial_1 U_3 \partial_2 U_3 \\ &\quad + \frac{a_{66}}{2}(\partial_2 U_3)^2 \Big\} dx, \end{aligned}$$

$$\begin{aligned}\mathcal{B}_1(\mathbf{k};U,U) &= \int_{-1}^{1} Q_\ell Q'_m dz \int_{\mathbf{k}} U_3\Big\{2a_{13}\partial_1 U_1 + 2a_{23}\partial_2 U_2 \\ &\qquad + \frac{3}{2}\, a_{34}\left(\partial_1 U_2 + \partial_2 U_1\right)\Big\}dx \\ &+ \int_{-1}^{1} Q'_\ell Q_m dz \int_{\mathbf{k}} \{a_{55}U_1\partial_1 U_3 + a_{56}\left(U_1\partial_2 U_3 + U_2\partial_1 U_3\right) \\ &\qquad + a_{66}U_2\partial_2 U_3\}\, dx, \\ \mathcal{B}_2(\mathbf{k};U,U) &= a_{33}b(q)\int_{\mathbf{k}}(U_3)^2 dx + \frac{a(q)}{2}\int_{\mathbf{k}}\{a_{55}(U_1)^2 + 2a_{56}U_1U_2 \\ &\qquad + a_{66}(U_2)^2\}\, dx\end{aligned}$$

with $a(q) = \int_{-1}^{1}(Q'_\ell)^2 dz$, $b(q) = \int_{-1}^{1}(Q'_m)^2 dz$. Furthermore we find, with

$$\overline{\Phi}^{I}_{\mathbf{n}}[u^I] = \begin{pmatrix} a(q)U_1(\bar{x}) \\ a(q)U_2(\bar{x}) \\ b(q)U_3(\bar{x}) \end{pmatrix}$$

and $\tilde{U} = (U_1, U_2)^\top$, that

$$c^{j\mathbf{n}}(u^I, u^I) = \theta^2 \mathcal{C}(U,U)$$

with the form $\mathcal{C}(\cdot,\cdot)$ given by

$$\mathcal{C}(U,U) = m_1(a(q))^2 \int_{\mathbf{k}} \left\{\tilde{U}^\top \left(\tilde{A}^{1/2}\right)^\top \tilde{A}^{1/2}\tilde{U}\right\} dx + m_2 a_{33} b(q) \int_{\mathbf{k}} (U_3)^2 dx.$$

Hence

$$\Lambda^I(M,\theta) = \inf_U \frac{\theta^{-2}\mathcal{B}_0(\mathbf{k};U,U) + \theta^{-1}\mathcal{B}_1(\mathbf{k};U,U) + \mathcal{B}_2(\mathbf{k};U,U)}{\mathcal{C}(U,U)}.$$

If $\Lambda^j(M,\theta)$ is to be uniformly bounded as $\theta \to 0$ then we must necessarily have $\mathcal{B}_0(\mathbf{k};U,U) \to 0$. Therefore, in the limit $\theta = 0$, the minimization is constrained to functions

$$U \in \mathcal{N} = \left\{U : \begin{pmatrix} U_1 \\ U_2 \end{pmatrix} \in \operatorname{span}\left\{\begin{pmatrix} 1 \\ 0 \end{pmatrix}, \begin{pmatrix} 0 \\ 1 \end{pmatrix}, \begin{pmatrix} -x_2 \\ x_1 \end{pmatrix}\right\}, U_3 = \text{const.}\right\}$$

If $U \in \mathcal{N}$ then it is easily verified that $\mathcal{B}_1(\mathbf{k};U,U) = 0$. Thus, since $\dim \mathcal{N} = 4$, we have in the limit $\theta \to 0$

$$\Lambda^I(M,0) = \inf_{U\in\mathcal{N}} \frac{\mathcal{B}_2(\mathbf{k};U,U)}{\mathcal{C}(U,U)}.$$

This is a symmetric, generalized eigenvalue problem in $\mathbb{R}^4$. Since

$$\mathcal{N} = \mathcal{N}_1 \times \{1\}, \quad \mathcal{N}_1 = \operatorname{span}\left\{ \begin{pmatrix} 1 \\ 0 \end{pmatrix}, \begin{pmatrix} 0 \\ 1 \end{pmatrix}, \begin{pmatrix} -x_2 \\ x_1 \end{pmatrix} \right\},$$

setting $\tilde{V} = (m_1)^{1/2} \tilde{A}^{1/2} \tilde{U}$ and $V_3 = (m_2 a_{33})^{1/2} U_3$, we have

$$\Lambda^I(M,0) = \min\left\{ \frac{1}{2m_1 a(q)}, \frac{1}{m_2 b(q)} \right\}.$$

as claimed. The proof for $j = II$ is analogous and yields the same expression for $\Lambda^{II}(M,0)$, with $a(q)$ and $b(q)$ exchanged. □

3.3.4 *Proof of Theorem 3.3*

We are now in a position to give the proof of Theorem 3.3. Recall that we have eqn. (3.36) with the constant $C_j(M,\theta)$ bounded by eqn. (3.43).

Let the matrix M be as in eqn. (3.48). We claim that for fixed $v \in \mathcal{H}^j(\Omega)$ and $d > 0$ there holds

$$\lim_{\theta \to 0+} \frac{\mathcal{B}(\Omega_{\theta d}; \tilde{v}, \tilde{v})}{\Lambda^j(M,\theta)} = \frac{\mathcal{B}(\Omega; v, v)}{\Lambda^j(M,0)}. \tag{3.53}$$

To verify the claim, note that $\tilde{v}\mid_\Omega = v$ and write

$$\frac{\mathcal{B}(\Omega_{\theta d}; \tilde{v}, \tilde{v})}{\Lambda^j(M,\theta)} - \frac{\mathcal{B}(\Omega; v, v)}{\Lambda^j(M,0)} = \frac{\mathcal{B}(\Omega_{\theta d}\backslash\Omega; \tilde{v}, \tilde{v})}{\Lambda^j(M,\theta)} + \mathcal{B}(\Omega; v, v) \frac{\Lambda^j(M,0) - \Lambda^j(M,\theta)}{\Lambda^j(M,0)\Lambda^j(M,\theta)}.$$

By eqn. (3.39) we have $|\Omega_{\theta d}\backslash\Omega| \le 2C(\gamma)\theta d^2$. Therefore $\mathcal{B}(\Omega_{\theta d}\backslash\Omega; \tilde{v}, \tilde{v}) \to 0$ as $\theta \to 0$ for fixed v and d. The statement (3.53) then follows from Lemma 3.9. Passing to the limit in eqn. (3.43) and using eqn. (3.36) gives

$$\left\| e^j(\mathbf{n}) \right\|^2_{E(\Omega)} \le \frac{d}{\Lambda^j(M,0)} \left\| M r^j_{\mathbf{n}} \right\|^2_{L^2(\omega)} \qquad j \in \{I, II\}$$

where the constants $\Lambda^j(M,0)$ are defined in eqs. (3.49) and (3.50). Thus, for $j = I$, we obtain the estimate

$$\left\| e^I(\mathbf{n}) \right\|^2_{E(\Omega)} \le d$$

$$\frac{\dfrac{1}{m_1} \displaystyle\int_\omega \begin{pmatrix} r^I_{\mathbf{n}1} \\ r^I_{\mathbf{n}2} \end{pmatrix}^{\mathrm{T}} \begin{pmatrix} a_{55} & a_{56} \\ a_{56} & a_{66} \end{pmatrix}^{-1} \begin{pmatrix} r^I_{\mathbf{n}1} \\ r^I_{\mathbf{n}2} \end{pmatrix} dx + \dfrac{1}{m_2 a_{33}} \displaystyle\int_\omega \left(r^I_{\mathbf{n}3}\right)^2 dx}{\min\left\{ \dfrac{1}{2m_1 a(q)}, \dfrac{1}{m_2 b(q)} \right\}}.$$

This estimate holds for *any* $m_1, m_2 > 0$. Therefore we may minimize the bound with respect to m_i. Choosing $m_2 = m_1\beta/\alpha$ gives the estimate

$$\min_{m \in \mathbb{R}^2_+} \left\{ \frac{a/m_1 + b/m_2}{\min\{\alpha/m_1, \beta/m_2\}} \right\} \le \frac{a}{\alpha} + \frac{b}{\beta}.$$

from where we obtain eqn. (3.26). This completes the proof of Theorem 3.3 for $j = I$. The proof for $j = II$ is analogous, with $a(q)$ and $b(q)$ interchanged.

3.4 Asymptotic exactness of the estimator

In this section we will prove that the modelling error estimator obtained in Theorem 3.3 is asymptotically, as $t \to 0$, and spectrally, as $q \to \infty$, exact provided some additional conditions on the data are satisfied. A similar result for the heat conduction problem in a plate was first obtained in [10].

To state the result, let EST^j be a computable quantity constituting an estimator for the modelling error $e^j(\mathbf{n})$ in energy norm. For instance, the right hand sides of eqs. (3.26) and (3.27) are suitable choices for $(EST^j)^2$, $j = I, II$. Then we define the *effectivity index* of the estimator EST^j by

$$\Theta^j_{\text{eff}} = \frac{EST^j}{\|e^j(\mathbf{n})\|_{E(\Omega)}} \qquad j \in \{I, II\}.$$

Theorem 3.10 *Suppose the assumptions of Theorem 3.3 hold and, in addition, assume that*

$$0 \neq r^j_{\mathbf{n}}(x) \in \left[H^1(\omega)\right]^3.$$

Let

$$D = \begin{pmatrix} \frac{1}{2}\begin{pmatrix} a_{55} & a_{56} \\ a_{56} & a_{66} \end{pmatrix} & 0 \\ 0 & a_{33} \end{pmatrix}^{-1}$$

and define the bilinear forms $\mathcal{B}_i(\cdot,\cdot)$ as eqn. (3.52). Then, for $j \in \{I, II\}$,

$$1 \leq \left(\Theta^j_{\text{eff}}\right)^2 \leq 1 + d\frac{\mathcal{B}_1(\omega; Dr^j_{\mathbf{n}}, Dr^j_{\mathbf{n}}) + d\mathcal{B}_0(\omega; Dr^j_{\mathbf{n}}, Dr^j_{\mathbf{n}})}{\mathcal{B}_2(\omega; Dr^j_{\mathbf{n}}, Dr^j_{\mathbf{n}})}. \tag{3.54}$$

Proof The lower bound for the effectivity index is evident from Theorem 3.3. Therefore, it only remains to prove the upper bound. Assume that $j = I$ and omit the index j throughout the rest of the proof. As in the proof of Theorem 3.3, we set $\ell = 2\lfloor\frac{q}{2}\rfloor + 1$, $m = 2\lfloor\frac{q+1}{2}\rfloor$ and again use $\tilde{A}$ given by eqn. (3.51). Since the volume forces are assumed to be polynomials of degree $\mathbf{m} \preceq \mathbf{n}$ in the transverse variable y, we have the variational characterization eqn. (3.29) of the modelling error with

$$\mathcal{R}_{\mathbf{n}}(v) = \int_\omega r_{\mathbf{n}}(x)^\top \Phi_{\mathbf{n}}[v](x)dx.$$

Now we select $v = v^*$ such that $\mathcal{R}_{\mathbf{n}}(v^*) = (EST)^2$ with EST denoting the computable modelling error estimator in Theorem 3.3. It is easily verified that this holds if

$$v^*(x,y) = dD \begin{pmatrix} Q_\ell\left(\frac{y}{d}\right) r_{\mathbf{n}1}(x) \\ Q_\ell\left(\frac{y}{d}\right) r_{\mathbf{n}2}(x) \\ Q_m\left(\frac{y}{d}\right) r_{\mathbf{n}3}(x) \end{pmatrix}.$$

This implies that

$$r_{\mathbf{n}}(x)^\top \Phi_{\mathbf{n}}[v^*](x) = d\left\{2a(q)\left(\tilde{r}_{\mathbf{n}}(x)\right)^\top \tilde{A}^{-1}\tilde{r}_{\mathbf{n}}(x) + b(q)\left(r_{\mathbf{n}3}(x)\right)^2/a_{33}\right\}$$

with $a(q)$ and $b(q)$ as in eqn. (3.28) and $\tilde{r}_{\mathbf{n}} = (r_{\mathbf{n}1}, r_{\mathbf{n}2})^\top$. Comparison with eqn. (3.26) shows that $\mathcal{R}(v^*) = (EST)^2$. Now, for any $\varepsilon > 0$, we estimate

$$\mathcal{R}_{\mathbf{n}}(v^*) = (EST)^2 = \mathcal{B}(e(\mathbf{n}), v^*) \leq \frac{\varepsilon}{2}\,\|e(\mathbf{n})\|^2_{E(\Omega)} + \frac{1}{2\varepsilon}\,\|v^*\|^2_{E(\Omega)},$$

and by selecting $\varepsilon = \varepsilon_0$ such that

$$\frac{1}{2\varepsilon_0}\,\|v^*\|^2_{E(\Omega)} \leq \frac{1}{2}(EST)^2,$$

we arrive at

$$(EST)^2 \leq \varepsilon_0\,\|e(\mathbf{n})\|^2_{E(\Omega)}.$$

We estimate ε_0 as follows. Thanks to the choice of v^*, we obtain

$$\varepsilon_0 = \frac{\|v^*\|^2_{E(\Omega)}}{\mathcal{R}_{\mathbf{n}}(v^*)}.$$

Now

$$\begin{aligned} \|v^*\|^2_{E(\Omega)} &= \mathcal{B}(v^*, v^*) \\ &= d^3\mathcal{B}_0(\omega; Dr_{\mathbf{n}}, Dr_{\mathbf{n}}) + d^2\mathcal{B}_1(\omega; Dr_{\mathbf{n}}, Dr_{\mathbf{n}}) + d\mathcal{B}_2(\omega; Dr_{\mathbf{n}}, Dr_{\mathbf{n}}) \end{aligned}$$

and from the definition of $\mathcal{B}_2(\omega; U, U)$,

$$\mathcal{B}_2(\omega; U, U) = a_{33}b(q)\int_\omega (U_3)^2\,dx + \frac{a(q)}{2}\int_\omega \begin{pmatrix} U_1 \\ U_2 \end{pmatrix}^\top \tilde{A} \begin{pmatrix} U_1 \\ U_2 \end{pmatrix} dx,$$

so that

$$d\mathcal{B}_2(\omega; Dr_{\mathbf{n}}, Dr_{\mathbf{n}}) = \mathcal{R}_{\mathbf{n}}(v^*) = (EST)^2.$$

The assertion then follows for $j = I$. The proof for $j = II$ is completely analogous. □

The main significance of this result is that it gives a computable upper bound on the effectivity index of the modelling error estimator. Moreover, since the bilinear forms $\mathcal{B}_i(\omega; U, U)$ are given explicitly, we may formulate sufficient conditions for the spectral exactness of the estimator. For this purpose, we introduce the class

$$T(\bar{A},\varepsilon) = \Big\{(g^+,g^-) : \text{ either } r^j_{\mathbf{n}} = 0 \text{ or } \\ |r^j_{\mathbf{n}}|_{H^1(\omega)} \,/\, \|r^j_{\mathbf{n}}\|_{L^2(\omega)} \le \bar{A}(q/t)^{1-\varepsilon}\Big\} \tag{3.55}$$

for some $\overline{A} > 0$ and $\varepsilon \geq 0$ independent of d. Then, we have

Corollary 3.11 *Let $(g^+,g^-) \in T(\bar{A},\varepsilon)$ for some $\bar{A} > 0$ and $\varepsilon > 0$. Then the modelling error estimator in Theorem 3.3 is asymptotically and spectrally exact.*

If $(g^+,g^-) \in T(\bar{A},0)$ for some $\bar{A} > 0$, the estimator in Theorem 3.3 is asymptotically and spectrally uniform. That is, the effectivity index is uniformly bounded with respect to t and q.

Proof Throughout the proof, let C denote a generic, positive constant depending only on the elastic moduli a_{ik}. We observe that

$$\int_{-1}^{1} (Q_\ell)^2\, dz = \frac{4}{(2\ell+1)((2\ell+1)^2-4)},$$

$$\int_{-1}^{1} Q_\ell Q'_m dz = \begin{cases} \dfrac{1}{(2\ell+1)(2\ell\pm 1)} & \text{if } m = \ell \pm 1, \\ 0 & \text{otherwise.} \end{cases}$$

Thus, letting

$$\frac{|r^j_{\mathbf{n}}|^2_{H^1(\omega)}}{\|r^j_{\mathbf{n}}\|^2_{L^2(\omega)}} \le \bar{A}^2 (q/t)^{2-2\varepsilon}$$

we find

$$\begin{aligned} \mathcal{B}_0(\omega; Dr^j_{\mathbf{n}}, Dr^j_{\mathbf{n}}) &\le Cq^{-3} |r^j_{\mathbf{n}}|^2_{H^1(\omega)} \\ &\le C\bar{A}^2 q^{-3+2-2\varepsilon} t^{-2+2\varepsilon} \|r^j_{\mathbf{n}}\|^2_{L^2(\omega)} \end{aligned}$$

and that

$$\begin{aligned} \mathcal{B}_1(\omega; Dr^j_{\mathbf{n}}, Dr^j_{\mathbf{n}}) &\le Cq^{-2}\left\{ \left\|r^j_{\mathbf{n}3}\right\|_{L^2(\omega)} |\tilde{r}^j_{\mathbf{n}}|_{H^1(\omega)} + \|\tilde{r}^j_{\mathbf{n}}\|_{L^2(\omega)} \left|r^j_{\mathbf{n}3}\right|_{H^1(\omega)} \right\} \\ &\le C\bar{A} q^{-2+1-\varepsilon} t^{-1+\varepsilon} \|r^j_{\mathbf{n}}\|^2_{L^2(\omega)} \end{aligned}$$

where we have set $\tilde{r}^j_{\mathbf{n}} = \left(r^j_{\mathbf{n}1}, r^j_{\mathbf{n}2}\right)^\top$. Furthermore,

$$\mathcal{B}_2(\omega; Dr^j_{\mathbf{n}}, Dr^j_{\mathbf{n}}) \ge Cq^{-1} \|r^j_{\mathbf{n}}\|^2_{L^2(\omega)}.$$

Consequently, we have

$$d\frac{\mathcal{B}_1(\omega; Dr^j_{\mathbf{n}}, Dr^j_{\mathbf{n}})}{\mathcal{B}_2(\omega; Dr^j_{\mathbf{n}}, Dr^j_{\mathbf{n}})} + d^2\frac{\mathcal{B}_0(\omega; Dr^j_{\mathbf{n}}, Dr^j_{\mathbf{n}})}{\mathcal{B}_2(\omega; Dr^j_{\mathbf{n}}, Dr^j_{\mathbf{n}})} \leq C\bar{A}(1+\bar{A})(t/q)^{\varepsilon}$$

and the assertion follows. □

The assumption $(g^+, g^-) \in T(\bar{A}, \varepsilon)$ implies, in particular, that the residual tractions $r^j_{\mathbf{n}}$ belong to $\left[H^1(\omega)\right]^3$. This assumption is true whenever the coefficient functions $X^{\mathbf{n}}_i$ belong to $H^2(\omega)$. This is the case, for example, if the data (g^+, g^-) belong to $H^1(\omega)$ componentwise and the edge $\partial\omega$ is smooth. This does not preclude $X^{\mathbf{n}}_i$ having layers so that the solution has components of the form $\exp(-a\mathrm{dist}(x, \partial\omega)/t)$. The constant $a > 0$ appearing here is independent of t but does depend on the model order and the elastic moduli [41]. Owing to the form of the boundary layers, we get

Corollary 3.12 *Suppose that $\partial\omega$ is smooth and that the surface tractions $g^{\pm} \in \left[H^1(\omega)\right]^3$. Then for fixed model order q the data belong to $T(\bar{A}, 0)$ for some $\bar{A} > 0$ independent of t.*

3.5 *A posteriori* control of discretization and modelling error

The present section is devoted to the investigation of the computational performance of the modelling error estimators derived in the foregoing sections. The estimators were derived under the assumption that the plate models in the hierarchy are solved exactly, and so the question arises of how well they will perform if only an approximate solution is available. Therefore, we will also consider the approximate solution of the models in the hierarchy using the *hp* finite element method.

The elliptic systems constituting the plate models are singularly perturbed so that the solutions exhibit boundary layers, containing components that are exponentially decaying away from the boundary $\partial\omega$. This feature, together with the phenomenon of shear locking, renders the accurate numerical solution of the plate models nontrivial. Nevertheless, the boundary layers can be resolved [38] and the shear locking overcome, by the use of high order p version finite element discretizations [37, 43]. In [38] it was shown that an *hp* finite element discretization with a single element of width $O(tp)$ near the edge of the plate is able to resolve the boundary layer with an exponential rate of convergence *uniformly* in t.

Consider the example of the pure bending of a clamped square plate, with midsurface $\omega = (-a/2, a/2)^2 \backslash \{x : |x| \leq a/10\}$ and thickness t, subjected to uniform unit normal loads on the faces. The $\mathbf{n}$ models, with $\mathbf{n}$ as in Table 1, are discretized using a p-hierarchic finite element method, with $1 \leq p \leq 8$, based on the mesh depicted in Fig. 1. If $q = 1$ the modification mentioned in Remark 20 is adopted so that model 1 is the Reissner–Mindlin model. The computations in this section were performed using the finite

j	$q=1$	$q=2$	$q=3$	$q=4$	$q=5$	$q=6$
I	$(0,0,1)$	$(2,2,1)$	$(2,2,3)$	$(4,4,3)$	$(4,4,5)$	$(6,6,5)$
II	$(1,1,0)$	$(1,1,2)$	$(3,3,2)$	$(3,3,4)$	$(5,5,4)$	$(5,5,6)$

Table 1 Model orders $\mathbf{n}$ for membrane and bending models depending on the maximal transverse degree q.

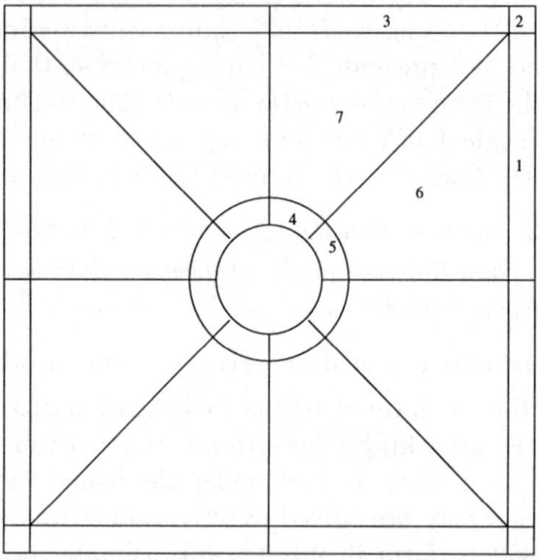

FIG. 1. Finite element mesh for the plate with central hole (not drawn to scale).

element code STRESSCHECK.[2] The software allows a separate computation of the bending and membrane models and also contains a unified implementation for homogeneous and laminated plates [44]. In STRESSCHECK all element mappings are performed exactly using transfinite blending maps which is an essential feature of p-hierarchic finite element methods in curvilinear geometries.

In particular, we considered a plate where $a = 2$, $d = 0.05$ and hence $a/t = 20$. The material was orthotropic with the constitutive parameters in eqn. (3.6) given, using engineering notation, by

$$\begin{aligned} E_1 &= 25 \cdot 10^6, & E_2 &= E_3 = 10^6, \\ G_1 &= G_3 = 0.5 \cdot 10^6, & G_2 &= 0.2 \cdot 10^6 \\ \nu_{12} &= \nu_{23} = 0.25, & \nu_{31} &= 0.01. \end{aligned} \tag{3.56}$$

[2]STRESSCHECK is a registered trademark of ESRD Inc., St Louis, Mo, USA

p	DOF	$\mathcal{G}_{3,p}$	α	$\left\| u_p^{\mathrm{n}} - u^{\mathrm{n}} \right\|_{E(\Omega)} / \left\| u^{\mathrm{n}} \right\|_{E(\Omega)}$ [%]
1	42	-4.3338026E-6	0.00	53.60
2	168	-5.7561015E-6	0.61	23.12
3	378	-6.0150995E-6	0.98	10.42
4	672	-6.0691653E-6	1.49	4.43
5	1050	-6.0774790E-6	1.34	2.44
6	1512	-6.0793402E-6	0.99	1.70
7	2058	-6.0802827E-6	1.24	1.16
8	2688	-6.0806806E-6	1.24	0.84
∞		-6.0811048023E-6		

Table 2 p-hierarchic solution of the $(3, 3, 2)$ model; $r1 = 0.04a$.

Thanks to symmetry, only the quarter of the plate located in the first quadrant need be discretized, and the data given below refer to this situation.

Since boundary layers will occur near the (free) perimeter of the hole and the (clamped) outer edge of the plate, thin elements parallel to these boundaries were inserted (elements 1 and 3 near the clamped edge and 4 and 5 near the perimeter of the hole). The perpendicular distance of the mesh lines defining elements 1 and 3, and elements 4 and 5, from the nearest edge is denoted by $r1$. The discrete potential energy corresponding to uniform polynomial degree p on the mesh shown in Fig. 1, and the model of order q, chosen as in eqn. (3.21), is denoted by $\mathcal{G}_{q,p}$. The finite element approximation of u^{n} is denoted by u_p^{n}. Table 2 shows data obtained from a computation using the $(3, 3, 2)$ model, corresponding to the choice $q = 3$ in eqn. (3.21), where p was increased hierarchically from 1 to 8. The discretization error was estimated by extrapolating the total potential energies $\mathcal{G}_{3,p}$ of the discrete solutions, based on fitting the data to the asymptotic convergence estimate $\mathcal{G}_{3,p} - \mathcal{G}_{3,\infty} = Cp^{-\alpha}$. The unknowns $\mathcal{G}_{3,\infty}$, C and the convergence rate α are determined from values of $\mathcal{G}_{q,p}$ obtained from three successive levels of p-refinement. At each stage, the mesh shown Fig. 1 was used along with a finite element space consisting of a full set of tensor product polynomials, since these prove to be advantageous for anisotropic materials.

Consider now the *a posteriori* estimation of the modelling error based on the residual tractions on the faces $R_{\pm}$. The element contributions to the estimator listed in Table 3 can be used as *modelling error indicators*. Thanks to symmetry, only data for those elements located in the first quadrant are shown. The convergence as the model order q is increased is clearly visible. Also observe the differing rates of convergence in the interior ele-

q	1	2	3	4	5	6
ω_1	1.95E-7	1.93E-7	4.46E-8	4.27E-8	1.40E-8	1.37E-8
ω_2	4.27E-10	3.20E-10	1.07E-10	1.01E-10	4.24E-11	4.07E-11
ω_3	8.95E-8	4.60E-8	1.08E-8	8.45E-9	4.38E-9	3.92E-9
ω_4	4.84E-10	2.51E-10	4.86E-11	4.31E-11	9.01E-13	6.13E-13
ω_5	3.03E-9	2.14E-9	1.76E-10	1.72E-10	1.82E-11	1.75E-11
ω_6	5.34E-7	5.09E-7	2.94E-9	2.68E-9	3.69E-11	2.55E-11
ω_7	1.13E-7	8.27E-8	2.01E-10	1.28E-11	6.03E-13	6.50E-13

Table 3 Elemental contributions to the error estimate ($p = 8$, $a/t = 20$).

q	2	3	4
$EST^2 / \|\|u\|\|^2_{E(\Omega)}$	6.8428E-2	4.8321E-3	4.4481E-3
$\|\|u^{\mathbf{n}}_8 - u\|\|^2_{E(\Omega)} / \|\|u\|\|^2_{E(\Omega)}$	5.5944E-2	1.6210E-3	1.3192E-3
$\Theta_{\text{eff}} = EST / \|\|u\|\|_E$	1.1059	1.7265	1.8363

Table 4 Residual and extrapolation based estimation of the modelling error.

ments ω_6 and ω_7 from those near the clamped edge of the plate. The error in the edge elements decreases at an inferior rate due to the fact that *edge singularities* in the three dimensional solution near the sets $\gamma \times \{\pm d\}$ are poorly approximated by a solution of form (3.19). It is therefore important to note that the hierarchic plate models used in elements ω_6 and ω_7 could be coupled to a fully three dimensional *hp* finite element discretization in the boundary layer subregions $\omega_j \times (-d, d)$, $j = 1, 3, 4, 5$ [32].

Finally, we compare the residual modelling error estimates given in Theorem 3.3 with those obtained by extrapolation in Table 4, based on

$$0 < 2(\mathcal{G}(u^{\mathbf{n}}) - \mathcal{G}(u)) = \mathcal{E}(u) - \mathcal{E}(u^{\mathbf{n}}) = ||u||^2 - ||u^{\mathbf{n}}||^2 = ||u^{\mathbf{n}} - u||^2 = ||e^{\mathbf{n}}||^2.$$

We see that the residual based estimators are guaranteed upper estimators and follow the modelling error accurately as the model order increases.

4 Incompressible fluid flow

In this section, we discuss the hierarchic modelling of Stokes flow in thin domains $\Omega = \omega \times (-d, d)$. Stokes and Navier–Stokes flows in thin domains occur in many areas such as atmospheric flow (since, viewed on a global scale, the atmosphere qualifies as a thin domain), injection molding, lakes, rivers and so on. Naturally, various effects not contained in the Stokes equations, such as convection, nonlinear constitutive laws and the like, become important in such cases. Therefore, the Stokes problem must be regarded as a first step in the direction of hierarchical dimensional reduction for general flow problems.

Formally, the results for linear elasticity obtained in the previous sections could be specialized to the case of an isotropic, incompressible material by choosing $\nu = 1/2$. One would thereby recover the Stokes case. However, it is easy to see this would impose special constraints on the dimensionally reduced solutions that are not easily satisfied.

Therefore, the three dimensional problem is, as usual, cast into saddle point form with the pressure as Lagrange multiplier and the hierarchic modelling starts from this formulation. The results presented here are due to O. Ricou [30].

4.1 Problem formulation

Let $\Omega = \omega \times (-d, d), \omega \subset \mathbb{R}^2$ be a bounded domain with Lipschitz boundary $\gamma = \partial\omega = \overline{\gamma}_D \cup \overline{\gamma}_N$. We consider the *Stokes problem*

$$\begin{aligned} -\nu\Delta u + \nabla p &= f \quad \text{in } \Omega \\ \nabla \cdot u &= 0 \\ u &= 0 \quad \text{on } R_\pm = \omega \times \{\pm d\} \\ u &= 0 \quad \text{on } \Gamma_D = \gamma_D \times (-d, d) \\ -\nu \frac{\partial u}{\partial n} + pn &= g \quad \text{on } \Gamma_N = \gamma_N \times (-d, d)\,. \end{aligned} \tag{4.1}$$

The variational form of eqn. (4.1) is: *Given* $f \in L^2(\Omega)$, *find* $(u, p) \in H_0^1(\Omega) \times L^2(\Omega)$ *such that*

$$\begin{aligned} \nu(\nabla u, \nabla v)_0 - (p, \nabla \cdot v)_0 &= (f, v)_0 + \langle g, v\rangle_\Gamma \quad \forall v \in H_0^1(\Omega) \\ (q, \nabla \cdot u)_0 &= 0 \quad \forall q \in L^2(\Omega) \end{aligned} \tag{4.2}$$

where $H_0^1(\Omega) = \{u \in H^1(\Omega) : u = 0 \text{ on } R_+ \cup R_- \cup \Gamma_D\}$.

Existence and uniqueness of a weak solution of eqn. (4.2) are assured by the *inf-sup condition*

$$\inf_{q\in L^2(\Omega)} \sup_{v\in H_0^1(\Omega)} \frac{(q, \nabla \cdot v)_0}{\|q\|_{L^2}\, \|\nabla v\|_0} \geq C(\omega, d) > 0\,. \tag{4.3}$$

4.2 Dimensional reduction

As in the previous sections, we obtain dimensionally reduced, two dimensional models of eqn. (4.2) by semidiscretization in the transverse direction.

Define

$$V_N = \left\{ v(x,y) \in H_0^1(\Omega) : v(x,y) = \sum_{j=0}^{N} X_j(x)\, \psi_j(y/d) \right\} \qquad (4.4)$$

$$Q_N = \left\{ p(x,y) \in L^2(\Omega) : p(x,y) = \sum_{j=0}^{N} P_j(x)\, \varphi_j(y/d) \right\} . \qquad (4.5)$$

Note that $X_j \in H^1(\omega)$ and $P_j \in L^2(\omega)$, and, due to the Dirichlet conditions on $y = \pm d$, we require

$$\psi_j \in H^1(-1,1), \quad \psi_j(\pm 1) = 0 . \qquad (4.6)$$

Once the basis functions ψ_j, φ_j in eqs. (4.4), (4.5) have been selected, the dimensionally reduced N-model of eqn. (4.2) is defined by: *find* $(u_N, p_N) \in V_N \times Q_N$ *such that*

$$\begin{array}{lcll} \nu(\nabla u_N, \nabla v)_0 - (p_N, \nabla \cdot v)_0 & = & (f,v)_0 + \langle g, v\rangle_\Gamma & \forall v \in V_N , \\ (q, \nabla \cdot u_N)_0 & = & 0 & \forall q \in P_N . \end{array} \qquad (4.7)$$

A necessary condition for the convergence of (u_N, p_N) to (u,p) as $N \to \infty$ is, as before, for the spaces $\{V_N\}_N$, $\{Q_N\}_N$ to be dense in $H_0^1(\Omega)$, $L^2(\Omega)$, respectively. This is the case provided that the transverse shape functions $\{\psi_j\}_{j=0}^N$, $\{\varphi_j\}_{j=0}^N$ are dense sets in $H_0^1(-1,1)$ and $L^2(-1,1)$, respectively. Accordingly, we select

$$\psi_j(z) = (2j+3) \int_0^z L_{j+1}(\xi)d\xi = L_{j+2}(z) - L_j(z), \quad j = 0,1,\ldots \qquad (4.8)$$

and

$$\varphi_j(z) = L_j(z), \quad j = 0,1,2,\ldots \qquad (4.9)$$

where L_j denotes the Legendre polynomial of degree j on $(-1,1)$, normalized such that $L_j(1) = 1$. Then we have

Proposition 4.1 *The sequences* $\{V_N\}, \{Q_N\}_N$ *of spaces defined by (4.4), (4.5), (4.10) and (4.9) are closed and dense in* $H_0^1(\Omega) \times L^2(\Omega)$.

For a detailed proof, we refer to [30].

Consider now the well-posedness of the reduced problem (4.7). As is well known, a sufficient condition for the well-posedness is the discrete analog of eqn. (4.3).

Proposition 4.2 *Suppose that the discrete inf-sup condition*

$$\inf_{0 \neq q \in Q_N} \sup_{0 \neq v \in V_N} \frac{(q, \nabla \cdot v)_0}{\|\nabla v\|_0 \, \|q\|_0} \geq \beta(N) > 0 \qquad (4.10)$$

holds. Then the hierarchic models are well defined and, for every N, admit a unique solution $(u_N, p_N) \in V_N \times Q_N$. Moreover, the modelling error estimates

$$\begin{aligned} \|u - u_N\|_{H^1(\Omega)} \leq\ & C\{(1 + 1/\beta(N)) \inf_{v \in V_N} \|u - v\|_{H^1(\Omega)} \\ & + \inf_{q \in Q_N} \|p - q\|_{L^2(\Omega)}\}, \end{aligned} \tag{4.11}$$

$$\|p - p_N\|_{L^2(\Omega)} \leq \frac{C}{\beta(N)} \{ \inf_{q \in Q_N} \|p - q\|_{L^2(\Omega)} + \|u - u_N\|_{"} H^1(\Omega)\} \tag{4.12}$$

hold.

For a proof, see, for example [11].

We now investigate the inf-sup condition (4.10). For general domains $\omega \subset \mathbb{R}^2$, this is still an open problem. However, the following is known.

Proposition 4.3 [30] *Assume that $\omega = (0,1) \subset \mathbb{R}$ and that we have the Dirichlet boundary condition*

$$u = 0 \text{ on } \Gamma_D = \{(x, y) : x = 0,\ |y| < 1\}. \tag{4.13}$$

Then we have (4.10) with $\beta(N) \geq \beta_0 > 0$ where β_0 is independent of N.

4.3 Hele–Shaw approximation

It will be shown that the lowest member of the hierarchy (u_N, p_N) based on the subspaces (4.4) and (4.5) is the classical Hele–Shaw approximation of Stokes flow in a thin domain. We briefly recall the physical reasoning leading to the Hele–Shaw model. First, one makes the assumption

$$p(x, y) = p(x), \tag{4.14}$$

so that the pressure is constant through the thickness. Then from momentum conservation we have

$$\frac{\partial p}{\partial x_\alpha} = \nu \frac{\partial^2 u_\alpha}{\partial y^2}, \quad \frac{\partial p}{\partial y} = 0. \tag{4.15}$$

We integrate these equations from $-d$ to d and use eqn. (4.14) to obtain $\partial^2 u_\alpha(x, y)/\partial y^2 = \nu^{-1} \partial p(x)/\partial x_\alpha$, which is independent of y. Therefore, u_α is parabolic through the thickness. Letting U_α be suitable coefficient functions to be determined and noting that $u_\alpha(x, \pm d) = 0$, we get

$$u_\alpha(x, y) = U_\alpha(x)\,(d^2 - y^2) \quad \alpha = 1, 2. \tag{4.16}$$

and

$$u_3(x, y) \equiv 0. \tag{4.17}$$

Inserting this into eqn. (4.15) gives

$$u_\alpha(x, y) = -\frac{1}{2\nu} \frac{\partial p}{\partial x_\alpha}(x)(d - y^2), \; u_3(x, y) = 0. \tag{4.18}$$

Evaluating the averages of the velocity through the thickness and using the incompressibility condition $\nabla \cdot u = 0$ we conclude

$$\left(\frac{\partial^2}{\partial x_1^2} + \frac{\partial^2}{\partial x_2^2}\right) p(x) = 0 \text{ in } \omega \,. \tag{4.19}$$

It may now be shown that the lowest member (u_0, p_0) in the hierarchy is the classical Hele–Shaw approximation. Note that $\varphi_0 = 1$ and $\psi_0(z) = L_2(z) - L_0(z) = 3(z^2-1)/2$. This, along with $\nabla \cdot u_0 = 0$, gives the relation

$$\left(\frac{\partial U_{01}}{\partial x_1} + \frac{\partial U_{02}}{\partial x_2}\right)(x)\,\psi_0\,(y/d) + U_{03}(x)\,d^{-1}\,\psi_0'\,(y/d) = 0 \tag{4.20}$$

for all $(x,y) \in \Omega$. Since ψ_0 and ψ_0' are linearly independent, eqn. (4.20) implies

$$\nabla_x \cdot (U_{01}, U_{02}) = 0, \;\; U_{03} = 0 \,. \tag{4.21}$$

Inserting u_0, p_0 into the first equation of (4.1) (with $f \equiv 0$) gives for the first two components

$$\nu \left\{\psi_0\,\Delta_x\,U_{0\alpha} + d^{-2}\,\psi_0''\,U_{0\alpha}\right\} = \frac{\partial p_0}{\partial x_\alpha}\,, \quad \alpha = 1,2 \tag{4.22}$$

and the third component equation is satisfied identically. Since ψ_0 and ψ_0'' are linearly independent and $p_0 = p_0(x)$, we find from eqn. (4.22) that

$$\Delta_x\,U_{0\alpha}(x) = 0 \text{ in } \omega, \;\; \alpha = 1,2 \,. \tag{4.23}$$

Substituting into eqn. (4.22) and integrating with respect to y shows that

$$U_{0\alpha} = \nu^{-1}\,\frac{\partial p_0}{\partial x_\alpha} \tag{4.24}$$

which is precisely the Hele–Shaw approximation (4.15).

Bibliography

1. R.L. Actis. *Hierarchic Models for Laminated Plates.* Doctoral Dissertation, Washington University, St. Louis, Mo (1991).
2. R.A. Adams. *Sobolev Spaces.* Academic Press, New York (1978).
3. M. Ainsworth and J.T. Oden. *A posteriori Error Analysis in the FEM.* Monographs on Comp. Mechanics, Elsevier Publ. (to appear).
4. S.S. Antman and R.S. Marlow. *Material Constraints, Lagrange Multipliers, and Compatibility. Applications to Rod and Shell Theories.* Arch. Rat. Mech. Anal. **116**, 257–299 (1991).
5. D.N. Arnold and R.S. Falk. *Asymptotic analysis of the boundary layer for the Reissner–Mindlin plate model.* To appear in SIAM J. Math. Anal.
6. I. Babuška and K. Aziz. *Survey lectures on the mathematical founda-*

tion of the finite element method, in *"The mathematical foundations of the finite element method with applications to partial differential equations"*, Ed. A.K. Aziz. Academic Press, New York, 3–343 (1972).

7. I. Babuška, I. Lee and C. Schwab. *On the a-posteriori estimation of the modeling error for the heat conduction in a plate and its use for adaptive hierarchical modeling.* Appl. Num. Math. **14**, 5–21 (1994).
8. I. Babuška and L. Li. *Hierarchic Modelling of Plates.* Computers and Structures **40**, 419–430. (1991)
9. I. Babuška and J. Pitkäranta. *The plate paradox for hard and soft simple support.* SIAM J. Math. Anal. **21**, 551–576 (1990).
10. I. Babuška and C. Schwab. *A-posteriori error estimation for hierarchic models of elliptic boundary value problems on thin domains.* SIAM J. Num. Anal. **33**, 221–246 (1996).
11. F. Brezzi and M. Fortin. *Mixed and Hybrid FEM.* Springer Series in Computational Mathematics, **15**, Springer-Verlag, New York (1991).
12. F. Brezzi and M. Fortin: *Mixed Finite Element Methods.* Springer Series in Computational Mathematics, Springer-Verlag, Berlin Heidelberg New York (1992).
13. P.G. Ciarlet. *Plates and junctions in elastic multistructures – an asymptotic analysis.* Masson, Paris and Springer-Verlag, Berlin Heidelberg New York (1990).
14. P.G. Ciarlet and P. Destuynder. *A justification of the two-dimensional plate model.* J. Mécanique **18**, 315–344 (1979).
15. P. Destuynder. *Comparaison entre les modèles tridimensionnels et bidimensionnels de plaques en élasticité.* RAIRO Anal. Numérique **15**, 331-369 (1981).
16. K. Eriksson, D. Estep, P. Hansbo and C. Johnson. *Adaptive Finite Element Methods.* Book manuscript, in preparation.
17. R.B. Kellogg. *Boundary layers and corner singularities for a self-adjoint problem.* In: Boundary value problems and integral equations in nonsmooth domains, eds. M. Costabel, M. Dauge, S. Nicaise. Lecture Notes in Pure and Applied Mathematics, vol. 167, Marcel Dekker, New York, (1995).
18. J.L. Lions. *Perturbations Singulières dans les Problèmes aux Limites et en Contrôle Optimal.* Lecture Notes in Mathematics No. 323, Springer-Verlag, Berlin Heidelberg New York (1973).
19. M. Melenk. *A note on robust exponential convergence of finite element methods for problems with boundary layers.* Report 96-06, Seminar für angewandte Mathematik, ETH-Zürich (1996) (submitted to IMA J. Numer. Analysis).
20. B. Miara. *Optimal spectral approximation in linearized plate theory.*

Applicable Analysis **31**, 291–307 (1989).

21. R.D. Mindlin. *Influence of rotatory inertia and shear on flexural motion of isotropic elastic plates.* J. Appl. Mech. **18**, 31–38 (1951).
22. S.A. Nazarov. *Vishik–Lyusternik method for elliptic boundary value problems in regions with conical points* parts I and II. Siberian Mathematical Journal **22**, 594–611 and 753–769 (1981).
23. J. Nečas. *Les Méthodes Directes en Théorie des Equations Elliptiques.* Masson, Paris (1967).
24. J. Nečas and I. Hlavaček. *Mathematical Theory of Elastic and Elasto-plastic Bodies: an Introduction.* Elsevier, Amsterdam New York (1981).
25. O.A. Oleinik, A.S. Shamaev and G.A. Yosifian. *Mathematical Problems in Elasticity and Homogenization.* Studies in Mathematics and its Applications Vol. 26, North-Holland Publ. Amsterdam, New York (1992).
26. W. Prager and J.L. Synge. *Approximations in elasticity based on the concept of the function space.* Quart. Appl. Math. **5**, 241–269 (1947).
27. M.H. Protter and H.F. Weinberger. *Maximum Principles in Differential Equations*, 2nd edition. Springer-Verlag, New York Heidelberg Berlin Tokyo (1984).
28. E. Reissner. *The effect of transverse shear deformation on the bending of elastic plates.* J. Appl. Mech. **67**, A69–A77 (1945).
29. E. Reissner. *On bending of elastic plates.* Quart. Appl. Math. **5**, 55–68 (1947).
30. O. Ricou. *A dimensional reduction for the Stokes problem.* Rapport Interne, Laboratoire d'Analyse Numérique, Univ. P. et M. Curie, Paris (1996).
31. C. Schwab. *Dimensional Reduction for elliptic boundary value problems.* Ph.D. thesis, Dept. of Mathematics, Univ. of Maryland College Park, MD 20742 USA (1989).
32. C. Schwab. *Boundary layer resolution in hierarchical models of laminated composites.* RAIRO Anal. Numér., Sér. rouge **28**, 517–537 (1994).
33. C. Schwab. *Boundary layer resolution in hierarchical plate modelling.* Mathematical Methods in the Applied Sciences **18**, 345–370 (1995).
34. C. Schwab. *A-posteriori modeling error estimation for hierarchic plate models.* Numer. Math., **74**, 221–259 (1996).
35. C. Schwab. *Hierarchic Models of Elliptic Boundary Value Problems on Thin Domains – a-posteriori Error Estimation and Fourier Analysis.* Book manuscript, to appear with Wiley Interscience, New York (1997).
36. C. Schwab and I. Babuška. *Subspace correction methods for the iterative solution of hierarchic plate models I: Heat transfer problems.*

Technical Note, Inst. Phys. Science and Technology, Univ. Maryland College Park, MD20742, USA (1994).

37. C. Schwab and M. Suri. *Locking and boundary layer effects in the finite element approximation of the Reissner–Mindlin plate model.* Proceedings of the 50th anniversary conference of MathComp, AMS-proceedings of Symposia in Applied Mathematics **48**, 367–371 (1994).
38. C. Schwab and M. Suri. *The p and hp versions of the finite element method for problems with boundary layers.* Math. Comp. **65**, 1403–1429 (1996).
39. C. Schwab, M. Suri and C.A. Xenophontos. *The hp version of the finite element method for problems with boundary layers.* Submitted to Comp. Meth. Appl. Mech. Engrg.
40. C. Schwab, M. Suri and C.A. Xenophontos. *The hp version of the finite element method for singularly perturbed problems in polygonal domains.* In preparation.
41. C. Schwab and S. Wright. *Boundary layers of hierarchical beam and plate models.* J. Elasticity **38**, 1–40 (1995).
42. E.M. Stein. *Singular Integrals and Differentiability Properties of Functions.* Princeton University Press, Princeton, NJ (1970).
43. M. Suri, I. Babuška C. and Schwab. *Locking effects in the finite element approximation of plate models.* Mathematics of Computation **64**, 461–482 (1995).
44. B.A. Szabo and R.L. Actis. *Hierarchic models for laminated plates.* In: Adaptive, Multilevel and Hierarchical Computational Strategies, ed. A.K. Noor. AMD-Vol.157, ASME NY, 69–94 (1992).
45. B.A. Szabo and I. Babuška. *Finite Element Analysis.* John Wiley Publ., New York (1991).
46. B.A. Szabo and G.J. Sahrmann. *Hierarchic plate and shell models based on p-extension.* Int. J. Num. Meth. Engg. **26**, 1855–1881 (1988).
47. S. Timoshenko and S. Woinowsky-Krieger. *Theory of Plates and Shells,* 2nd edition. McGraw-Hill, New York (1959).
48. C. Truesdell (Ed.): *Mechanics of Solids, Volume II.* Springer-Verlag, Berlin Heidelberg New York (1984).
49. M. Vogelius and I. Babuška. *On a dimensional reduction method I.* Math. Comp. **37**, 31–68 (1981).
50. R. Verfürth. *A review of a posteriori error estimation and adaptive mesh-refinement techniques.* Wiley-Teubner Series in Advances in Numerical Mathematics, Teubner Verlag, Stuttgart, Germany (1996).
51. R. Verfürth. *Robust a-posteriori error estimation for singularly perturbed problem.* To appear in Num. Math.
52. C.A. Xenophontos. *The hp version of the finite element method for*

singularly perturbed problems. Ph.D. thesis Univ. Maryland, Baltimore County, Baltimore MD21228 USA (1996).

53. J. Xu. *Iterative methods by space decomposition and subspace correction.* SIAM Rev. **34**, 581–613 (1992).

Wavelets from filter banks

Gilbert Strang

Department of Mathematics, Massachusetts Institute of Technology

1 Introduction

This subject has two parts. One part is discrete, the other is continuous. In discrete time we develop the idea and applications of **filter banks**. In continuous time we have **scaling functions** $\varphi(t)$ and **wavelets** $w(t)$. By a natural limiting process, iteration of the lowpass filter leads to the scaling function. One highpass filter then produces a wavelet. Our goal is to make this connection clear. We find the conditions on the discrete coefficients that lead to good filter banks and good wavelets.

Historically and mathematically, the filters come first. Perfect reconstruction filter banks were developed in the early 1980's. The excitement around wavelets started later (and grew quickly). This excitement was not universal — designers of filter banks naturally asked what was new. Part of the answer is precisely in that process of **iteration**. For a filter to behave well in practice, when it is combined with subsampling and repeated five times, it must have an extra property — not built into earlier designs. This property expresses itself in the frequency domain by a sufficient number of "zeros at π". Then the frequency band can be successfully separated into five octaves.

The underlying problem is to choose a good basis. We want to represent a signal well, by a small number of basic signals. These can be sinusoids and they can be wavelets. On a discrete grid, $\omega = \pi$ is the highest frequency at which a signal can oscillate. Those oscillations $\boldsymbol{x}(n) = e^{i\pi n} = (-1)^n$ are stopped by the lowpass filter with a "zero at π". The highpass filter lets fast oscillations through, and the synthesis filters can reconstruct the exact input. But **compression** may come between analysis and synthesis. Frequencies that are barely represented will be intentionally lost. That mostly means high frequencies but the filter bank is impartial — it keeps the basis functions that are important to the specific signal. We want to show when, and why, filter banks and wavelets are effective in reconstruction and signal representation and compression.

Filter banks

For filter banks, we identify the two conditions for perfect reconstruction (in the absence of lossy compression). One condition removes distortion, the other condition removes aliasing. The anti-distortion condition applies to the products $\boldsymbol{F}_0\boldsymbol{H}_0$ and $\boldsymbol{F}_1\boldsymbol{H}_1$ along the channels of the filter bank. Then the anti-aliasing condition controls how those products can be separated into the four filters.

The design of a perfect reconstruction filter bank is a choice of $\boldsymbol{F}_0\boldsymbol{H}_0$ and then a factorization. To understand the conditions on distortion and aliasing, we apply the techniques of multirate filtering. The algebra is moved into the "z-domain". We are selecting and factoring a polynomial $P_0(z) = F_0(z)H_0(z)$. We go forward now, to illustrate a filter bank that gives perfect reconstruction.

The analysis bank is on the left. It has a lowpass filter $\boldsymbol{H}_0$, and a high-pass filter $\boldsymbol{H}_1$, and decimation by $(\downarrow 2)$ which removes the odd-numbered components after filtering. The analysis bank yields two "half-length" outputs. Then the synthesis bank on the right begins with the upsampling operation $(\uparrow 2)$ which inserts zeros in those odd components:

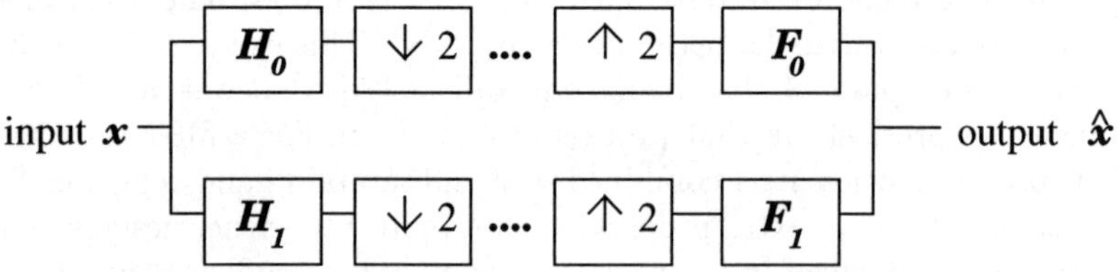

Two-channel filter bank: Separate the input into frequency bands (filter and downsample). Then reassemble (upsample and filter).

The gap in the center indicates where the subband signals are compressed or enhanced. The applications of this structure are extremely widespread. We believe that any reader interested in signal processing (and image processing) will find that filter bank analysis is extremely useful.

The filters $\boldsymbol{H}_0$, $\boldsymbol{H}_1$, $\boldsymbol{F}_0$ and $\boldsymbol{F}_1$ are linear and time-invariant. The operators $(\downarrow 2)$ and $(\uparrow 2)$ are **not** time-invariant. These multirate operations are responsible for *aliasing* and for *imaging* – they create undesirable and extraneous signals that the filters must cancel. The structure of an **orthogonal** bank is very special, and the next figure shows how the filters are related. For length 4 all filters use the four coefficients, a, b, c, d that Daubechies derived:

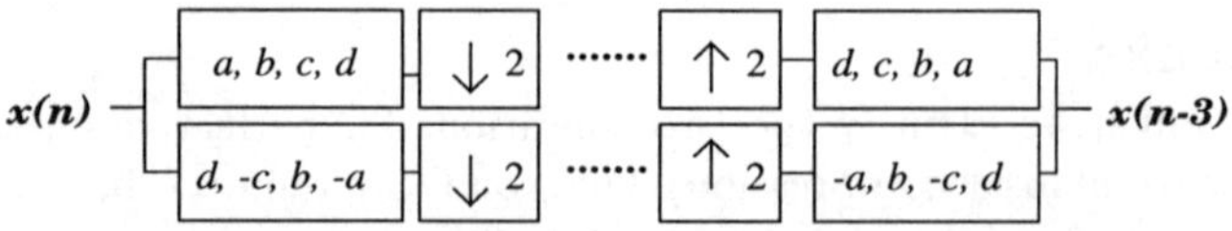

The form of an orthogonal filter bank with four coefficients.

How did she choose a, b, c, d? Part of the answer will have to wait, but here is the essential idea. The product along the top channel gives a particular "halfband filter" as $\boldsymbol{P}_0$:

$$(a,\ b,\ c,\ d)\ *\ (d,\ c,\ b,\ a)\ =\ (-1,\ 0,\ 9,\ 16,\ 9,\ 0,\ -1)/16.$$

This convolution is a multiplication of two polynomials, when a, b, c, d are the coefficients:

$$(a + bz^{-1} + cz^{-2} + dz^{-3})\quad(d + cz^{-1} + bz^{-2} + az^{-3}) =$$

$$(-1 + 9z^{-2} + 16z^{-3} + 9z^{-4} - z^{-6})/16.$$

The four coefficients are pleasant to calculate. The serious job is to explain what is special about that 6th degree polynomial in which z^{-1} and z^{-5} are missing.

A filter bank also gives perfect reconstruction if it is **biorthogonal**. This design is less restricted. The product $\boldsymbol{F}_0\boldsymbol{H}_0$ must skip the same odd powers of z^{-1}, but $\boldsymbol{F}_0$ does not have to be the transpose (the flip) of $\boldsymbol{H}_0$. Here are specific numbers for the filter coefficients – not the only choice and maybe not the best. They show how the filters $\boldsymbol{F}_0$ and $\boldsymbol{F}_1$ on the synthesis side are related to the analysis filters $\boldsymbol{H}_1$ and $\boldsymbol{H}_0$ (by alternating signs):

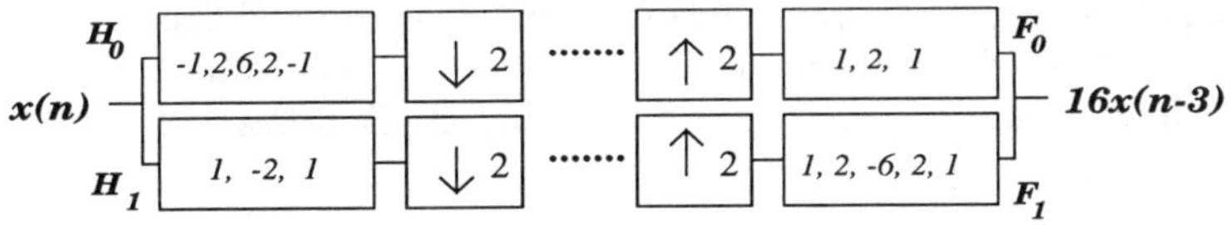

A biorthogonal filter bank: Perfect reconstruction with 3 delays.

For filters, we stop here. This time $F_0(z) = 1 + 2z^{-1} + z^{-2}$. Multiplied by $H_0(z)$ it gives the same important sixth degree polynomial as before. To understand why the zero coefficients are necessary in that polynomial, and why $-\frac{1}{16}$ and $\frac{9}{16}$ are desirable, we refer to the new book [7].

Our discussion went this far so as to make a basic and encouraging point: *the construction of new filter banks need not be complicated.* This subject is accessible to new ideas and experiments.

Wavelets

Wavelets are localized waves. Instead of oscillating forever, they drop to zero. They come from the **iteration** of filters (with rescaling). The link between discrete-time filters and continuous-time wavelets is in the limit of a logarithmic filter tree:

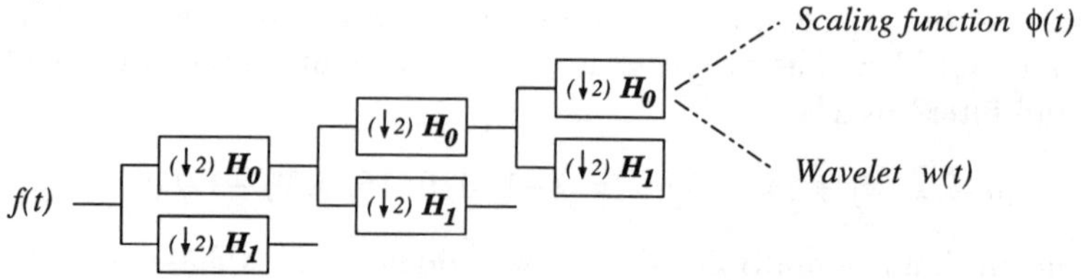

Scaling function and wavelets from iteration of the lowpass filter.

Scaling functions and wavelets have remarkable properties. They inherit orthogonality, or biorthogonality, from the filter bank. Because of the repeated rescaling that produces them, wavelets decompose a signal into details at all scales. The wavelet $w(t)$ and its shifts $w(t-k)$ are at unit scale. The wavelets $w(2^j t)$ and $w(2^j t - k)$ are at scale 2^{-j}. The biorthogonal functions $\tilde{\varphi}(t)$ and $\tilde{w}(t)$ come from iterating the synthesis bank.

Wavelets produces a natural "**multiresolution**" of every image, including the all-important edges. Where the low frequency part of the Fourier transform is often a blur, the output from the lowpass channel is a useful compression.

The wavelets created by Ingrid Daubechies are orthogonal, with the advantages and limitations that this property brings. The biorthogonal alternatives come from different factorizations of the same polynomial (as above). This polynomial corresponds to a "maxflat halfband filter", and we hope you will like the connections. This subject is a beautiful combination of mathematical analysis and signal processing applications.

Summary of the theory

There are four conditions that play a central part in the analysis. Because of their importance we highlight them here. They apply directly to the coefficients in the filter banks and the consequences are felt (after iteration!) in the scaling functions and wavelets. Here are the four conditions — some might say in decreasing order of importance:

PR condition *Perfect reconstruction*
The synthesis bank inverts the analysis bank, with ℓ delays.
Biorthogonal banks with no aliasing and no distortion.

Condition O *Orthogonality.*
The analysis bank is inverted by its transpose.
The wavelets are orthogonal to their dilates and translates.

Condition A_p *Accuracy of order p from the scaling functions.*
p vanishing moments in the wavelets.
pth order decay of wavelet coefficients for smooth $f(t)$.

Condition E *Accuracy of order p from the scaling functions.*
Determines convergence to $\varphi(t)$ and smoothness of wavelets.
Equivalent to stability of the wavelet basis.

The four fundamental conditions will be stated explicitly for a two-channel filter bank. We continue to use the polynomials $H_0(z)$, $H_1(z)$, $F_0(z)$, and $F_1(z)$, whose coefficients come directly from the filters. By convention, these are polynomials in z^{-1}, and the lowpass analysis filter is represented by $H_0(z) = \boldsymbol{h}(0) + \boldsymbol{h}(1)z^{-1} + \ldots + \boldsymbol{h}(N)z^{-N}$. Here are the conditions that give filters and wavelets their good properties:

1. Perfect reconstruction (**PR condition**)

$$F_0(z)H_0(z)+F_1(z)H_1(z) = 2z^{-\ell} \quad \text{and} \quad F_0(z)H_0(-z)+F_1(z)H_1(-z) = 0.$$

The second equation gives the anti-aliasing choices $F_0(z) = H_1(-z)$ and $F_1(z) = -H_0(-z)$.

2. Orthogonality (**Condition O**)
The filter coefficients are reversed by $F_0(z) = z^{-N}H_0(z^{-1})$ and $F_1(z) = z^{-N}H_1(z^{-1})$. Then perfect reconstruction depends on the "double-shift orthogonality" of the lowpass coefficients $\boldsymbol{h}(k)$:

$$\sum \boldsymbol{h}(k)\, \boldsymbol{h}(k+2n) = \boldsymbol{\delta}(n).$$

In terms of the polynomials this is $H_0(z)H_0(z^{-1}) + H_0(-z)H_0(-z^{-1}) = 2$.

3. Accuracy of order p (**Condition $\boldsymbol{A_p}$**)
The lowpass filter has a zero of order p at $z = -1$:

$$H_0(z) = \left(\frac{1+z^{-1}}{2}\right)^p Q(z).$$

4. Convergence and stability (**Condition E**)
The transition matrix $\boldsymbol{T}$ has $\lambda = 1$ as simple eigenvalue and all other $|\lambda(\boldsymbol{T})| < 1$.

Final note: The sixth degree polynomial in the examples above has *four* *zeros* at $z = -1$:

$$-1 + 9z^{-2} + 16z^{-3} + 9z^{-4} - z^{-6} = (1 + z^{-1})^4 (-1 + 4z^{-1} - z^{-2}).$$

These zeros give flat responses near $\omega = \pi$ and also $\omega = 0$. The absence of z^{-1} and z^{-5} is the key to perfect reconstruction. Polynomials of higher degree, also with zeros at $z = -1$ and also with only one odd power, factor into $F_0(z)H_0(z)$ and give the best filters for iteration. In the limit of the iterations, these filters give the best wavelets.

Note to the reader: The four sections of these notes are adapted from recent papers and also from our new textbook. The lectures could not cover the subject in full, and it may be helpful to indicate the sources from which these notes were drawn. The Guide to the Book and the journal articles are available by e-mail from `gs@math.mit.edu`.

I. Introduction (from the Guide to the Book: *Wavelets and Filter Banks* by Gilbert Strang and Truong Nguyen, Wellesley-Cambridge Press)

II. Filter banks and perfect reconstruction (from Chapters 4 and 5 of *Wavelets and Filter Banks*, [7])

III. Eigenvalues of $(\downarrow 2)H$ and convergence of the cascade algorithm (from [6])

IV. Zeros of the Daubechies polynomials (from a paper with Jianhong Shen [4]; a later paper is also available by email, and an earlier paper by Kateb and Lemarie [3] is highly important)

The cascade algorithm studied in Section 3 of these notes connects filter banks to wavelets. May I emphasize that the new textbook [7] studies wavelets in full!

2 Filter banks and perfect reconstruction

2.1 Overview and notation

We begin with an overview of **filters**, **filter banks** and **wavelets**. We want to indicate, first in rough outline and then in detail, the connections between these three topics. Our immediate purpose is to open up the problem and the language, starting with the filter coefficients $\boldsymbol{h}(n)$. The choice of those coefficients is the crucial decision. Their properties govern all that follows.

Each step is a natural development from the one before:

(1) A filter is a linear time-invariant operator. It acts on input vectors $\boldsymbol{x}$. The output vector $\boldsymbol{y}$ is the convolution of $\boldsymbol{x}$ with a fixed vector $\boldsymbol{h}$. The vector $\boldsymbol{h}$ contains the filter coefficients $\boldsymbol{h}(0), \boldsymbol{h}(1), \boldsymbol{h}(2), \ldots$. Our filters are digital, not analog, so the coefficients $\boldsymbol{h}(n)$ come at discrete times $t = nT$. The sampling period T is assumed to be 1 here. The inputs $\boldsymbol{x}(n)$ and outputs $\boldsymbol{y}(n)$ come at all times $t = 0, \pm 1, \pm 2, \ldots$:

$$\boldsymbol{y}(n) \;=\; \sum_k \boldsymbol{h}(k)\, \boldsymbol{x}(n-k) \;=\; \text{convolution } \boldsymbol{h} * \boldsymbol{x} \text{ in the time domain.}$$

One input $\boldsymbol{x} = (\ldots, 0, 1, 0, \ldots)$ has special importance — a unit impulse at time zero. The input has $\boldsymbol{x}(n-k) = 0$ except when $n = k$. The sum in the convolution has only one term, and that term is $\boldsymbol{h}(n)$. This output $\boldsymbol{y}(n) = \boldsymbol{h}(n)$ is the response at time n to the unit impulse $\boldsymbol{x}(0) = 1$. It is the *impulse response* $\boldsymbol{h}(0)$, $\boldsymbol{h}(1)$, $, \ldots,$ $\boldsymbol{h}(N)$.

In a moment the same filter will be described in the frequency domain. Convolution with the vector $\boldsymbol{h}$ will become *multiplication* by a function H. It is the simplicity of multiplication that makes this subject a success. The action of a filter in time and frequency is the foundation on which signal processing is built.

(2) A **filter bank** is a set of filters. The analysis bank often has two filters, lowpass and highpass. They separate the input signal into frequency bands. Those subsignals can be compressed much more efficiently than the original signal. Then they can be transmitted or stored. We are describing "subband coding" and its applications. At any time the signals can be recombined (by the *synthesis bank*).

It is not necessary to preserve the full outputs from the analysis filters. Normally they are *downsampled* by the operator $(\downarrow 2)$. **We keep only the even components of the lowpass and highpass filter outputs.** If there are M filters, then keeping every Mth component of each output gives a total of the same length as the input. Critical sampling is the key to subband coding.

This article explains how two or more filters, with downsampling, can jointly achieve properties that are impossible for a single filter. We are particularly interested in "perfect reconstruction FIR filter banks". In this case the reconstructed output $\hat{\boldsymbol{x}}(n)$ from the synthesis bank is identical to the original input $\boldsymbol{x}$ to the analysis bank (with only a time delay). In matrix language, a banded matrix (for the analysis bank) has a banded inverse (the synthesis bank).

In the frequency domain, each filter leads to a multiplication. But downsampling is *not a time-invariant operation.* If we delay all components of $\boldsymbol{y}$ by one time unit, the output from downsampling is totally different. The new samples $\boldsymbol{y}(-1)$, $\boldsymbol{y}(1)$, $\boldsymbol{y}(3)$ are entirely separate and independent from the original samples $\boldsymbol{y}(0)$, $\boldsymbol{y}(2)$, $\boldsymbol{y}(4)$. Those two subsampled signals are two "phases" of $\boldsymbol{y}$, not connected. Therefore downsampling alters the multiplication picture in the frequency domain. In fact it introduces *aliasing.*

The simplicity of multiplication can be rescued by looking at each phase separately. Each phase of $\boldsymbol{y}$ comes from filtering the phases of $\boldsymbol{x}$ (using phases of $\boldsymbol{h}$). These separate pieces are multiplications in the frequency domain. The whole operation together, filtering followed by downsampling, becomes a matrix multiplication — by the *polyphase matrix.*

This is the foundation of filter bank theory (still to be explained in detail!). The analysis polyphase matrix $\boldsymbol{H}_p$ will reveal the correct synthesis bank for perfect reconstruction. That synthesis filter bank uses $\boldsymbol{H}_p^{-1}$.

(3) Wavelets are basis functions $w_{jk}(t)$ in continuous time. A basis is a set of linearly independent functions that can be used to produce all admissible functions $f(t)$:

$$f(t) = \text{ combination of basis functions } = \sum_{j,k} b_{jk}\, w_{jk}(t). \tag{2.1}$$

The special feature of the wavelet basis is that all functions $w_{jk}(t)$ are constructed from a single mother wavelet $w(t)$. This wavelet is a small wave (a pulse). Normally it starts at time $t = 0$ and ends at time $t = N$.

The shifted wavelets w_{0k} start at time $t = k$ and end at time $t = k+N$. The rescaled wavelets w_{j0} start at time $t = 0$ and end at time $t = N/2^j$. Their graphs are compressed by the factor 2^j, where the graphs of w_{0k} are translated (shifted to the right) by k:

$$\textit{compressed:}\quad w_{j0} = w\left(2^j t\right) \qquad\qquad \textit{shifted:}\quad w_{0k}(t) = w(t-k).$$

A typical wavelet w_{jk} is compressed j times and shifted k times. Its formula is

$$w_{jk}(t) = w\left(2^j t - k\right).$$

The remarkable property that is achieved by many wavelets is *orthogonality*. The wavelets are orthogonal when their "inner products" are zero:

$$\int_{-\infty}^{\infty} w_{jk}(t)\, w_{JK}(t)\, dt = \text{ inner product of } w_{jk} \text{ and } w_{JK} = 0. \tag{2.2}$$

In this case the wavelets form an *orthogonal basis* for the space of admissible functions. This basis corresponds to a set of axes that meet at 90° angles — as most good axes do. Orthogonality leads to a simple formula for each coefficient b_{JK} in the expansion for $f(t)$. Multiply the expansion displayed in eqn. (2.1) by $w_{JK}(t)$ and integrate:

$$\int_{-\infty}^{\infty} f(t)\, w_{JK}(t)\, dt = b_{JK} \int_{-\infty}^{\infty} \left(w_{JK}(t)\right)^2 dt. \tag{2.3}$$

All other terms in the sum disappear because of orthogonality. Eqn. (2.2) eliminates all integrals of w_{jk} times w_{JK}, except the one term that has

$j = J$ and $k = K$. That term produces $(w_{JK}(t))^2$. Then b_{JK} is the ratio of the two integrals in eqn. (2.3).

As we describe the connection between filter banks and wavelets, you will see that it is the "*highpass filter*" that leads to $w(t)$. The "*lowpass filter*" leads to a scaling function $\varphi(t)$. In most constructions the lowpass filter comes first — **the scaling function is obtained before the wavelet**. In fact the scaling function (in continuous time) comes from infinite repetition $\boldsymbol{L}\,\boldsymbol{L}\ldots\boldsymbol{L}$ of the lowpass filter, with rescaling at each iteration. The wavelet follows from $\varphi(t)$ by just *one* application of the highpass filter.

2.2 Multiresolution

At a given resolution of a signal or image, the scaling functions $\varphi\left(2^j t - k\right)$ are a basis for the set of signals. The level is set by j, and the time steps at that level are 2^{-j}. The new details at level j are represented by the wavelets $w\left(2^j t - k\right)$. Then the smooth signal plus the details, the φ's plus the w's, combine into a **multiresolution** of the signal at the finer level $j + 1$. Averages come from the scaling functions, details come from the wavelets:

signal at level j (local averages) ↘
+ signal at level $j+1$
details at level j (local differences) ↗

That is multiresolution for one signal. When we apply it to all signals, we have multiresolution for *spaces* of functions:

V_j = scaling space at level j ↘
$\oplus$ V_{j+1} = scaling space at level $j+1$
W_j = wavelet space at level j ↗

This idea of multiresolution is absolutely basic to wavelet analysis. Again, we are only introducing it. We are sending a coarse signal to the reader, not the details. You only have the input at level 1.

Thus the signal is divided into different **scales** of resolution, rather than different frequencies. The "time-scale plane" takes the place for wavelets that the "time-frequency plane" takes for filters. Multiresolution divides the frequencies into *octave bands*, from ω to 2ω, instead of uniform bands from ω to $\omega + \Delta\omega$. The compression of a graph, when $f(t)$ is replaced by $f(2t)$, means expansion of its Fourier transform from $F(\omega)$ to $\frac{1}{2}F\left(\frac{\omega}{2}\right)$. Frequencies shift upward by an octave, when time is rescaled by two. The time-frequency plane is partitioned naturally into *rectangles of constant area*.

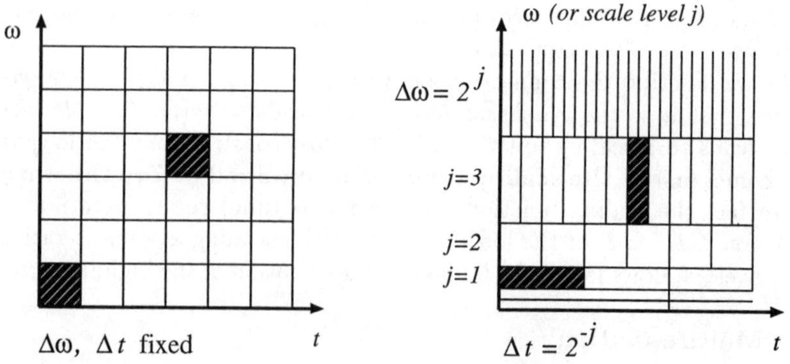

Time-frequency squares for Fourier decompositions become rectangles for wavelets.
Short time intervals are natural for high frequencies.

This matching of long time with low frequency and short time with high frequency occurs in a natural way for wavelets. It is one of the attractions of a wavelet decomposition.

2.3 Frequency domain and notation

To see a filter as a multiplication, we must take Fourier transforms. This will be the *discrete-time Fourier transform*, since the vectors $\boldsymbol{x}(n)$, $\boldsymbol{h}(n)$ and $\boldsymbol{y}(n)$ are discrete. The time index n goes from $-\infty$ to ∞. (A vector with zero components at all negative times is called *causal*.) The transform of $\boldsymbol{x}$ has two reasonable notations. They both stand for the same transform, which we denote by X:

$$X(e^{j\omega}) = \sum_{-\infty}^{\infty} \boldsymbol{x}(n)\, e^{-jn\omega} \quad \text{(signal processing notation)}$$

$$X(\omega) = \sum_{-\infty}^{\infty} \boldsymbol{x}(n)\, e^{-in\omega} \quad \text{(reduced notation)}.$$

The standard notation allows a direction conversion of the Fourier transform to the *z-transform*. The transform is still X but the variable becomes z:

$$X(z) = \sum_{-\infty}^{\infty} \boldsymbol{x}(n)\, z^{-n}.$$

We simply replace $e^{j\omega}$ by z, extending the formal definition of X from $e^{j\omega}$ on the unit circle to z in the whole complex plane. (Remember: $e^{j\omega}$ has magnitude 1.) The Fourier transform will dominate the first part of the article, but the z-transform appears more frequently towards the end.

Convolution by h in time becomes multiplication by H in frequency:

$$\begin{aligned} Y(e^{j\omega}) &= H(e^{j\omega})\,X(e^{j\omega}) && \text{in signal processing notation} \\ Y(\omega) &= H(\omega)\,X(\omega) && \text{in reduced notation.} \end{aligned}$$

This is the transform of $\boldsymbol{y}(n) = \sum \boldsymbol{h}(k)\,\boldsymbol{x}(n-k)$. It is the "convolution rule". In the z-domain it becomes $Y(z) = H(z)\,X(z)$.

2.4 Perfect reconstruction

A filter bank is a set of filters, linked by sampling operators and sometimes by delays. The downsampling operators are decimators, the upsampling operators are expanders. In a two-channel filter bank, the analysis filters are normally lowpass and highpass. Those are the filters $\boldsymbol{H}_0$ and $\boldsymbol{H}_1$ at the start of the following filter bank:

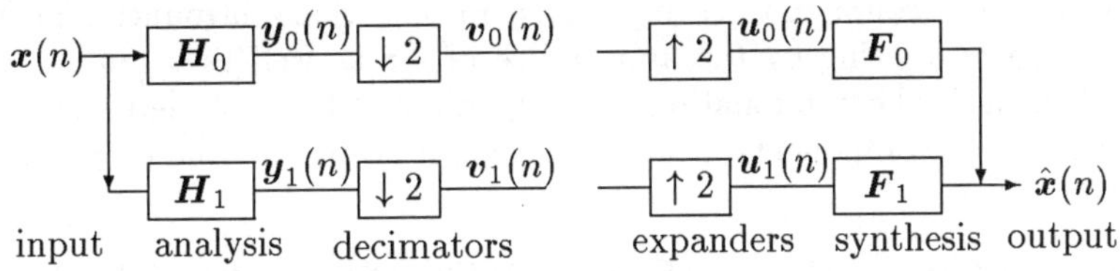

This structure was introduced in the 1980's. It gradually became clear how to choose $\boldsymbol{H}_0$, $\boldsymbol{H}_1$, $\boldsymbol{F}_0$, $\boldsymbol{F}_1$ to get perfect reconstruction: $\hat{\boldsymbol{x}}(n) = \boldsymbol{x}(n-l)$. The gap in the figure indicates where the downsampled signals might be coded for storage or transmission. At that point we may compress the signal and destroy information. Perfect reconstruction assumes no compression, so the gap is closed.

To indicate that $\boldsymbol{H}_0$ is lowpass and $\boldsymbol{H}_1$ is highpass, we often sketch the frequency responses. The drawing shows that they are not ideal brick wall filters. The responses overlap. *There is aliasing in each channel.* There is also amplitude distortion and phase distortion (our drawing does not show the phase). The synthesis filters $\boldsymbol{F}_0$ and $\boldsymbol{F}_1$ must be specially adapted to the analysis filters $\boldsymbol{H}_0$ and $\boldsymbol{H}_1$, in order to cancel the errors in this analysis bank.

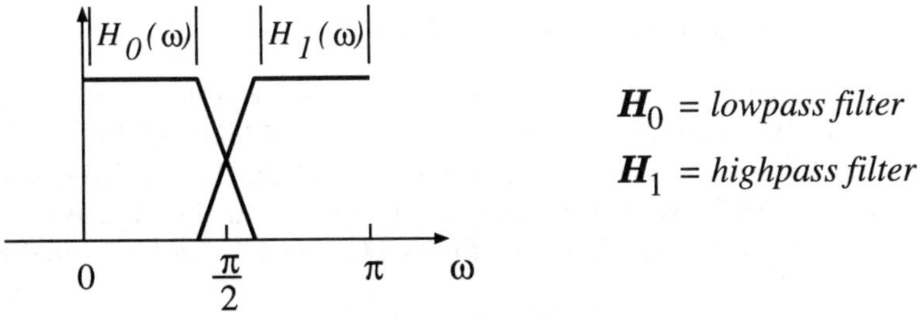

Rough sketch of frequency responses. Do not expect $|H_0(\omega)| + |H_1(\omega)| = 1$.

The goal of this section is to discover the conditions for perfect reconstruction. This means that the filter bank is **biorthogonal**. The synthesis bank, from $\boldsymbol{F}_0$ and $\boldsymbol{F}_1$ and $\uparrow 2$, is the inverse of the analysis bank. Inverse matrices automatically involve biorthogonality. (The rows of $\boldsymbol{T}$ and the columns of $\boldsymbol{T}^{-1}$ are by definition biorthogonal.) When the analysis bank leads to scaling functions $\varphi(t-k)$, and the synthesis bank leads to $\widetilde{\varphi}(t-j)$, those are biorthogonal. So are the wavelets.

Perfect reconstruction is a crucial property. If the sampling operators $(\downarrow 2)$ and $(\uparrow 2)$ were not present, a reconstruction without delay would mean that $\boldsymbol{F}_0\boldsymbol{H}_0 + \boldsymbol{F}_1\boldsymbol{H}_1 = \boldsymbol{I}$. A perfect reconstruction with an l-step delay would mean (in the z-domain) that

$$\textit{without } (\downarrow 2) \text{ and } (\uparrow 2): \quad F_0(z)H_0(z) + F_1(z)H_1(z) = z^{-l}. \tag{2.4}$$

We do expect an overall delay z^{-l}, because each individual filter is causal.

Now take account of the sampling operators, which introduce *aliasing*. We recognize aliasing by the appearance of $-z$ as well as z (and $\omega + \pi$ as well as ω). The combination of $(\downarrow 2)$ followed by $(\uparrow 2)$ zeros out the odd-numbered components. In the z-domain it keeps only the *even powers* of $H_0(z)X(z)$:

$$\text{The transform of } (\uparrow 2)(\downarrow 2)\boldsymbol{H}_0\boldsymbol{x} \text{ is } \tfrac{1}{2}(H_0(z)X(z) + H_0(-z)X(-z)).$$

This is an even function, because the odd components are gone. The aliasing term $H_0(-z)X(-z)$ is multiplied by $F_0(z)$ at the synthesis step. This alias has to cancel the alias $F_1(z)H_1(-z)X(-z)$ from the other channel. So there is an alias cancellation condition in addition to a reconstruction condition:

$$\textit{Alias cancellation} \quad F_0(z)H_0(-z) + F_1(z)H_1(-z) = 0. \tag{2.5}$$

Correction The sampling operators also produce a change in eqn. (2.4). *The right side has an extra factor* 2. You can see this by considering a simple set of filters: $H_0(z) = 1$ and $H_1(z) = z^{-1}$, $F_0(z) = z^{-1}$ and $F_1(z) = 1$. This satisfies (2.5) and cancels aliasing. The left side of eqn. (2.4) equals $2z^{-1}$ rather than z^{-1}. The overall delay is $l = 1$ for this filter bank, and its perfect reconstruction comes from

$$\textit{No distortion} \quad F_0(z)H_0(z) + F_1(z)H_1(z) = 2z^{-l}. \tag{2.6}$$

The next page establishes these two conditions for perfect reconstruction.

One further point. In a genuine filter bank, the highpass filter has $H_1 = 0$ at $z = 1$ (or $\omega = 0$). Eqn. (2.6) becomes $F_0(1)H_0(1) = 2$. That equation is more natural if we include an extra factor $\sqrt{2}$ in the filter coefficients. For a similar reason the highpass filters $\boldsymbol{H}_1$ and $\boldsymbol{F}_1$ can be normalized by an extra $\sqrt{2}$.

No aliasing and no distortion Conditions (2.4) and (2.5) for perfect reconstruction come directly from following a signal through the filter bank. The original signal is $\boldsymbol{x}(n)$. The lowpass analysis filter is $\boldsymbol{H}_0$. In the z-domain this produces $H_0(z)X(z)$. Now downsample and upsample:

$$\text{First } (\downarrow 2) \text{ produces} \quad \tfrac{1}{2}[H_0(z^{\frac{1}{2}})X(z^{\frac{1}{2}}) + H_0(-z^{\frac{1}{2}})X(-z^{\frac{1}{2}})]$$

$$\text{Then } (\uparrow 2) \text{ produces} \quad \tfrac{1}{2}[H_0(z)X(z) + H_0(-z)X(-z)].$$

The filtered signal $\boldsymbol{H}_0\boldsymbol{x}$ now has zeros in its odd-numbered components.

Those zeros are produced by averaging $\boldsymbol{H}_0\boldsymbol{x}(n)$ with its alternating alias $(-1)^n\boldsymbol{H}_0\boldsymbol{x}(n)$. In the z-domain this is the average of $H_0(z)X(z)$ with $H_0(-z)X(-z)$. The aliasing term has entered the filter bank.

The final filter multiplies by $F_0(z)$. This yields the output from the lowpass channel. Below it we write the corresponding output from the highpass channel (same formula with subscripts changed to 1):

$$\text{lowpass output} = \tfrac{1}{2}F_0(z) \;\; [H_0(z)X(z) + H_0(-z)\; X(-z)]$$

$$\text{highpass output} = \tfrac{1}{2}F_1(z) \;\; [H_1(z)X(z) + H_1(-z)\; X(-z)].$$

Now add. The filter bank combines the channels to get $\hat{\boldsymbol{x}}(n)$. In the z-domain this is $\hat{X}(z)$. Half the terms involve $X(z)$ and half involve $X(-z)$:

$$\begin{aligned}\hat{X}(z) &= \tfrac{1}{2}[F_0(z)H_0(z) + F_1(z)H_1(z)]X(z) \\ &+ \tfrac{1}{2}[F_0(z)H_0(-z) + F_1(z)H_1(-z)]X(-z).\end{aligned}$$

For perfect reconstruction with l time delays, $\hat{X}(z)$ must be $z^{-l}X(z)$. So the "distortion term" must be z^{-l} and the "alias term" must be zero:

Theorem 2.1 *A 2-channel filter bank gives perfect reconstruction when*

$$F_0(z)H_0(z) + F_1(z)H_1(z) = 2z^{-l} \tag{2.7}$$

$$F_0(z)H_0(-z) + F_1(z)H_1(-z) = 0. \tag{2.8}$$

In vector-matrix form these two conditions involve the "*modulation matrix*" $\boldsymbol{H}_m(z)$ (the 2×2 matrix on the left hand side below):

$$[F_0(z) \quad F_1(z)]\begin{bmatrix} H_0(z) & H_0(-z) \\ H_1(z) & H_1(-z) \end{bmatrix} = [2z^{-l} \quad 0]. \tag{2.9}$$

This matrix $\boldsymbol{H}_m(z)$ will play a very important role. It involves the responses $H_k(z)$ and the alias terms $H_k(-z)$. For an M-channel bank the matrix will be $M \times M$. But the real problem is clearly identified by the separate conditions (2.7) and (2.8) — *how to design filters that meet those conditions.*

2.5 Alias cancellation and the product filter $\boldsymbol{P}_0 = \boldsymbol{F}_0\boldsymbol{H}_0$

At this point we have four filters $\boldsymbol{H}_0, \boldsymbol{H}_1, \boldsymbol{F}_0, \boldsymbol{F}_1$ to design. They must satisfy (2.7) and (2.8). It is almost irresistible to determine some of the filters from the others:

$$\textit{for alias cancellation choose} \quad \boxed{F_0(z) = H_1(-z) \quad \text{and} \quad F_1(z) = -H_0(-z)} \tag{2.10}$$

Important: This choice automatically satisfies

$$F_0(z)H_0(-z) + F_1(z)H_1(-z) = 0.$$

Aliasing is removed; it cancels itself! This relation of $\boldsymbol{F}_0$ to $\boldsymbol{H}_1$ and of $\boldsymbol{F}_1$ to $\boldsymbol{H}_0$ gives the *alternating signs* pattern of a 2-channel filter bank:

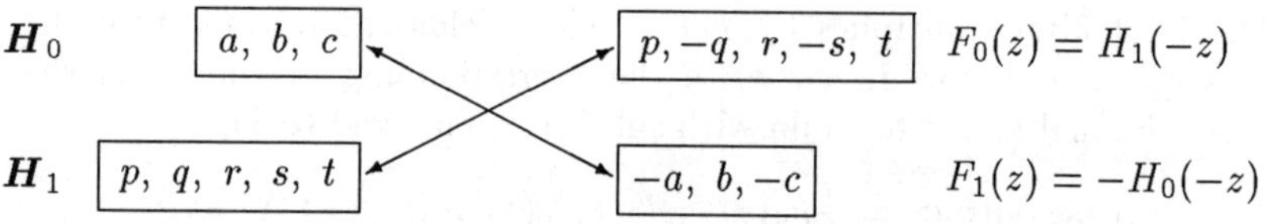

Now comes a definition that allows us to rewrite eqn. (2.7) for no distortion:

Define the "**product filter**" *by* $P_0(z) = F_0(z)H_0(z)$.

This is a lowpass filter. The highpass product filter is $P_1(z) = F_1(z)H_1(z)$. These products P_0 and P_1 are exactly the terms in (2.7). The crucial point is the relation between $P_0(z)$ and $P_1(z)$, when the synthesis filters are determined by $F_0(z) = H_1(-z)$ and $F_1(z) = -H_0(-z)$. We substitute directly to find that $P_1(z) = -P_0(-z)$:

$$P_1(z) = -H_0(-z)H_1(z) = -H_0(-z)F_0(-z) = -P_0(-z).$$

The reconstruction equation $F_0(z)H_0(z) + F_1(z)H_1(z) = 2z^{-l}$ simplifies to

$$\boxed{P_0(z) - P_0(-z) = 2z^{-l}.} \tag{2.11}$$

The design of a 2-channel PR filter bank is reduced to two steps:

Step 1. Design a lowpass filter $\boldsymbol{P}_0$ satisfying (2.11).

Step 2. Factor $\boldsymbol{P}_0$ into $\boldsymbol{F}_0\boldsymbol{H}_0$. Then use (2.10) **to find $\boldsymbol{F}_1$ and $\boldsymbol{H}_1$.**

The length of $\boldsymbol{P}_0$ determines the sum of the lengths of $\boldsymbol{F}_0$ and $\boldsymbol{H}_0$. There are many ways to design $\boldsymbol{P}_0$ in Step 1. And there are many ways to factor it in Step 2. Experiments are going on as this article is written, and undoubtedly they are going on as the article is read, to find the best factors $\boldsymbol{F}_0$ and $\boldsymbol{H}_0$ of the best product filter $\boldsymbol{P}_0$.

Note that (2.11) is a condition on the *odd powers* in $P_0(z)$. Those odd powers must have coefficient zero, except z^{-l} has coefficient one.

A look forward To help the reader find the specific filters that are coming, we point to an outstanding choice for the product filter:

$$P_0(z) = (1 + z^{-1})^{2p} Q(z).$$

The polynomial $Q(z)$ of degree $2p - 2$ is chosen so that (2.11) is satisfied. There are $2p - 1$ odd powers in $P_0(z)$, and $2p - 1$ coefficients to choose in $Q(z)$. *Then* $Q(z)$ *is unique.* This is the Daubechies construction. Since the construction starts with the special factor $(1 + z^{-1})^{2p}$, these filters are called *binomial* or *maxflat*. The binomial factor gives a maximum number p of zeros at $z = -1$, which means that the frequency response is maximally flat at $\omega = \pi$. The binomial by itself, without $Q(z)$, represents a "spline filter". $Q(z)$ is needed to give perfect reconstruction.

Splitting P_0 into $F_0 H_0$ can give linear phase filters (symmetry in F_0 and H_0 separately). It can give orthogonal filters (symmetry between F_0 and H_0). *It cannot give both*, except in the Haar case $p = 1$.

Simplification The equation $P_0(z) - P_0(-z) = 2z^{-l}$ can be made a little more convenient. The left side is an odd function, so l is odd. Normalize $P_0(z)$ by z^l to center it:

$$\textit{the normalized product filter is} \quad P(z) = z^l P_0(z).$$

Then $P(-z) = (-z)^l P_0(-z)$. Since l is odd, this is $-z^l P_0(-z)$. The reconstruction equation $P_0(z) - P_0(-z) = 2z^{-l}$ takes an extremely simple form when we multiply by z^l. The factor z^{-l} disappears and the minus sign becomes plus, giving us the following note.

Note 1: perfect reconstruction condition $P(z)$ must be a "halfband filter"

$$\boxed{P(z) + P(-z) = 2.} \tag{2.12}$$

This means that *all even powers in* $P(z)$ *are zero*, except the constant term (which is 1). The odd powers cancel when $P(z)$ combines with $P(-z)$ — so the coefficients of odd powers in $P(z)$ are design variables in 2-channel PR filter banks.

2.6 Modulation matrices

The conditions for perfect reconstruction are expressed in (2.9) by two equations:

$$[F_0(z) \quad F_1(z)] \begin{bmatrix} H_0(z) & H_0(-z) \\ H_1(z) & H_1(-z) \end{bmatrix} = [2z^{-l} \quad 0]. \tag{2.13}$$

This displays the **analysis modulation matrix** $\boldsymbol{H}_m(z)$ which is central to filter bank theory. With no extra effort we can also produce the synthesis modulation matrix $\boldsymbol{F}_m(z)$. The two matrices should play matching (and even reversible) roles. This balance between $\boldsymbol{F}_m$ and $\boldsymbol{H}_m$ is achieved by expanding (2.13) into a matrix equation:

$$\begin{bmatrix} F_0(z) & F_1(z) \\ F_0(-z) & F_1(-z) \end{bmatrix} \begin{bmatrix} H_0(z) & H_0(-z) \\ H_1(z) & H_1(-z) \end{bmatrix} = \begin{bmatrix} 2z^{-l} & 0 \\ 0 & 2(-z)^{-l} \end{bmatrix}. \quad (2.14)$$

The second row of equations follows from the first, when $-z$ replaces z. Note that the synthesis matrix $\boldsymbol{F}_m$ has $F_1(z)$ in the (1,2) position, while the analysis matrix $\boldsymbol{H}_m$ has $H_1(z)$ in the (2,1) position. This *transpose convention* between analysis and synthesis will appear again for polyphase matrices. The reader sees why it is necessary.

The reconstruction condition (2.14) is now a statement about the matrix product $\boldsymbol{F}_m(z)\boldsymbol{H}_m(z)$. If we "center" the filter coefficients around the zero position, the right side becomes the identity matrix! This is so desirable and memorable that we do it. It is the same normalization that centered $P_0(z)$ into $P(z)$, and it is especially clear when the filters are linear phase (and l is odd).

Theorem 2.2 *If all filters are symmetric (or antisymmetric) around zero, as in $H(z) = H(z^{-1})$ and $\boldsymbol{h}(k) = \boldsymbol{h}(-k)$, then the condition for perfect reconstruction becomes a statement about inverse matrices:*

$$\boxed{\mathbf{F}_m(z)\,\mathbf{H}_m(z) = 2\mathbf{I}.} \quad (2.15)$$

The H's determine the F's. The analysis bank is inverted by the synthesis bank. When we express it that way, eqn. (2.15) becomes almost obvious.

2.7 A brief history of $\boldsymbol{H}_1$

The reader understands that the filters $\boldsymbol{H}_0$ and $\boldsymbol{H}_1$ are still to be chosen. These choices are connected. Historically, designers chose the lowpass filter coefficients $\boldsymbol{h}(0), \ldots \boldsymbol{h}(N)$ and then constructed $\boldsymbol{H}_1$ from $\boldsymbol{H}_0$. Here are two possibilities that produce *equal length filters.* $\boldsymbol{H}_1$ will be highpass whenever $\boldsymbol{H}_0$ is lowpass.

Alternating signs:

$$H_1(z) = H_0(-z) \text{ comes from } (\boldsymbol{h}(0), -\boldsymbol{h}(1), \boldsymbol{h}(2), -\boldsymbol{h}(3), \ldots)$$

Alternating flip:

$$H_1(z) = -z^{-N} H_0(-z^{-1}) \text{ comes from } (\boldsymbol{h}(N), -\boldsymbol{h}(N-1), \ldots).$$

For convenience we are assuming real coefficients. The number N is odd in the alternating flip. The perfect reconstruction condition is *still to be imposed.* When that is satisfied, the overall system delay is $l = N$.

Note 2: early choice Estaban and Galand [2] chose alternating signs $H_1(z) = H_0(-z)$. The resulting filter bank was called QMF (Quadrature Mirror Filter). The highpass response $|H_1(e^{j\omega})|$ is a mirror image of the lowpass magnitude $|H_0(e^{j\omega})|$ with respect to the middle frequency $\frac{\pi}{2}$ — the quadrature frequency. Note that IIR filters $\boldsymbol{H}_0$ and $\boldsymbol{H}_1$ are allowed (and needed for PR, except for Haar!). This name QMF has since been extended to a larger class of filter banks, allowing M channels.

Note 3: better choice Smith and Barnwell [5] and Mintzer chose the alternating flip $H_1(z) = -z^{-N}H_0(-z^{-1})$. This leads to orthogonal filter banks, when $\boldsymbol{H}_0$ is correctly chosen. The Daubechies filters will fit this pattern.

Note 4: general choice The product $F_0(z)H_0(z)$ is a halfband filter. This leads to biorthogonal filter banks, when aliasing is cancelled by the relation of $\boldsymbol{F}_0$ to $\boldsymbol{H}_1$ and $\boldsymbol{F}_1$ to $\boldsymbol{H}_0$.

Actually the synthesis bank has little freedom. Alias cancellation requires $F_0(z)H_0(-z) + F_1(z)H_1(-z) = 0$. Croisier, Estaban and Galand wrote each F_k directly in terms of H_k by

$$F_0(z) = H_0(z) \quad \text{and} \quad F_1(z) = -H_1(z).$$

With alternating signs $H_1(z) = H_0(-z)$ inside the analysis bank, their synthesis construction agrees (as it must) with the anti-aliasing equations

$$F_0(z) = H_1(-z) \quad \text{and} \quad F_1(z) = -H_0(-z). \tag{2.16}$$

Smith and Barnwell also made the anti-aliasing choice (2.16). With the alternating flip in $H_1(z)$, their synthesis filters (remembering that N is odd) are

$$F_0(z) = H_1(-z) = z^{-N}H_0(z^{-1}),$$

coming from $(\boldsymbol{h}(N), \boldsymbol{h}(N-1), \ldots, \boldsymbol{h}(0))$, and

$$F_1(z) = -H_0(-z) = z^{-N}H_1(z^{-1}),$$

coming from $(-\boldsymbol{h}(0), \boldsymbol{h}(1), -\boldsymbol{h}(2), \ldots, \boldsymbol{h}(N))$. **Notice!** Each $\boldsymbol{F}_k$ has become the ordinary flip of the corresponding $\boldsymbol{H}_k$. In matrix language the synthesis matrices are the *transposes* of the analysis matrices. A shift by N delays makes them causal. When we flip to get $\boldsymbol{F}_0$ and then alternate signs to get $\boldsymbol{H}_1$, we have the alternating flip from $\boldsymbol{H}_0$ to $\boldsymbol{H}_1$.

The alternating flip automatically gives us double-shift orthogonality between highpass and lowpass (to be explained). *Conclusion*: when the design of $\boldsymbol{H}_0$ leads to perfect reconstruction in the alternating flip filter bank, it also leads to orthogonality.

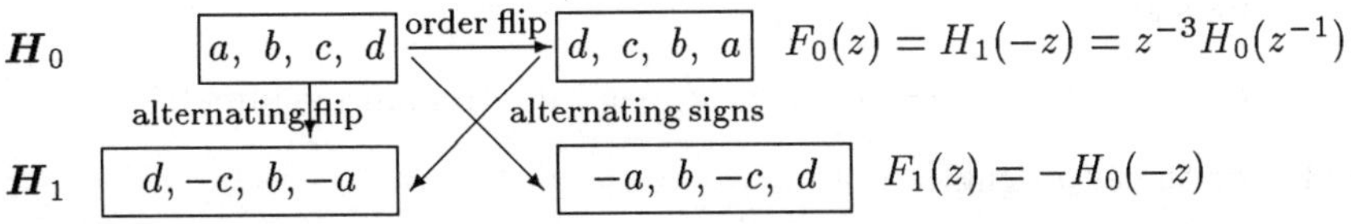

Relations between the filters allowing orthogonality when $N = 3$.

With aliasing cancelled, we now look at the PR condition

$$F_0(z)H_0(z) + F_1(z)H_1(z) = 2z^{-l}.$$

The early choice was alternating signs $H_1(z) = H_0(-z)$. With $F_0(z) = H_1(-z)$ and $F_1(z) = -H_0(-z)$, PR requires

$$H_0^2(z) - H_1^2(z) = H_0^2(z) - H_0^2(-z) = 2z^{-l}.$$

Therefore $H_0^2(z)$ has exactly one odd power z^{-l}. This is not easy for the square of a polynomial. An FIR filter is restricted to *two coefficients* (not good). This forces the filters to be IIR.

The better choice is the alternating flip. Perfect reconstruction is possible. The product filters $\boldsymbol{F}_0\boldsymbol{H}_0$ and $\boldsymbol{F}_1\boldsymbol{H}_1$ become

$$P_0(z) = z^{-N}H_0(z^{-1})H_0(z), \quad P_1(z) = -z^{-N}H_0(-z^{-1})H_0(-z).$$

Multiply by $z^l = z^N$ to center these filters. The normalized product filter is $P(z)$ and the reconstruction condition is (2.12):

$$P(z) + P(-z) = 2 \text{ with } P(z) = H_0(z^{-1})H_0(z).$$

This is spectral factorization of a halfband filter! On the unit circle $z = e^{j\omega}$, the product $H_0(e^{-j\omega})H_0(e^{j\omega})$ is a magnitude squared:

$$P(e^{j\omega}) = \sum_{-N}^{N} \boldsymbol{p}(n)e^{-jn\omega} = \left|\sum_{0}^{N} \boldsymbol{h}(n)e^{-jn\omega}\right|^2.$$

The halfband coefficients are $\boldsymbol{p}(n) = \boldsymbol{p}(-n)$ for odd n and $\boldsymbol{p}(n) = 0$ for even n (except $\boldsymbol{p}(0) = 1$). *We design $P(z)$ and factor to find $H_0(z)$.* This symmetric factorization coincides with the Smith-Barnwell alternating flip. *It yields orthogonal banks with perfect reconstruction.* The flattest $P(z)$ will lead us to the Daubechies wavelets.

2.8 A note on biorthogonality (= PR) with linear phase

Theorem 2.3 *In a biorthogonal linear-phase filter bank with two channels, the filter lengths are all odd or all even. The analysis filters can be*

(a) both symmetric, of odd length;
(b) one symmetric and the other antisymmetric, of even length.

Proof: The difference between odd and even lengths comes when we alternate signs. For the odd length,

$$a \quad b \quad c \quad b \quad a \quad \rightarrow \quad a \quad -b \quad c \quad -b \quad a \quad \text{(remains symmetric)},$$

while for the even length

$$a \quad b \quad b \quad a \quad \rightarrow \quad a \quad -b \quad b \quad -a \quad \text{(becomes antisymmetric)}.$$

To cancel aliasing, there is sign alternation in $F_0(z) = H_1(-z)$. There is also alternation in $F_1(z) = -H_0(-z)$. The extra minus sign does not change the symmetry type. The two successful combinations are

$$\begin{array}{ccc|ccc} H_0 = \text{symm} & & F_0 = \text{symm} & H_0 = \text{symm} & & F_0 = \text{symm} \\ & \times & & & \times & \\ H_1 = \text{symm} & & F_1 = \text{symm} & H_1 = \text{anti} & & F_1 = \text{anti} \\ & & \textit{odd lengths} & & & \textit{even lengths} \end{array}$$

The other possibilities are excluded by the PR condition: $F_0(z)\,H_0(z)$ has to be a **halfband filter**. It must have an odd number of coefficients, and the center coefficient must be 1. For $F_0(z)\,H_0(z)$ to have odd length (which means even degree), the factors $F_0(z)$ and $H_0(z)$ must be both odd length or both even length. If one is symmetric and the other antisymmetric, the product $F_0(z)\,H_0(z)$ will be antisymmetric with zero at the center — not allowed. We conclude that $F_0(z)$ and $H_0(z)$ must match: both odd length or both even length, both symmetric or both antisymmetric.

This leaves the two successful possibilities shown above, and two more: H_0 and F_0 both antisymmetric. But the sum of lowpass coefficients cannot be zero. So antisymmetry of H_0 is ruled out.

Perfect reconstruction with M channels In reality a filter bank can have M channels. Although $M = 2$ is standard in a range of applications, we often see $M > 2$. There are M analysis filters $\boldsymbol{H}_0, \boldsymbol{H}_1, \ldots, \boldsymbol{H}_{M-1}$. The sampling is done at the critical rate by $(\downarrow M)$ and $(\uparrow M)$. There are M filters $\boldsymbol{F}_0, \boldsymbol{F}_1, \ldots, \boldsymbol{F}_{M-1}$ in the synthesis bank. The outputs from all channels are combined into a single output $\hat{\boldsymbol{x}}$. Our standard picture of this implementation is

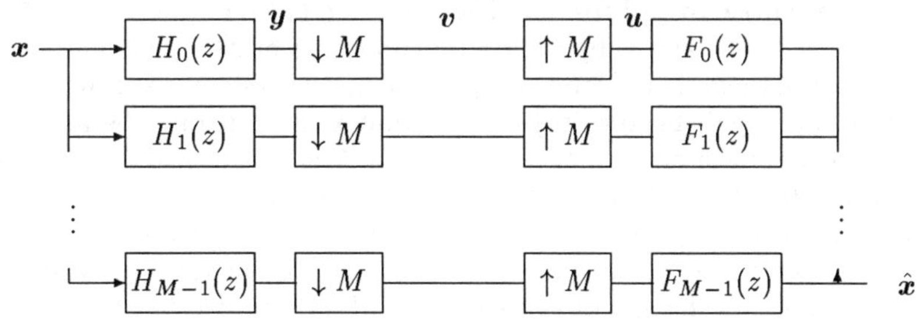

Several perfect reconstruction filter banks deserve special mention. The simplest is the average-difference pair from Haar. This is a useful example but a poor filter. That is the first in a family of "maxflat filters", corresponding to the Daubechies wavelets. The others in the family are orthogonal but not linear phase, since those two properties conflict.

Different factorizations of the product $P_0(z)$ lead to linear phase (not orthogonality). Those filters have become favorites for compression.

For $M > 2$, the design of separate filters $\boldsymbol{H}_0, \boldsymbol{H}_1, \ldots, \boldsymbol{H}_{M-1}$ can become unwieldy. We look for constructions in which these all come from one prototype filter. A particular class consists of the *cosine-modulated filter banks*. A phase change (= modulation) is the key to their construction. Those are efficient in every way.

2.9 The polyphase matrix

This section establishes a key idea and a valuable notation. The word "*polyphase*" has gained a certain mystique in the theory of multirate filters. Perhaps we can begin by explaining the meaning of the word, and also the purpose of the idea. Then the notation and applications will come naturally.

Note 5: meaning of polyphase When a vector is downsampled by 2, its even-numbered components are kept. Its odd-numbered components are lost. Those are the two phases, *even* and *odd*. It is natural to follow the two phases of the input vector, $\boldsymbol{x}_{\text{even}}$ and $\boldsymbol{x}_{\text{odd}}$, as they go through the filter bank. They are acted on by the two phases $\boldsymbol{H}_{\text{even}}$ and $\boldsymbol{H}_{\text{odd}}$ of the filter.

For downsampling by M there are M phases. The ideas still apply to this "several-phase" or "polyphase" decomposition. Instead of even and odd inputs we will have M vectors (phases of $\boldsymbol{x}$). Instead of even and odd filters we will have M filters (phases of $\boldsymbol{H}$). The vector of filter coefficients $\boldsymbol{h}(n)$ is separated into phases, exactly as $\boldsymbol{x}(n)$ is separated. Then we watch those phases during downsampling.

The word "phase" is applied because the even filter with coefficients $\boldsymbol{h}(0)$, $\boldsymbol{h}(2)$ has a different delay (phase shift) from the odd phase with coefficients $\boldsymbol{h}(1)$, $\boldsymbol{h}(3)$.

Note 6: purpose of polyphase The operation $(\downarrow 2)\boldsymbol{Hx}$, taken literally, is not efficient. We are computing all components of $\boldsymbol{Hx}$ and then destroying half of them. If we don't compute them, the system is still working at a fast rate (high bandwidth). The output is at half rate, because of downsampling. Each output component needs N additions and $N+1$ multiplications, to apply all the coefficients $\boldsymbol{h}(0), \ldots, \boldsymbol{h}(N)$.

The polyphase implementation works on the different phases separately. The input vector is separated into $\boldsymbol{x}_{\text{even}}$ and $\boldsymbol{x}_{\text{odd}}$. The operator $(\downarrow 2)$ comes *before the filter*! It changes one input at a high rate to two (or M) inputs at a lower rate. Then the separate phases of the filters act simultaneously (*in parallel*) on separate phases of the input.

The notation has to keep track of each phase. Often we find that "even multiplies even" and "odd multiplies odd". The *Noble identities* justify an interchange of filtering and sampling. For the whole filter this interchange is forbidden, but it is allowed for each phase.

Note 7: polyphase in the time domain = block Toeplitz matrix: We can display the infinite matrix for a 2-channel analysis bank. The two filters $\sqrt{2}\boldsymbol{H}_0 = \boldsymbol{C}$ and $\sqrt{2}\boldsymbol{H}_1 = \boldsymbol{D}$ are downsampled by $(\downarrow 2)$. This removes the odd-numbered rows. Then we *interleave the rows* of $\boldsymbol{L} = (\downarrow 2)\boldsymbol{C}$ and $\boldsymbol{B} = (\downarrow 2)\boldsymbol{D}$ to see the analysis bank as a *block-Toeplitz matrix*:

$$\textbf{Block}\quad \boldsymbol{H}_b = \begin{bmatrix} \cdot & \cdot & & & & & \\ \cdot & \cdot & & & & & \\ c(3) & c(2) & c(1) & c(0) & & & \\ d(3) & d(2) & d(1) & d(0) & & & \\ & & c(3) & c(2) & c(1) & c(0) & \\ & & d(3) & d(2) & d(1) & d(0) & \\ & & & & \cdot & \cdot & \cdot\ \cdot \\ & & & & \cdot & \cdot & \cdot\ \cdot \end{bmatrix}.$$

This takes the input in blocks (two samples at a time). It gives the output in blocks. It is time-invariant in blocks! By block z-transform, multiplication by the infinite matrix $\boldsymbol{H}_b$ (which is block convolution) becomes multiplication by the *polyphase matrix*:

$$\textbf{Polyphase matrix}\ \boldsymbol{H}_p(z) = \begin{bmatrix} c(0) & c(1) \\ d(0) & d(1) \end{bmatrix} + z^{-1} \begin{bmatrix} c(2) & c(3) \\ d(2) & d(3) \end{bmatrix}. \tag{2.17}$$

The polyphase matrix is nothing but *the z-transform of a block of filters.* There are 2^2 or M^2 filters, from M phases of M original filters. Here those filters have four coefficients and their phases have two coefficients.

Notice especially how the block matrix $\boldsymbol{H}_b$ relates to the two separate downsampled filters $(\downarrow 2)\boldsymbol{C}$ and $(\downarrow 2)\boldsymbol{D}$:

- the efficient form downsamples the input *first* (to make blocks for $\boldsymbol{H}_b$);
- the inefficient form downsamples *last* (after the filters $\boldsymbol{C}$ and $\boldsymbol{D}$).

The Noble Identities prove the equivalence. It is just a removal of useless odd-numbered rows and an interleaving of the remaining rows. Next we discuss the algebra and the implementation.

Note 8: key identity in the z-domain The even part of $X(z)$ is $\frac{1}{2}(X(z) + X(-z))$. The odd part is $\frac{1}{2}(X(z) - X(-z))$. The first has even powers $1, z^2, z^4$; the second has z, z^3, z^5. The original X is the sum of even plus odd (obviously). The same splitting holds for $C(z)$, and furthermore for $C(z)X(z)$. *The key is to find the even part of* $C(z)X(z)$. It is the even coefficients of $\boldsymbol{Cx}$ that survive downsampling and appear in $(\downarrow 2)\boldsymbol{Cx}$. In most of this section the lowpass filter is denoted by $\boldsymbol{C}$, to avoid the subscripts on $\boldsymbol{H}$.

A simple and important identity shows how the even part of $C(z)X(z)$ comes from even times even plus odd times odd:

$$\begin{aligned}\tfrac{1}{2}[C(z)X(z)+C(-z)X(-z)] &= \tfrac{1}{4}[C(z)+C(-z)][X(z)+X(-z)]\\ &+ \tfrac{1}{4}[C(z)-C(-z)][X(z)-X(-z)].\end{aligned}$$

In multiplying numbers, odd times odd is odd. But we are *adding exponents*, as in $(z^3)(z^5)=z^8$. So it is really odd *plus* odd, and even *plus* even, that yield the even part of the z-transform. This is the part that downsampling picks out, when $\boldsymbol{Cx}$ is decimated.

The importance of the key identity is this. The left side involves *all* coefficients of $C(z)$ and $X(z)$. Each product on the right involves only *half* the coefficients. The multiplication in the z-domain, which is $(\downarrow 2)\boldsymbol{Cx}$ in the time domain, becomes computationally efficient. We don't want all of $C(z)X(z)$, only the even half. The right side shows how to do half the work. Better still, it shows how even-even and odd-odd can be executed in parallel at half the rate.

Downsampling an even function effectively replaces z by $z^{1/2}$. It "closes the gaps" in $1, z^2, z^4$ by changing to $1, z, z^2$. For an odd function we will need a delay or an advance. We cannot change z and z^3 to $z^{1/2}$ and $z^{3/2}$. You will see how the coefficients of z^{-1}, z^{-3}, z^{-5} in $C(z)$ become coefficients of $1, z^{-1}, z^{-2}$ in the odd phase $C_{\text{odd}}(z)$. There is a delay for the odd phase and a "delay chain" when there are multiple phases.

This section works out the polyphase notation. We concentrate most on $M=2$; the phases are even and odd. Then the polyphase forms of the analysis and synthesis banks lead quickly to a main goal of the theory. We find the *perfect reconstruction condition* on the polyphase matrices, when the filters are centered:

$$\boxed{\boldsymbol{F}_p(z)\,\boldsymbol{H}_p(z)=\mathbf{I}.}$$

This tells us, clearly and directly, what is required:

1. At a minimum, $\boldsymbol{H}_p(z)$ must be **invertible**. (biorthogonality)
2. Better than that, its inverse $\boldsymbol{F}_p(z)$ should be a **polynomial**. (FIR)
3. Better still, $\boldsymbol{F}_p(z)$ might be the **transpose** of $\boldsymbol{H}_p(z)$. (orthogonality)

In case **3**, the polyphase matrices are "paraunitary". The analysis and synthesis banks are orthogonal. In the more general case **1**, the banks are "biorthogonal". In case **2**, the synthesis bank is biorthogonal and also FIR.

The rows of a matrix are always biorthogonal to the columns of its inverse. When the rows of one are identical to the columns of the other, the matrix is self-orthogonal. Then it is an *orthogonal* matrix if real, a *unitary* matrix if complex, and a *paraunitary* matrix if it is a function of a complex parameter z.

2.10 Polyphase for vectors

Any input vector **x** and any filter vector $\boldsymbol{c}$ or $\boldsymbol{h}$ can be separated into even and odd:

$$\boldsymbol{x} = (\ldots, \boldsymbol{x}(0), 0, \boldsymbol{x}(2), 0, \ldots) + (\ldots, 0, \boldsymbol{x}(1), 0, \boldsymbol{x}(3), 0, \ldots).$$

The z-transform is separated into even powers and odd powers, as in

$$X(z) \quad = \quad [\boldsymbol{x}(0) + \boldsymbol{x}(2)z^{-2} + \ldots] + z^{-1}[\boldsymbol{x}(1) + \boldsymbol{x}(3)z^{-2} + \ldots]. \tag{2.18}$$

The even part has powers of z^2. So has the odd part, when we factor out z^{-1}. This is the polyphase decomposition of $\boldsymbol{x}$ in the z-domain:

$$\boxed{X(z) = X_{\text{even}}(z^2) + z^{-1}X_{\text{odd}}(z^2)} \tag{2.19}$$

Each phase has its own z-transform:

$$\boldsymbol{x}_{\text{even}} \quad = \quad \begin{bmatrix} \boldsymbol{x}(0) \\ \boldsymbol{x}(2) \\ \cdot \end{bmatrix} \leftrightarrow X_0(z) = \sum \boldsymbol{x}(2k)z^{-k}$$

$$\boldsymbol{x}_{\text{odd}} \quad = \quad \begin{bmatrix} \boldsymbol{x}(1) \\ \boldsymbol{x}(3) \\ \cdot \end{bmatrix} \leftrightarrow X_1(z) = \sum \boldsymbol{x}(2k+1)z^{-k}.$$

Because of the z^2 in the definition, the in-between zeros are gone from $X_0(z)$ and $X_1(z)$. Please verify that the phases of $X(z) = z^{-1} + z^{-2} + z^{-3}$ are $X_{\text{even}} = z^{-1}$ and $X_{\text{odd}} = 1 + z^{-1}$.

Now reverse the process, to recover $\boldsymbol{x}$. Upsampling puts zeros back into $\boldsymbol{x}_{\text{even}}$ and $\boldsymbol{x}_{\text{odd}}$. *Those zeros change z to z^2.* The odd phase is delayed by z^{-1}, to move $\boldsymbol{x}(1)$ from position 0 to position 1. Then addition reconstructs eqn. (2.19).

Here is the splitting and the reconstruction in block form. Notice that so far the filters are not included, and $(\downarrow 2)\boldsymbol{x}$ is exactly $\boldsymbol{x}_{\text{even}}$:

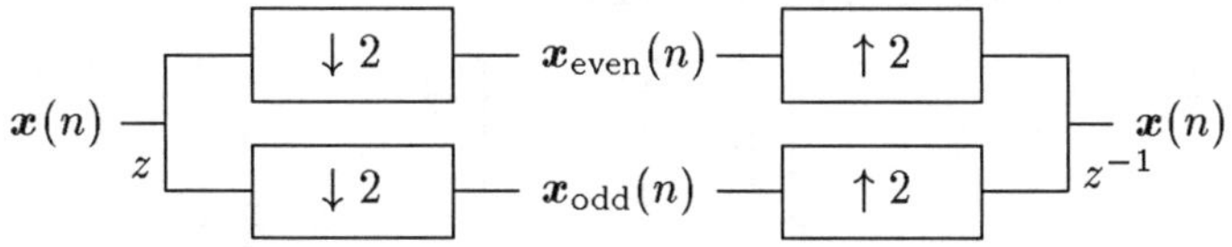

Important! The z at the start of the odd channel is because the odd phase has $\boldsymbol{x}(1)$ in its zeroth position. We have to advance the signal to achieve that. But advances look bad in our flow diagram. So the advance can be replaced by a delay, if we make up for it at the end by delaying the even part too.

Here is the "delay form" that we use in later sections. Please go through that form:

Polyphase with delay

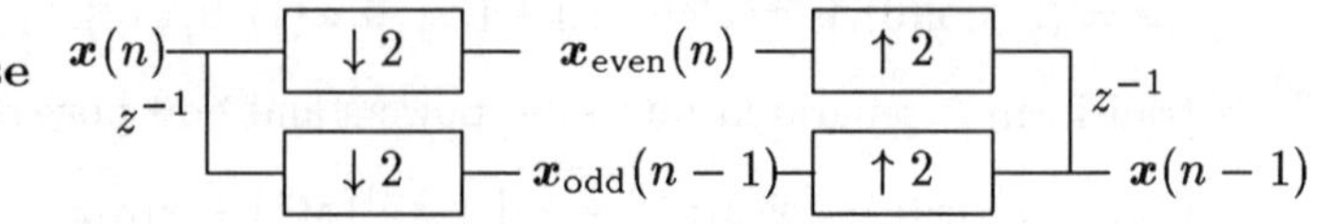

2.11 Polyphase matrices for filters

The polyphase form of a filter $\boldsymbol{C}$ comes directly from the polyphase form of $\boldsymbol{c}$ (the vector of filter coefficients). That vector separates into $\boldsymbol{c}_{\text{even}}$ and $\boldsymbol{c}_{\text{odd}}$. Its z-transform $C(z)$ separates into phases exactly as $X(z)$ did:

$$C(z) = C_0(z^2) + z^{-1}C_1(z^2).$$

The filtering step is $C(z)X(z)$. This is ordinary filtering $\boldsymbol{Cx}$, where even mixes with odd. But when downsampling picks out the even part of the product $C(z)X(z)$, it comes from even times even plus odd times odd. The transform of $(\downarrow 2)\boldsymbol{Cx}$ is

$$\boxed{(C(z)X(z))_{\text{even}} = C_0(z)X_0(z) + z^{-1}C_1(z)X_1(z).} \tag{2.20}$$

The direct multiplication of $C(z)$ times $X(z)$ will have even parts from $C_0(z^2)X_0(z^2)$ and from $z^{-2}C_1(z^2)X_1(z^2)$. Those give the even part of $C(z)X(z)$. Downsampling picks out those terms. It changes z^2 to z in their transform. The result is the z-transform of $(\downarrow 2)\boldsymbol{Cx}$.

Example 2.4 The moving average filter, downsampled.

Section 1 introduced the lowpass filter $\frac{1}{2}\boldsymbol{x}(n) + \frac{1}{2}\boldsymbol{x}(n-1)$. The coefficients are $\boldsymbol{h}(0) = \frac{1}{2}$ and $\boldsymbol{h}(1) = \frac{1}{2}$. Thus $\frac{1}{2}$ is the leading and only coefficient in H_{even} and H_{odd}. Both matrices have $\frac{1}{2}$ on *one diagonal* – the main diagonal. Here is the two-phase form of $(\downarrow 2)\boldsymbol{Hx}$:

$$\begin{bmatrix} \frac{1}{2} & \frac{1}{2} & & & \\ & & \frac{1}{2} & \frac{1}{2} & \\ & & & & \cdot\;\cdot \end{bmatrix}\begin{bmatrix} \\ \boldsymbol{x} \\ \\ \end{bmatrix} = \begin{bmatrix} \frac{1}{2} & & \\ & \frac{1}{2} & \\ & & \cdot \end{bmatrix}\begin{bmatrix} \boldsymbol{x}(0) \\ \boldsymbol{x}(2) \\ \cdot \end{bmatrix} + \begin{bmatrix} \frac{1}{2} & & \\ & \frac{1}{2} & \\ & & \cdot \end{bmatrix}\begin{bmatrix} \boldsymbol{x}(-1) \\ \boldsymbol{x}(1) \\ \cdot \end{bmatrix}.$$

The polyphase components of $H(z) = \frac{1}{2} + \frac{1}{2}z^{-1}$ are constants: $H_{\text{even}}(z) = \frac{1}{2}$ and $H_{\text{odd}}(z) = \frac{1}{2}$.

The same splitting occurs for the highpass filter $\boldsymbol{D}$. Its even and odd phases are represented by $D_0(z)$ and $D_1(z)$. Those go into the 1 by 2 polyphase matrix $\mathbf{D}_p(z)$. Then the whole analysis bank comes together when we combine the polyphase matrices for $\boldsymbol{C} = \boldsymbol{H}_0$ and $\boldsymbol{D} = \boldsymbol{H}_1$ into a single polyphase matrix $\boldsymbol{H}_p(z)$:

$$\begin{aligned}\boldsymbol{H}_p(z) &= \begin{bmatrix} \boldsymbol{C}_p(z) \\ \boldsymbol{D}_p(z) \end{bmatrix} = \begin{bmatrix} C_0(z) & C_1(z) \\ D_0(z) & D_1(z) \end{bmatrix} \\ &= \begin{bmatrix} H_{0,\text{even}}(z) & H_{0,\text{odd}}(z) \\ H_{1,\text{even}}(z) & H_{1,\text{odd}}(z) \end{bmatrix}.\end{aligned}$$

This shows the matrix that we are aiming for. Now we go back for the close look at $(\downarrow 2)\boldsymbol{C}$. This operator is fundamental in the theory of multirate filters and wavelets.

Polyphase in the time domain When downsampling follows the filter $\boldsymbol{C}$, we get the crucial matrix $\boldsymbol{L} = (\downarrow 2)\boldsymbol{C}$. This has to display the separation of even and odd, and I would like to show how this happens. Most of polyphase theory is developed in the z-domain, and we will do that too. But first, look at the filter matrix as it produces $\boldsymbol{y} = \boldsymbol{C}\boldsymbol{x}$:

$$\begin{bmatrix} \cdot \\ \boldsymbol{y}(0) \\ \boldsymbol{y}(1) \\ \boldsymbol{y}(2) \\ \boldsymbol{y}(3) \\ \cdot \end{bmatrix} = \begin{bmatrix} \cdot & & & & & \\ \cdot & \boldsymbol{c}(0) & & & & \\ \cdot & \boldsymbol{c}(1) & \boldsymbol{c}(0) & & & \\ \cdot & \boldsymbol{c}(2) & \boldsymbol{c}(1) & \boldsymbol{c}(0) & & \\ \cdot & \boldsymbol{c}(3) & \boldsymbol{c}(2) & \boldsymbol{c}(1) & \boldsymbol{c}(0) & \\ \cdot & \cdot & \cdot & \cdot & \cdot & \cdot \end{bmatrix} \begin{bmatrix} \cdot \\ \boldsymbol{x}(0) \\ \boldsymbol{x}(1) \\ \boldsymbol{x}(2) \\ \boldsymbol{x}(3) \\ \cdot \end{bmatrix}.$$

Downsampling leaves only the even-numbered components $\boldsymbol{y}(2n)$. To reach $\boldsymbol{v} = (\downarrow 2)\boldsymbol{y}$, we throw away every other row (the odd-numbered rows). This leaves the matrix $\boldsymbol{L} = (\downarrow 2)\boldsymbol{C}$:

$$\begin{bmatrix} \cdot \\ \boldsymbol{y}(0) \\ \boldsymbol{y}(2) \\ \boldsymbol{y}(4) \\ \cdot \end{bmatrix} = \begin{bmatrix} \cdot & & & & & & \\ \cdot & \boldsymbol{c}(1) & \boldsymbol{c}(0) & & & & \\ \cdot & \boldsymbol{c}(3) & \boldsymbol{c}(2) & \boldsymbol{c}(1) & \boldsymbol{c}(0) & & \\ \cdot & \boldsymbol{c}(5) & \boldsymbol{c}(4) & \boldsymbol{c}(3) & \boldsymbol{c}(2) & \boldsymbol{c}(1) & \boldsymbol{c}(0) \\ \cdot & \cdot & \cdot & \cdot & \cdot & \cdot & \cdot \end{bmatrix} \begin{bmatrix} \cdot \\ \boldsymbol{x}(-1) \\ \boldsymbol{x}(0) \\ \boldsymbol{x}(1) \\ \boldsymbol{x}(2) \\ \cdot \end{bmatrix}.$$

For polyphase here is the important point. *Only the even-numbered coefficients* $\boldsymbol{c}(2n)$ *are multiplying the even-numbered coefficients* $\boldsymbol{x}(2n)$. The even and odd $\boldsymbol{c}$'s are in separate columns. The even-numbered $\boldsymbol{x}(0)$ is multiplying the column that starts with $\boldsymbol{c}(0)$. The odd-numbered component $\boldsymbol{x}(1)$ is multiplying the column containing $\boldsymbol{c}(1), \boldsymbol{c}(3), \ldots$. We can separate the matrix multiplication $(\downarrow 2)\boldsymbol{Cx}$ into *even times even and odd times odd*:

$$\begin{bmatrix} \cdot \\ \boldsymbol{y}(0) \\ \boldsymbol{y}(2) \\ \boldsymbol{y}(4) \\ \cdot \end{bmatrix} = \begin{bmatrix} \cdot & & & \\ \cdot & \boldsymbol{c}(0) & & \\ \cdot & \boldsymbol{c}(2) & \boldsymbol{c}(0) & \\ \cdot & \boldsymbol{c}(4) & \boldsymbol{c}(2) & \boldsymbol{c}(0) \\ \cdot & \cdot & \cdot & \cdot \end{bmatrix} \begin{bmatrix} \cdot \\ \boldsymbol{x}(0) \\ \boldsymbol{x}(2) \\ \cdot \end{bmatrix}$$

$$+ \begin{bmatrix} \cdot & & & \\ \cdot & \boldsymbol{c}(1) & & \\ \cdot & \boldsymbol{c}(3) & \boldsymbol{c}(1) & \\ \cdot & \boldsymbol{c}(5) & \boldsymbol{c}(3) & \boldsymbol{c}(1) \\ \cdot & \cdot & \cdot & \cdot \end{bmatrix} \begin{bmatrix} \cdot \\ \boldsymbol{x}(-1) \\ \boldsymbol{x}(1) \\ \cdot \end{bmatrix}.$$

This is a matrix display of eqn. (2.20):

$$(\downarrow 2)\boldsymbol{Cx} = \boldsymbol{C}_{\text{even}}\ \boldsymbol{x}_{\text{even}} + (\text{delay})\ \boldsymbol{C}_{\text{odd}}\ \boldsymbol{x}_{\text{odd}}. \tag{2.21}$$

The two phases $\boldsymbol{x}_{\text{even}}$ and $\boldsymbol{x}_{\text{odd}}$ are filtered by the two polyphase components $\boldsymbol{C}_{\text{even}}$ and $\boldsymbol{C}_{\text{odd}}$. We need a delay in the odd phase, because $\boldsymbol{c}(1)\boldsymbol{x}(1)$ contributes to $\boldsymbol{y}(2)$ and not to $\boldsymbol{y}(0)$. Then $(\downarrow 2)\boldsymbol{Cx}$ is the sum from the two phases in (2.21):

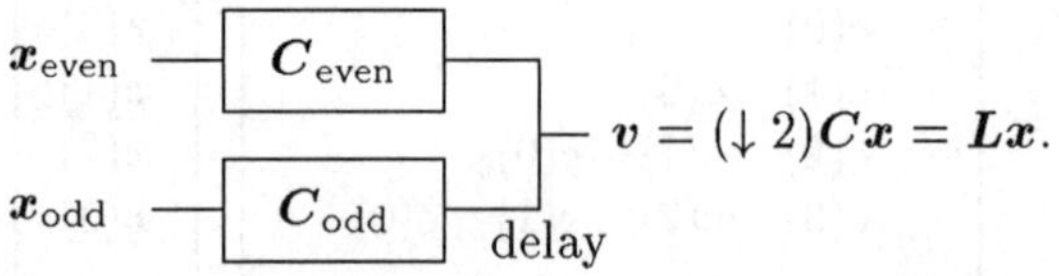

Notice something nice. The two matrices in eqn. (2.21) have constant diagonals. *The two operators* $\boldsymbol{C}_{\text{even}}$ *and* $\boldsymbol{C}_{\text{odd}}$ *are time-invariant filters.* They have frequency responses $C_0(z) = C_{\text{even}}(z)$ and $C_1(z) = C_{\text{odd}}(z)$. The delay in the odd channel can go before or after C_1, because it commutes with C_1. (C_1 is time-invariant!) The two filters involve even coefficients $\boldsymbol{c}(2n)$ and odd coefficients $\boldsymbol{c}(2n+1)$, without zeros in between. They can operate in parallel, more efficiently.

Note 9: Key point The polyphase form puts $(\downarrow 2)$ *before* the filters. This order is more efficient. It was possible to use the Noble identity on each phase separately, because $C_{\text{even}}(z^2)$ and also $C_{\text{odd}}(z^2)$ appeared in the right place:

$$\textbf{(direct)}\quad (\downarrow 2)C_{\text{even}}(z) = C_{\text{even}}(z^2)(\downarrow 2)\quad \textbf{(polyphase)}.$$

Example 2.5 Four-tap filters yield two taps for each phase. The even phase $\boldsymbol{C}_{\text{even}}$ has two coefficients $\boldsymbol{c}(0)$ and $\boldsymbol{c}(2)$. The odd phase has $\boldsymbol{C}_{\text{odd}} = \boldsymbol{c}(1) + \boldsymbol{c}(3)z^{-1}$. The same pattern holds for $\boldsymbol{D}$. The polyphase matrix for the filter bank is

$$\boldsymbol{H}_p(z) = \begin{bmatrix} \boldsymbol{c}(0)+\boldsymbol{c}(2)z^{-1} & \boldsymbol{c}(1)+\boldsymbol{c}(3)z^{-1} \\ \boldsymbol{d}(0)+\boldsymbol{d}(2)z^{-1} & \boldsymbol{d}(1)+\boldsymbol{d}(3)z^{-1} \end{bmatrix}.$$

Even is separated from odd. This reflects what happens in the filter bank, when $\boldsymbol{Cx}$ and $\boldsymbol{Dx}$ are downsampled:

$$\begin{bmatrix} \boldsymbol{v}_0 \\ \boldsymbol{v}_1 \end{bmatrix} = \begin{bmatrix} (\downarrow 2)\boldsymbol{Cx} \\ (\downarrow 2)\boldsymbol{Dx} \end{bmatrix} = \begin{bmatrix} \boldsymbol{C}_{\text{even}} & \boldsymbol{C}_{\text{odd}} \\ \boldsymbol{D}_{\text{even}} & \boldsymbol{D}_{\text{odd}} \end{bmatrix} \begin{bmatrix} 1 & \\ & \text{delay} \end{bmatrix} \begin{bmatrix} \boldsymbol{x}_{\text{even}} \\ \boldsymbol{x}_{\text{odd}} \end{bmatrix}.$$

In case you like matrices, I am going to write the time-domain filter bank matrix in three ways. Downsampling is included in all three! First comes the matrix $\boldsymbol{H}_d = \boldsymbol{H}_{\text{direct}}$ that multiplies the input vector $\boldsymbol{x}$ in the direct form:

$$\textbf{Direct} \qquad \boldsymbol{H}_d = \begin{bmatrix} \cdot & \cdot & & & & & \\ \boldsymbol{c}(3) & \boldsymbol{c}(2) & \boldsymbol{c}(1) & \boldsymbol{c}(0) & & & \\ & & \boldsymbol{c}(3) & \boldsymbol{c}(2) & \boldsymbol{c}(1) & \boldsymbol{c}(0) & \\ & & & & \cdot & \cdot & \cdot\,\cdot \\ \cdot & \cdot & & & & & \\ \boldsymbol{d}(3) & \boldsymbol{d}(2) & \boldsymbol{d}(1) & \boldsymbol{d}(0) & & & \\ & & \boldsymbol{d}(3) & \boldsymbol{d}(2) & \boldsymbol{d}(1) & \boldsymbol{d}(0) & \\ & & & & \cdot & \cdot & \cdot\,\cdot \end{bmatrix}.$$

Downsampling has removed every other row. That leaves this "square" infinite matrix. Each column is completely odd or completely even.

For the second form I rearrange the *rows* of $\boldsymbol{H}_d$. The highpass outputs are interleaved with the lowpass outputs, both downsampled by 2. This produces the block-diagonal form (or *block-Toeplitz form*) $\boldsymbol{H}_b = \boldsymbol{H}_{\text{block}}$:

$$\textbf{Block} \qquad \boldsymbol{H}_b = \begin{bmatrix} \cdot & \cdot & & & & & \\ \cdot & \cdot & & & & & \\ \boldsymbol{c}(3) & \boldsymbol{c}(2) & \boldsymbol{c}(1) & \boldsymbol{c}(0) & & & \\ \boldsymbol{d}(3) & \boldsymbol{d}(2) & \boldsymbol{d}(1) & \boldsymbol{d}(0) & & & \\ & & \boldsymbol{c}(3) & \boldsymbol{c}(2) & \boldsymbol{c}(1) & \boldsymbol{c}(0) & \\ & & \boldsymbol{d}(3) & \boldsymbol{d}(2) & \boldsymbol{d}(1) & \boldsymbol{d}(0) & \\ & & & & \cdot & \cdot & \cdot\,\cdot \\ & & & & \cdot & \cdot & \cdot\,\cdot \end{bmatrix}.$$

Your eye will divide that matrix into 2 by 2 blocks. It is like an ordinary time-invariant constant-diagonal matrix, but the entries are blocks instead of scalars. The main diagonal block corresponds to the constants in the polyphase matrix. The subdiagonal block produces the z^{-1} terms. There are only two diagonals because the phases of C and D have *two* coefficients. The original C and D had four coefficients.

The third form is the polyphase form $\boldsymbol{H}_p$. We are still in the time domain. For this third form I rearrange the *columns* of the direct form:

$$\textbf{Polyphase} \quad \boldsymbol{H}_p = \begin{bmatrix} \cdot & & & \cdot & & \\ \boldsymbol{c}(2) & \boldsymbol{c}(0) & & \boldsymbol{c}(3) & \boldsymbol{c}(1) & \\ & \boldsymbol{c}(2) & \boldsymbol{c}(0) & & \boldsymbol{c}(3) & \boldsymbol{c}(1) \\ & & \cdot & \cdot & & \cdot \quad \cdot \\ \cdot & & & \cdot & & \\ \boldsymbol{d}(2) & \boldsymbol{d}(0) & & \boldsymbol{d}(3) & \boldsymbol{d}(1) & \\ & \boldsymbol{d}(2) & \boldsymbol{d}(0) & & \boldsymbol{d}(3) & \boldsymbol{d}(1) \\ & & \cdot & \cdot & & \cdot \quad \cdot \end{bmatrix} = \begin{bmatrix} C_0 & C_1 \\ D_0 & D_1 \end{bmatrix}.$$

When the columns are rearranged, the vector $\boldsymbol{x}$ must be rearranged. Here $\boldsymbol{x}_{\text{even}}$ comes above $\boldsymbol{x}_{\text{odd}}$(delayed). The transform of the time-domain matrix $\boldsymbol{H}_p$ is the z-domain polyphase matrix $\boldsymbol{H}_p(z)$. *This 2 by 2 matrix of filters becomes a 2 by 2 matrix of functions.*

The block form $\boldsymbol{H}_b$ is an infinite matrix of 2 by 2 blocks. The polyphase form $\boldsymbol{H}_p$ is a 2 by 2 matrix of infinite blocks. Each block is a time-invariant filter. Either of those matrices leads directly by z-transform, to the 2×2 polyphase matrix $\boldsymbol{h}_p(0) + z^{-1}\boldsymbol{h}_p(1)$:

$$\textbf{Polyphase matrix} \quad \boldsymbol{H}_p(z) = \begin{bmatrix} \boldsymbol{c}(0) & \boldsymbol{c}(1) \\ \boldsymbol{d}(0) & \boldsymbol{d}(1) \end{bmatrix} + z^{-1} \begin{bmatrix} \boldsymbol{c}(2) & \boldsymbol{c}(3) \\ \boldsymbol{d}(2) & \boldsymbol{d}(3) \end{bmatrix}. \tag{2.22}$$

2.12 Paraunitary matrices

Definition 2.6 *The matrix $\boldsymbol{H}(z)$ is paraunitary if it is unitary for all $|z| = 1$:*

$$\boxed{\boldsymbol{H}^T(e^{-j\omega})\boldsymbol{H}(e^{j\omega}) = \mathbf{I} \quad \textit{for all } \omega.} \tag{2.23}$$

This extends to all $z \neq 0$ by $\widetilde{\boldsymbol{H}}(z) = \boldsymbol{H}^T(z^{-1})$. Then a paraunitary matrix has

$$\boldsymbol{H}^T(z^{-1})\boldsymbol{H}(z) = \widetilde{\boldsymbol{H}}(z)\ \boldsymbol{H}(z) = \boldsymbol{I} \quad \textit{for all } z. \tag{2.24}$$

When the coefficients $\boldsymbol{h}(k)$ are complex, they are conjugated in $\widetilde{\boldsymbol{H}}(z)$.

The matrix $\boldsymbol{H}$ need not be 2 by 2. If it is 1 by 1, then $|\boldsymbol{H}(e^{j\omega})| = 1$. The corresponding filter is *allpass*. The best allpass examples are ratios of polynomials coming from IIR filters – since only trivial polynomials z^{-l} can have $|\boldsymbol{H}(e^{j\omega})| = 1$.

If $\boldsymbol{H}(z)$ is $M \times M$, it could come from an M-channel filter bank. It might be the polyphase matrix $\boldsymbol{H}_p(z)$ or the modulation matrix $\boldsymbol{H}_m(z)$ (divided by $\sqrt{2}$). We will show that the filter bank is orthogonal if these matrices are paraunitary. That is the important connection for this article.

Eqn. (2.23) gives the inverse matrix by transposing and conjugating the original. The synthesis bank comes by "reversing" the analysis bank. Note that for a square matrix, $\boldsymbol{H}(z)$ is paraunitary when $\boldsymbol{H}^{-1}(z)$ and $\boldsymbol{H}^T(z)$ and $\widetilde{\boldsymbol{H}}(z)$ are paraunitary. And notice especially what eqn. (2.24) says about the *determinants* of these matrices:

$$\left(\det \widetilde{\boldsymbol{H}}(z)\right)(\det \boldsymbol{H}(z)) = 1. \tag{2.25}$$

The determinants are 1 by 1 allpass!

Theorem 2.7 *If a square paraunitary matrix* $\boldsymbol{H}(z)$ *is FIR (= polynomial), then its determinant must be a delay:*

$$\det \boldsymbol{H}(z) = \pm z^{-l}. \tag{2.26}$$

The determinant of $\boldsymbol{H}_p(z)$ is also a delay for any *bi*orthogonal filter bank. Orthogonality requires more; the polyphase matrix $\boldsymbol{H}_p(z)$ must be paraunitary.

2.13 Orthonormal filter banks

This section brings together the requirements for an orthonormal filter bank. We will see those requirements in the *time domain* and the *polyphase domain* and the *modulation domain*. These requirements are conditions on the filter coefficients $\boldsymbol{c}(k)$ and $\boldsymbol{d}(k)$. Equation (2.39) below indicates a simple choice of the $\boldsymbol{d}$'s coming from the $\boldsymbol{c}$'s. If the lowpass filter meets the orthogonality requirements, it is easy to construct a highpass filter to go with it.

The discussion is in terms of a 2-channel FIR filter bank, $M = 2$. But the conditions extend immediately to any M. *The polyphase matrix and the modulation matrix must be paraunitary.* In the M-channel case, the lowpass filter does not immediately determine the $M - 1$ remaining filters.

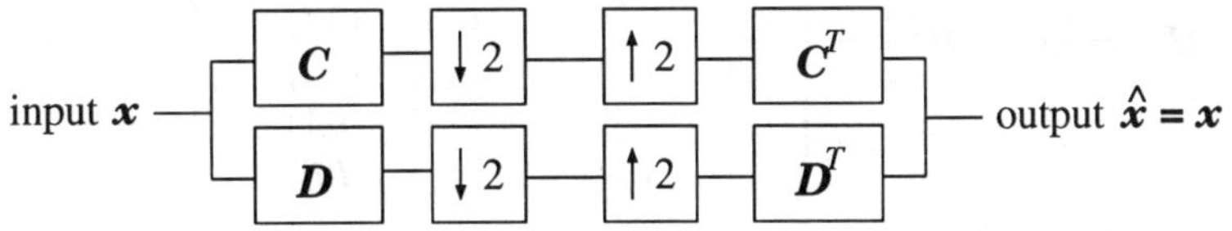

An orthogonal filter bank has synthesis bank = transpose of analysis bank.

The figure shows the structure of an orthogonal filter bank. We intend to achieve $\hat{\boldsymbol{x}} = \boldsymbol{x}$, with synthesis filters $\boldsymbol{C}^T$ and $\boldsymbol{D}^T$ that are time-reversals of the analysis filters:

$$\begin{aligned} \widetilde{\boldsymbol{C}} &= \boldsymbol{C}^T \quad \text{and} \quad \tilde{\boldsymbol{c}}(n) = \boldsymbol{c}(-n) && (2.27)\\ \widetilde{\boldsymbol{D}} &= \boldsymbol{D}^T \quad \text{and} \quad \tilde{\boldsymbol{d}}(n) = \boldsymbol{d}(-n). && (2.28)\end{aligned}$$

As it stands, $\widetilde{\boldsymbol{C}}$ and $\widetilde{\boldsymbol{D}}$ are anticausal. At the end we make them causal by N delays. The output $\hat{\boldsymbol{x}}(n)$ is equally delayed; it is $\boldsymbol{x}(n-N)$. But the algebra is easiest for $\boldsymbol{C}^T$ and $\boldsymbol{D}^T$ with no delays.

This special structure imposes special conditions on the $\boldsymbol{c}$'s and $\boldsymbol{d}$'s for perfect reconstruction. We will call the requirements **Condition O** (for orthogonality). This section finds four equivalent forms: Condition O on the infinite matrix, on the lowpass coefficients, and on the polyphase and modulation matrices $\boldsymbol{H}_p$ and $\boldsymbol{H}_m$. We look first at the infinite matrices in the time domain.

Time domain: condition O and the alternating flip

The key matrix in the time domain is $\boldsymbol{H}_t$. It represents the direct form of the analysis bank, with downsampling. The lowpass part $\boldsymbol{L} = (\downarrow 2)\boldsymbol{C}$ comes above the highpass part $\boldsymbol{B} = (\downarrow 2)\boldsymbol{D}$. We display this infinite matrix for filters of length four:

$$\boldsymbol{H}_t = \begin{bmatrix} \boldsymbol{L} \\ \boldsymbol{B} \end{bmatrix} = \begin{bmatrix} \boldsymbol{c}(3) & \boldsymbol{c}(2) & \boldsymbol{c}(1) & \boldsymbol{c}(0) & & \\ & & \boldsymbol{c}(3) & \boldsymbol{c}(2) & \boldsymbol{c}(1) & \boldsymbol{c}(0) \\ & & & & \cdot & \cdot \\ \boldsymbol{d}(3) & \boldsymbol{d}(2) & \boldsymbol{d}(1) & \boldsymbol{d}(0) & & \\ & & \boldsymbol{d}(3) & \boldsymbol{d}(2) & \boldsymbol{d}(1) & \boldsymbol{d}(0) \\ & & & & \cdot & \cdot \end{bmatrix}. \tag{2.29}$$

The shifts by 2 were created by downsampling, which removed the odd-numbered rows.

With orthogonality, the synthesis filters are to be time-reversals of the analysis filters. The infinite synthesis matrix contains the transposes of $(\downarrow 2)\boldsymbol{C}$ and $(\downarrow 2)\boldsymbol{D}$. Those transposes are $\boldsymbol{C}^T(\uparrow 2)$ and $\boldsymbol{D}^T(\uparrow 2)$, since upsampling is the transpose of downsampling:

$$\boldsymbol{H}_t^T = \begin{bmatrix} \boldsymbol{L}^T & \boldsymbol{B}^T \end{bmatrix} = \begin{bmatrix} \boldsymbol{c}(3) & & & \boldsymbol{d}(3) & & \\ \boldsymbol{c}(2) & & & \boldsymbol{d}(2) & & \\ \boldsymbol{c}(1) & \boldsymbol{c}(3) & & \boldsymbol{d}(1) & \boldsymbol{d}(3) & \\ \boldsymbol{c}(0) & \boldsymbol{c}(2) & & \boldsymbol{d}(0) & \boldsymbol{d}(2) & \\ & \boldsymbol{c}(1) & \cdot & & \boldsymbol{d}(1) & \cdot \\ & \boldsymbol{c}(0) & \cdot & & \boldsymbol{d}(0) & \cdot \end{bmatrix}. \tag{2.30}$$

The shifts by 2 in the columns are created by upsampling, which removes every other column. We require $\hat{\boldsymbol{x}} = \boldsymbol{x}$. This means that $\boldsymbol{H}_t^T\boldsymbol{H}_t = \boldsymbol{I}$. The matrix $\boldsymbol{H}_t$ is required to be an *orthogonal matrix*. Its columns are orthonormal and so are its rows: $\boldsymbol{H}_t\boldsymbol{H}_t^T = \boldsymbol{I}$. We can express this Condition O in matrix form, in block form, and in coefficient form.

Condition O An orthogonal filter bank comes from an orthogonal matrix:

$$\boldsymbol{H}_t^T\boldsymbol{H}_t = \boldsymbol{I} \quad \text{and} \quad \boldsymbol{H}_t\boldsymbol{H}_t^T = \boldsymbol{I}. \tag{2.31}$$

In block form this means that

$$\begin{bmatrix} \boldsymbol{L}^T & \boldsymbol{B}^T \end{bmatrix} \begin{bmatrix} \boldsymbol{L} \\ \boldsymbol{B} \end{bmatrix} = \boldsymbol{L}^T\boldsymbol{L} + \boldsymbol{B}^T\boldsymbol{B} = \boldsymbol{I} \tag{2.32}$$

and

$$\begin{bmatrix} \boldsymbol{L} \\ \boldsymbol{B} \end{bmatrix} \begin{bmatrix} \boldsymbol{L}^T & \boldsymbol{B}^T \end{bmatrix} = \begin{bmatrix} \boldsymbol{L}\boldsymbol{L}^T & \boldsymbol{L}\boldsymbol{B}^T \\ \boldsymbol{B}\boldsymbol{L}^T & \boldsymbol{B}\boldsymbol{B}^T \end{bmatrix} = \begin{bmatrix} \boldsymbol{I} & 0 \\ 0 & \boldsymbol{I} \end{bmatrix}. \tag{2.33}$$

For the coefficients $\boldsymbol{c}(k)$ and $\boldsymbol{d}(k)$, eqn. (2.33) becomes *orthogonality to double shifts*:

$$\boldsymbol{L}\boldsymbol{L}^T = \boldsymbol{I}: \quad \sum \boldsymbol{c}(n)\boldsymbol{c}(n-2k) = \boldsymbol{\delta}(k) \tag{2.34}$$

$$\boldsymbol{L}\boldsymbol{B}^T = 0: \quad \sum \boldsymbol{c}(n)\boldsymbol{d}(n-2k) = 0 \tag{2.35}$$

$$\boldsymbol{B}\boldsymbol{B}^T = \boldsymbol{I}: \quad \sum \boldsymbol{d}(n)\boldsymbol{d}(n-2k) = \boldsymbol{\delta}(k). \tag{2.36}$$

Because of (2.34)–(2.36), we refer to Condition O as **"double-shift orthogonality"**. Its equivalent in the frequency domain is presented below. This double-shift orthogonality immediately rules out odd length filters! If the length is $N+1=5$, a shift by 4 gives an inner product that cannot be zero:

$$(\boldsymbol{c}(0), \boldsymbol{c}(1), \boldsymbol{c}(2), \boldsymbol{c}(3), \boldsymbol{c}(4)) \cdot (0,0,0,0,\boldsymbol{c}(0)) = \boldsymbol{c}(0)\boldsymbol{c}(4) \neq 0.$$

N cannot be even (the filter length cannot be odd) because $\frac{N}{2}$ double shifts would give a shift by N – and the inner product $\boldsymbol{c}(0)\boldsymbol{c}(N)$ is not zero. So N is odd. The degree $2N$ of the halfband product filter $\boldsymbol{P}$ is $2, 6, 10, \ldots$ The clearest examples have $N=3$ and $2N=6$.

Example Condition O in (2.34) imposes two constraints on four coefficients:

$$\boldsymbol{c}(0)^2 + \boldsymbol{c}(1)^2 + \boldsymbol{c}(2)^2 + \boldsymbol{c}(3)^2 = 1 \quad \text{and} \quad \boldsymbol{c}(0)\boldsymbol{c}(2) + \boldsymbol{c}(1)\boldsymbol{c}(3) = 0. \tag{2.37}$$

Eqn. (2.36) is an identical condition on the $\boldsymbol{d}$'s, from $\boldsymbol{B}\boldsymbol{B}^T = \boldsymbol{I}$. Eqn. (2.35) is the orthogonality of the rows of $\boldsymbol{L}$ to the rows of $\boldsymbol{B}$. Those are the fundamental design constraints on an orthogonal filter bank.

The conditions on the $\boldsymbol{c}$'s and $\boldsymbol{d}$'s are independent, but something good happens. *If the* $\boldsymbol{c}$*'s satisfy eqn.* (2.34), *it is easy to choose the* $\boldsymbol{d}$*'s.* We display a choice of $\boldsymbol{d}$'s that automatically gives orthogonality:

$$\begin{bmatrix} \boldsymbol{L} \\ \boldsymbol{B} \end{bmatrix} = \begin{bmatrix} \boldsymbol{c}(3) & \boldsymbol{c}(2) & \boldsymbol{c}(1) & \boldsymbol{c}(0) & & \\ & & \boldsymbol{c}(3) & \boldsymbol{c}(2) & \boldsymbol{c}(1) & \boldsymbol{c}(0) \\ & & & & \cdot & \cdot \\ -\boldsymbol{c}(0) & \boldsymbol{c}(1) & -\boldsymbol{c}(2) & \boldsymbol{c}(3) & & \\ & & -\boldsymbol{c}(0) & \boldsymbol{c}(1) & -\boldsymbol{c}(2) & \boldsymbol{c}(3) \\ & & & & \cdot & \cdot \end{bmatrix}. \tag{2.38}$$

This is the **alternating flip**. The $\boldsymbol{c}$'s are reversed in order and alternated in sign, to produce the $\boldsymbol{d}$'s. We start with lowpass coefficients $\boldsymbol{c}(0), \ldots, \boldsymbol{c}(N)$, where N is odd. The highpass coefficients are

$$\boxed{\boldsymbol{d}(k) = (-1)^k \boldsymbol{c}(N-k).} \tag{2.39}$$

The essential point can be checked by eye, in the infinite matrix (2.38). *If the* **top** *rows are orthogonal to each other, then* **all** *rows are orthogonal.* The zero dot products in $\boldsymbol{LB}^T$ are

$$\begin{aligned} -\boldsymbol{c}(3)\boldsymbol{c}(0) + \boldsymbol{c}(2)\boldsymbol{c}(1) - \boldsymbol{c}(1)\boldsymbol{c}(2) + \boldsymbol{c}(0)\boldsymbol{c}(3) &= 0, \\ -\boldsymbol{c}(1)\boldsymbol{c}(0) + \boldsymbol{c}(0)\boldsymbol{c}(1) &= 0. \end{aligned} \tag{2.40}$$

Furthermore, the $\boldsymbol{d}$'s are orthonormal within themselves ($\boldsymbol{BB}^T = \boldsymbol{I}$) because the $\boldsymbol{c}$'s are. Equations (2.37) hold for the $\boldsymbol{d}$'s, when they are constructed by flipping the $\boldsymbol{c}$'s. Minus signs cancel in $\boldsymbol{d}(3)\boldsymbol{d}(1)$ which is $(-\boldsymbol{c}(0))(-\boldsymbol{c}(2))$.

Our four-tap example has $N = 3$. The alternating flip gives $\boldsymbol{LB}^T = 0$ for every odd N. The top rows of $\boldsymbol{H}_t$ in (2.29) are *always* orthogonal to the bottom rows. Also (2.36) for the $\boldsymbol{d}$'s follows from (2.34) for the $\boldsymbol{c}$'s. Thus the alternating flip reduces orthogonality to (2.34):

$$\textbf{Condition O} \text{ on the coefficients: } \sum \boldsymbol{c}(n)\boldsymbol{c}(n-2k) = \boldsymbol{\delta}(k).$$

Polyphase domain: condition O and the alternating flip

The polyphase form separates $(\downarrow 2)\boldsymbol{C}$ and $(\downarrow 2)\boldsymbol{D}$ into even phase and odd phase. In the time domain, we are rearranging the columns of the matrix $\boldsymbol{H}_t$. All even columns come before the odd columns. The matrix goes into this 2 by 2 block form with time-invariant filters as the blocks:

$$\boldsymbol{H}_{\text{block}} = \begin{bmatrix} \boldsymbol{C}_{\text{even}} & \boldsymbol{C}_{\text{odd}} \\ \boldsymbol{D}_{\text{even}} & \boldsymbol{D}_{\text{odd}} \end{bmatrix}.$$

In the z-domain, we are rearranging the response functions $C(z)$ and $D(z)$:

$$\sum \boldsymbol{c}(k) z^{-k} = C_{\text{even}}(z^{-2}) + z^{-1} C_{\text{odd}}(z^{-2}). \tag{2.41}$$

Those phase responses are written $H_{00}(z)$ and $H_{01}(z)$ when $C(z)$ is $H_0(z)$. The highpass response $D(z) = H_1(z)$ decomposes in the same way. The polyphase matrix is

$$\boldsymbol{H}_p(z) = \begin{bmatrix} C_{\text{even}}(z) & C_{\text{odd}}(z) \\ D_{\text{even}}(z) & D_{\text{odd}}(z) \end{bmatrix} = \begin{bmatrix} H_{00}(z) & H_{01}(z) \\ H_{10}(z) & H_{11}(z) \end{bmatrix}. \tag{2.42}$$

Now Condition O translates directly into a requirement on $\boldsymbol{H}_p(z)$. A filter bank is **orthogonal** when its polyphase matrix is **paraunitary**:

$$\boldsymbol{H}_p^T(e^{-j\omega})\boldsymbol{H}_p(e^{j\omega}) = \boldsymbol{I} \text{ for all } \omega \text{ and } \widetilde{\boldsymbol{H}}_p(z)\boldsymbol{H}_p(z) = \boldsymbol{I} \text{ for all } z. \tag{2.43}$$

The inverse of $\boldsymbol{H}_p(z)$ is the synthesis polyphase matrix. The matrix and its inverse can be multiplied in either order — here is analysis times synthesis:

$$\begin{bmatrix} C_{\text{even}}(z) & C_{\text{odd}}(z) \\ D_{\text{even}}(z) & D_{\text{odd}}(z) \end{bmatrix} \begin{bmatrix} C_{\text{even}}(z^{-1}) & D_{\text{even}}(z^{-1}) \\ C_{\text{odd}}(z^{-1}) & D_{\text{odd}}(z^{-1}) \end{bmatrix} = \begin{bmatrix} 1 & 0 \\ 0 & 1 \end{bmatrix}. \quad (2.44)$$

On the unit circle, where z^{-1} is $\bar{z}$, row 1 times column 1 becomes

$$\boxed{|C_{\text{even}}(z)|^2 + |C_{\text{odd}}(z)|^2 = 1 \text{ when } |z| = 1.} \quad (2.45)$$

This is the essence of Condition O in the polyphase domain.

For the example with four coefficients, it is helpful to multiply out eqn. (2.45):

$$\left(c(0) + c(2)z^{-1}\right)\left(c(0) + c(2)z\right) + \left(c(1) + c(3)z^{-1}\right)\left(c(1) + c(3)z\right) =$$

$$c(0)^2 + c(2)^2 + c(1)^2 + c(3)^2 + [c(0)c(2) + c(1)c(3)]\left(z^{-1} + z\right) = 1.$$

Thus (2.45) is equivalent to the explicit statement (2.37). The sum of squares is 1 and the dot product $c(0)c(2) + c(1)c(3)$ is zero.

The other multiplications in (2.44) give answers 1 or 0 in the same way. All these requirements on the $\boldsymbol{d}$'s are automatically satisfied by the flip construction! We write that choice $d(k) = (-1)^k c(N-k)$ in the z-domain:

$$\sum d(k)z^{-k} = \sum c(N-k)(-z)^{-k} = \sum c(n)(-z)^{n-N}.$$

This relation between highpass and lowpass is an *alternating flip*:

$$D(z) = (-z)^{-N} C(-z^{-1}). \quad (2.46)$$

The number N is odd. Because of $(-z)^{-N}$, the even and odd phases in $\boldsymbol{C}$ are reversed to odd and even phases in $\boldsymbol{D}$. We take $N = 3$ as typical. An alternating flip of $c(0), c(1), c(2), c(3)$ yields

$$\begin{aligned} D(z) &= c(3) - c(2)z^{-1} + c(1)z^{-2} - c(0)z^{-3} \qquad (2.47) \\ &= \left(c(3) + c(1)z^{-2}\right) - z^{-1}\left(c(2) + c(0)z^{-2}\right). \end{aligned}$$

With this flip in row 2, the multiplication $\boldsymbol{H}_p(z)\boldsymbol{H}_p^T(z^{-1})$ becomes

$$\begin{bmatrix} c(0) + c(2)z^{-1} & c(1) + c(3)z^{-1} \\ c(3) + c(1)z^{-1} & -c(2) - c(0)z^{-1} \end{bmatrix} \begin{bmatrix} c(0) + c(2)z & c(3) + c(1)z \\ c(1) + c(3)z & -c(2) - c(0)z \end{bmatrix}$$

$$= \boldsymbol{I}.$$

The off-diagonal entries of the product are *automatically* zero. *The alternating flip achieves* $\boldsymbol{LB}^T = \boldsymbol{0}$ *with or without orthogonality*. The 2, 2 entry of the product is the same as the 1, 1 entry, when $|z| = 1$. The orthogonality requirement (2.45) makes the 1, 1 entry equal to 1.

Modulation domain: condition O and the alternating flip

The function that arises from modulating $C(z)$ is $C(-z)$. The frequency in $z = e^{j\omega}$ changes by π to produce $-z = e^{j(\omega+\pi)}$. This modulation takes $C(\omega)$ to $C(\omega+\pi)$. The frequency response graph is shifted by π.

Our goal is to relate $C(z)$ to $C(-z)$ and $D(z)$ to $D(-z)$ for an orthogonal filter bank. We already know Condition O on the coefficients:

$$c(0)^2 + c(1)^2 + c(2)^2 + c(3)^2 = 1 \quad \text{and} \quad c(0)c(2) + c(1)c(3) = 0.$$

Watch how 1 and 0 appear in $|C(z)|^2$. Stay on the unit circle where $\bar{z}^{-1} = z$:

$$\left(c(0) + c(1)z^{-1} + c(2)z^{-2} + c(3)z^{-3}\right)\left(c(0) + c(1)z + c(2)z^2 + c(3)z^3\right) =$$
$$1 + [c(0)c(1) + c(1)c(2) + c(2)c(3)]\left(z^{-1} + z\right) + c(0)c(3)\left(z^{-3} + z^3\right).$$

Now change z to $-z$. The odd powers z and z^3 change sign. When we add, those odd powers cancel:

$$\boxed{|C(z)|^2 + |C(-z)|^2 = |C(\omega)|^2 + |C(\omega+\pi)|^2 = 2 \quad \text{for all } z = e^{j\omega}.} \tag{2.48}$$

This is the *halfband condition*, also called the *Nyquist condition*: $|C(z)|^2$ is a halfband filter. Those filters we can design! In other words, the lowpass analysis filter $C(z)$ is a spectral factor of a halfband filter.

Note 10: Condition O on $\boldsymbol{H}_m(z)$ *The modulation matrix of an orthogonal filter bank is a paraunitary matrix times* $\sqrt{2}$:

$$\boldsymbol{H}_m(z)\widetilde{\boldsymbol{H}}_m(z) = 2\boldsymbol{I} \quad \text{for all } z. \tag{2.49}$$

On the circle $z = e^{j\omega}$, the modulation matrix is a unitary matrix times $\sqrt{2}$:

$$\begin{bmatrix} C(\omega) & C(\omega+\pi) \\ D(\omega) & D(\omega+\pi) \end{bmatrix} \begin{bmatrix} \overline{C(\omega)} & \overline{D(\omega)} \\ \overline{C(\omega+\pi)} & \overline{D(\omega+\pi)} \end{bmatrix} = \begin{bmatrix} 2 & 0 \\ 0 & 2 \end{bmatrix}. \tag{2.50}$$

The $1,1$ entry of this matrix product is $|C(\omega)|^2 + |C(\omega+\pi)|^2 = 2$ by (2.48). The other entries, when we multiply them out, follow immediately from (2.35) and (2.36). Thus Condition O on the coefficients is equivalent to Condition O on the modulation matrix $\boldsymbol{H}_m$. It is also equivalent to Condition O on $\boldsymbol{H}_p$. It is the statement that the analysis bank followed by its transpose gives perfect reconstruction. We summarize.

Theorem 2.8 *For an orthogonal filter bank the lowpass coefficients must satisfy* Condition O; *we give four equivalent forms.*

Matrix form $\boldsymbol{L}\boldsymbol{L}^T = (\downarrow 2)\boldsymbol{C}\boldsymbol{C}^T(\uparrow 2) = \boldsymbol{I}$

Coefficient form $\sum \boldsymbol{c}(n)\boldsymbol{c}(n-2k) = \boldsymbol{\delta}(k)$

Polyphase form $|C_{\text{even}}(e^{j\omega})|^2 + |C_{\text{odd}}(e^{j\omega})|^2 = 1$

Modulation form $|C(\omega)|^2 + |C(\omega+\pi)|^2 = 2.$

By an alternating flip, $\boldsymbol{L}\boldsymbol{B}^T = \boldsymbol{0}$ and $\boldsymbol{B}\boldsymbol{B}^T = \boldsymbol{I}$ follow immediately from the lowpass part $\boldsymbol{L}\boldsymbol{L}^T = \boldsymbol{I}$. The real problem is the design of the lowpass filter.

Symmetry prevents orthogonality

It is natural to want two good properties at once. Symmetry is good for the eye, and orthogonality is good for the algorithm. But the only filters with both properties are averaging filters (Haar filters) with two coefficients. We are forced to use IIR filters, or M channels, or multifilters with matrix coefficients. Extra computation is unavoidable, because the next theorem rules out a perfect filter.

Theorem 2.9 *A symmetric orthogonal FIR filter can only have two nonzero coefficients.*

Proof N is odd for orthogonality. The filter length must be even. By convention $\boldsymbol{c}(0)$ is the first nonzero coefficient. Shift the filter if necessary to achieve this. With $N = 5$ a symmetric filter of length 6 has the form

$$(\boldsymbol{c}(0), \boldsymbol{c}(1), \boldsymbol{c}(2), \boldsymbol{c}(2), \boldsymbol{c}(1), \boldsymbol{c}(0)).$$

This vector must be orthogonal to all its double shifts. The inner product with its shift by *four* must be $2\,\boldsymbol{c}(0)\boldsymbol{c}(1) = 0$. Therefore $\boldsymbol{c}(1) = 0$. Then the inner product with its shift by *two* gives $2\,\boldsymbol{c}(0)\boldsymbol{c}(2) = 0$. The only nonzero coefficient is $\boldsymbol{c}(0)$ at both ends of the filter. This completes the proof for $N = 5$.

Now consider general N. The only symmetric orthogonal possibilities are $\boldsymbol{c} = (1,1)/\sqrt{2}$, $(1,0,0,1)/\sqrt{2}$ and $(1,0,\ldots,0,1)/\sqrt{2}$. Only the Haar coefficients $(1,1)/\sqrt{2}$ will lead to orthogonal wavelets. Symmetry really conflicts with orthogonality.

A second proof observes that the odd phase is the flip of the even phase:

$$(\boldsymbol{c}(0), \boldsymbol{c}(4), \boldsymbol{c}(2), \boldsymbol{c}(2), \boldsymbol{c}(4), \boldsymbol{c}(0)) \quad \text{has} \quad |C_{\text{even}}(z)|^2 = |C_{\text{odd}}(z)|^2.$$

Condition O is $|C_{\text{even}}(z)|^2 + |C_{\text{odd}}(z)|^2 = 2$. With symmetry this separates into $|C_{\text{even}}(z)|^2 = 1$ and $|C_{\text{odd}}(z)|^2 = 1$. *The even phase is an allpass filter!* So is the odd phase. But FIR allpass filters can only have one nonzero coefficient, which completes the second proof.

2.14 Spectral factorization

Out of all these equations we would like to emphasize one. It came at the end. It applied first of all to the lowpass filter $C(z)$, and then by the alternating flip also to $D(z)$. It was eqn. (2.48), that the frequency response $C(\omega) = \sum c(k)e^{-jk\omega}$ satisfies

$$|C(\omega)|^2 + |C(\omega+\pi)|^2 = 2. \tag{2.51}$$

The key question is, *what does eqn.* (2.51) *say about* $|C(\omega)|^2$ *itself?*

We assign the symbol $P(\omega)$ to this important quantity $|C(\omega)|^2$. It is the **power spectral response**. Because $C(\omega)$ multiplies $\overline{C(\omega)}$, the filter with this response $P(\omega)$ is *symmetric*. It is the "autocorrelation filter". When $\sum c(k)e^{-jk\omega}$ multiplies its conjugate $\sum c(l)e^{jl\omega}$, we watch for $n = k - l$ which is $l = k - n$. The coefficient $p(n)$ is the sum of $c(k)$ times $c(k-n)$:

$$p(n) = \sum c(k)c(k-n) = \text{autocorrelation of the sequence } c(k). \tag{2.52}$$

Autocorrelation is $p = c * c^T$. This is the convolution of

$$c = (c(0), c(1), c(2), \ldots)$$

with its time reversal

$$c^T = (\ldots, c(2), c(1), c(0)).$$

Replacing $-n$ by n in (2.52) brings no change in p.

The reason for the surprising notation c^T is that multiplying $C(\omega)$ by $\overline{C(\omega)}$ corresponds exactly to multiplying the infinite filter matrix C by C^T. $P = CC^T$ is a symmetric positive definite (or semidefinite) Toeplitz matrix.

What does eqn. (2.51) say about the other coefficients $p(n)$? In a word, it says *nothing* about the odd coefficients and it assigns *zero* to the even coefficients. C is the start of an orthogonal filter bank if and only if the autocorrelation filter P is a **halfband filter**:

$$P(z) + P(-z) = 2. \tag{2.53}$$

The *even coefficients* with $n = 2m$ must be $\delta(m)$:

$$\textbf{Halfband filter} \quad p(2m) = \sum c(k)c(k-2m) = \begin{cases} 1 & \text{if } m = 0 \\ 0 & \text{if } m \neq 0. \end{cases} \tag{2.54}$$

In an orthonormal filter bank, $C(z) = \sqrt{2}\,H(z)$ is a *spectral factor* of a symmetric halfband filter $P(z)$. The factorization is $P(z) = C(z^{-1})C(z)$ and the halfband property is $P(z) + P(-z) = 2$. In frequency, $P(\omega) = |C(\omega)|^2$ achieves the orthogonality condition $|C(\omega)|^2 + |C(\omega+\pi)|^2 = 2$. In the reverse direction, $P(z)$ is the *autocorrelation* of $C(z)$. This intimate relation of spectral factor $C(z)$ and its autocorrelation $P(z)$ is fundamental throughout signal processing.

Two questions arise immediately:

1. (Theory) Can every polynomial with $P(\omega) \geq 0$ be factored into $|C(\omega)|^2$?
2. (Practice) How is this spectral factorization actually done?

The answer to Question 1 is *yes*. This is the Féjer–Riesz Theorem. The answer to Question 2 is not so quick. There are many competing algorithms for spectral factorization. Short filters offer no serious difficulty, but with 100 or even 50 coefficients the weaker algorithms become slow and/or unreliable. When $C(\omega)$ is only approximate, the reconstruction is not perfect.

The trigonometric polynomials $P(\omega)$ and $C(\omega)$ are both of degree N:

$$\sum_{-N}^{N} \boldsymbol{p}(n)e^{-in\omega} = |C(e^{i\omega})|^2 = \left|\sum_{0}^{N} \boldsymbol{c}(n)e^{-in\omega}\right|^2 .$$

$P(z)$ has symmetric coefficients $\boldsymbol{p}(n) = \boldsymbol{p}(-n)$. There are $N+1$ independent coefficients in P and the same number in C. They are linked by quadratic equations, when we solve $P(\omega) = |C(\omega)|^2$. Those equations are solvable if and only if $P(\omega) \geq 0$ for all ω.

As an aside, note that *matrix* spectral factorization is also possible where $P(\omega)$ is symmetric positive definite. Both 1 and 2, theory and practice, are nontrivial. The Riccati equation is involved.

With real symmetric coefficients $\boldsymbol{p}(n)$, we have $P(z) = P(1/z)$. **If** z_i **is a root, so is** $1/z_i$. When z_i is inside the unit circle, $1/z_i$ is outside. The roots z_j *on* the unit circle must have even multiplicity, by the crucial assumption that $P(\omega) \geq 0$. Therefore the polynomial $z^N P(z)$ of degree $2N$, with leading coefficient $\boldsymbol{p}(N) \neq 0$, must have these $2N$ factors:

$$z^N P(z) = \boldsymbol{p}(N) \prod_{i=1}^{M} (z - z_i)\left(z - \frac{1}{z_i}\right) \prod_{j=1}^{N-M} (z - z_{j+M})^2 . \tag{2.55}$$

This contains the key point, but we know more. Real coefficients ensure that the complex conjugate $\bar{z}$ is a root when z is a root. The complex roots off the unit circle actually come *four at a time*: z_i and $\bar{z}_i$ inside, $1/z_i$ and $1/\bar{z}_i$ outside. The complex roots on the circle also come four at a time: z_j twice and $\bar{z}_j$ twice. Real roots on the circle come two at a time (even multiplicity).

Now construct $C(z)$ by taking *all* the roots z_i (including $\bar{z}_i$) inside the circle, and also take one out of every double root z_j on the circle:

$$z^N C(z) = |\boldsymbol{p}(N)|^{1/2} \prod_{i=1}^{M} (z - z_i) \prod_{j=1}^{N-M} (z - z_{j+M}) . \tag{2.56}$$

This is the "minimum phase spectral factor." It has no roots outside the circle. The coefficients of $C(z)$ are still real, because the complex roots are automatically in conjugate pairs: $\bar{z}_i$ and $\bar{z}_j$ came with z_i and z_j.

Example 2.10 The 4-tap Daubechies filter has zeros at $z_i = 2-\sqrt{3}$ and $z_i^{-1} = 2+\sqrt{3}$. The other four roots of $z^3P(z)$ are at $z_j = -1$ (on the unit circle and again real). Two of those roots go into the spectral factor (2.56). Thus $z^3C(z)$ is a cubic polynomial with roots $2-\sqrt{3}$, -1, and -1. It is minimum phase.

Every factorization of $P(z)$ into $F(z)\,H(z)$ must put some roots into $H(z)$ and the remaining roots into $F(z)$. The rules for this separation of roots of $P(z)$ are as follows.

- For F and H to be **real** filters, z and $\bar{z}$ must stay together.
- For F and H to be **symmetric** filters, z and z^{-1} must stay together.
- For F to be the **transpose** of H, z and z^{-1} must go separately. (This factorization $C(z)\,C(z^{-1})$ gives an orthogonal filter bank when P is halfband.)

To achieve the first two properties at the same time, all zeros of $P(z)$ — not just the zeros on the unit circle — must be of even multiplicity. This gives another proof that orthogonality conflicts with symmetry.

2.15 Maxflat (Daubechies) filters

This section is about an important family of filters, which lead to an outstanding family of wavelets. The same construction yields both.

1. These particular filters (and wavelets) are **orthogonal**.
2. The frequency responses have **maximum flatness** at $\omega = 0$ and $\omega = \pi$.

The lowpass filters will have $p = 1, 2, 3, 4, \ldots$ zeros at π. They have $2p = 2, 4, 6, 8, \ldots$ coefficients, so that $N = 2p-1$. We use **boldface $\mathbf{p}$** for the coefficients of $P(\omega) = |C(\omega)|^2$ and *lightface* p to count the zeros of $C(\omega)$ at $\omega = \pi$. The highpass coefficients $\mathbf{d}(k)$ come from an alternating flip. The first member of this family was $\mathbf{c}(0) = \mathbf{c}(1) = 1/\sqrt{2}$. Note the normalization $\mathbf{c}(0)^2 + \mathbf{c}(1)^2 = 1$. These numbers go into a *unitary matrix*. For each $p = 1, 2, 3, 4, \ldots$ the filter bank is orthonormal. The product filters have degree $2N = 4p-2$:

$$P_0(z) = \left(\frac{1+z^{-1}}{2}\right)^{2p} Q_{2p-2}(z) \quad \textit{will be halfband by special choice of } Q.$$

In the literature on filters, this family is described as *maxflat*. The coefficients were given by Herrmann. They were already in formulas for interpolation, described below. In the history of wavelets, we are reproducing the great 1988 discovery by Ingrid Daubechies [1]. The filters are FIR with $2p$

coefficients. The wavelets are supported on the interval $[0, N] = [0, 2p-1]$. As p increases, the filters are increasingly "regular" and the wavelets are increasingly "smooth."

Condition O and condition $\mathbf{A}_p$

Before starting, it is helpful to count the requirements we must impose. There are $2p$ numbers to be chosen. These can be the coefficients in the lowpass filter, $\boldsymbol{c}(0), \ldots, \boldsymbol{c}(2p-1)$, with frequency response $C(\omega)$. They could equally well be the coefficients $\boldsymbol{p}(0), \ldots, \boldsymbol{p}(2p-1)$ of the centered (even) polynomial $P(\omega) = |C(\omega)|^2$. The $\boldsymbol{c}$'s come from the $\boldsymbol{p}$'s by spectral factorization. The nonnegative polynomial $P(\omega)$ is factored by the methods of the previous section:

$$P(\omega) = \sum_{1-2p}^{2p-1} \boldsymbol{p}(n)e^{-in\omega} \text{ equals } |C(\omega)|^2 = \left|\sum_{0}^{2p-1} \boldsymbol{c}(n)e^{-in\omega}\right|^2. \tag{2.57}$$

Our formulas yield the numbers $\boldsymbol{p}(n) = \boldsymbol{p}(-n)$. Except for the first few filters in the family, there are no simple formulas for $\boldsymbol{c}(n)$.

These $2p$ numbers are determined by p conditions for orthogonality from Condition O, and p conditions for a flat response from Condition A. More precisely, the requirement is "Condition A_p" — the subscript indicates the order of flatness at $\omega = \pi$ (and $\omega = 0$). Here are the $p + p$ conditions:

Condition O $\quad P = |C|^2$ is a normalized halfband filter:

$$\boxed{\boldsymbol{p}(0) = 1 \text{ and } \boldsymbol{p}(2) = \boldsymbol{p}(4) = \cdots = \boldsymbol{p}(2p-2) = 0.} \tag{2.58}$$

Condition A_p $\quad C(\omega)$ has a zero of order p at $\omega = \pi$:

$$\boxed{C(\pi) = C'(\pi) = \cdots = C^{(p-1)}(\pi) = 0.} \tag{2.59}$$

The equation $C(\pi) = 0$ says that $\sum \boldsymbol{c}(n)(-1)^n = 0$. The odd-numbered coefficients have the same sum as the even-numbered coefficients:

$$\text{Condition } A_1 \text{ on } \boldsymbol{c}(n): \quad \sum_{\text{odd } n} \boldsymbol{c}(n) = \sum_{\text{even } n} \boldsymbol{c}(n). \tag{2.60}$$

This is the first of the "sum rules". Altogether we can impose the pth order zero in (2.59) as p sum rules on the coefficients:

$$\text{Condition } A_p \text{ on } \boldsymbol{c}(n): \quad \sum_{n=0}^{2p-1} (-1)^n n^k \boldsymbol{c}(n) = 0 \quad \text{for } k = 0, 1, \ldots, p-1. \tag{2.61}$$

The factor n^k comes from the kth derivative of $\sum \boldsymbol{c}(n)e^{-in\omega}$. Then $(-1)^n$ comes from substituting $\omega = \pi$. The convention for n^0 is 1.

The p zeros at π mean that $C(\omega)$ has a factor $\left(1+e^{-i\omega}\right)^p$:

$$\text{Condition } A_p \text{ on } C(\omega): \quad C(\omega) = \left(\frac{1+e^{-i\omega}}{2}\right)^p R(\omega). \tag{2.62}$$

$R(\omega)$ has degree $p-1$, to bring the total degree of $C(\omega)$ to $2p-1$. You could say that the pth order flatness is accounted for by $\left(1+e^{-i\omega}\right)^p$. Then the p coefficients in $R(\omega)$ are chosen to satisfy the p equations of Condition O.

Formulas for $P(\omega)$

We intend to give two formulas for $P(\omega) = |C(\omega)|^2$. The one associated with Ingrid Daubechies has $(1+\cos\omega)^p$ times a sum of p terms. The formula associated with Yves Meyer gives the derivative of $P(\omega)$ as $-c(\sin\omega)^{2p-1}$. Then integration determines c and $P(\omega)$.

The best starting point is the ordinary polynomial $B_p(y)$. This has degree $p-1$, with p coefficients. It is the binomial series for $(1-y)^{-p}$, truncated after p terms:

$$B_p(y) = 1 + py + \frac{p(p+1)}{2}y^2 + \ldots + \binom{2p-2}{p-1} y^{p-1} = (1-y)^{-p} + O(y^p).$$

The coefficient of y^k is $\binom{p+k-1}{k}$. The remainder has order y^p because this is the first term to be dropped. The complex zeros of this polynomial $B_p(y)$ will be all-important for the Daubechies filters.

We combine $B_p(y)$ with the factor $(1-y)^p$ that has p zeros at $y=1$. The variable y on $[0,1]$ will correspond to the frequency ω on $[0,\pi]$. The product $\widetilde{P}(y) = 2(1-y)^p B_p(y)$ has exactly the flatness we want at $y=0$:

$$2(1-y)^p B_p(y) = 2(1-y)^p[(1-y)^{-p} + O(y^p)] = 2 + O(y^p).$$

This is a polynomial of degree $2p-1$. It is the unique polynomial with $2p$ coefficients that satisfies p conditions at each endpoint:

$\widetilde{P}(y)$ and its first $p-1$ derivatives are zero at $y=0$ and $y=1$, except $\widetilde{P}(0)=2$.

Two more properties follow quickly. First, the derivative has $p-1$ zeros at both end points. It is a polynomial of degree $2p-2$ and with those zeros it must be

$$\widetilde{P}'(y) = -Cy^{p-1}(1-y)^{p-1} \quad \text{for some } C. \tag{2.63}$$

The second property comes when we add $\widetilde{P}(y)$ to $\widetilde{P}(1-y)$. The sum equals 2 at both ends and is still flat. Its $2p$ coefficients are uniquely determined;

it must be the constant polynomial 2:

$$\widetilde{P}(y) + \widetilde{P}(1-y) \equiv 2. \tag{2.64}$$

At $y = \frac{1}{2}$ this gives $\widetilde{P}(\frac{1}{2}) = 1$. $\widetilde{P}(y)$ is odd around its middle value. This "Hermite interpolating polynomial" drops from 2 to 0 with flatness at the ends. Here are the polynomials for $p = 2$ and $p = 3$:

$$B_2(y) = 1 + 2y \quad \text{and} \quad \widetilde{P}(y) = 2(1-y)^2(1+2y) = 2 - 6y^2 + 4y^3,$$

$$\begin{aligned} B_3(y) = 1 + 3y + 6y^2 \quad \text{and} \quad \widetilde{P}(y) &= 2(1-y)^3 B_3(y) \\ &= 2 - 20y^3 + 30y^4 - 12y^5. \end{aligned}$$

Now we go from ordinary polynomials in y to trigonometric polynomials in ω. The degree stays at $2p-1$. The change that takes $0 \le y \le 1$ into $0 \le \omega \le \pi$ is

$$y = \frac{1-\cos\omega}{2} \quad \text{and} \quad 1 - y = \frac{1+\cos\omega}{2}.$$

The polynomial $\widetilde{P}(y)$ becomes our desired $P(\omega)$. We summarize its properties.

Theorem 2.11 *The polynomial $2(1-y)^p B_p(y)$ becomes the halfband response*

$$\boxed{P(\omega) = 2\ \left(\tfrac{1+\cos\omega}{2}\right)^p\ \sum_{k=0}^{p-1} \binom{p+k-1}{k} \left(\tfrac{1-\cos\omega}{2}\right)^k.} \tag{2.65}$$

This satisfies Conditions O and A_p. Its Meyer form, by integrating $P'(\omega)$ and choosing c to give $P(\pi) = 0$, is

$$\boxed{P(\omega) = 2 - c\int_0^\omega (\sin\omega)^{2p-1} d\omega.}$$

For $p = 1, 2, 3$ the Daubechies and Meyer forms are

$$P(\omega) = \quad 1 + \cos\omega = 2 - \int_0^\omega \sin\omega\, d\omega$$

$$P(\omega) = \quad (1+\cos\omega)^2(1 - \tfrac{1}{2}\cos\omega) = 2 - \tfrac{3}{2}\int_0^\omega \sin^3\omega\, d\omega$$

$$P(\omega) = \quad (1+\cos\omega)^3(1 - \tfrac{9}{8}\cos\omega + \tfrac{3}{8}\cos^2\omega) = 2 - \tfrac{15}{4}\int_0^\omega \sin^5\omega\, d\omega$$

Most authors emphasize the Daubechies form, with its highly visible factor $(1+\cos\omega)^p$. That immediately ensures a pth order zero for the factors at $\omega = \pi$. Spectral factorization is speeded up, because only a lower-degree polynomial remains. It may not be so clear that (2.65) is a halfband filter. The even powers like $\cos^2\omega$ and $\cos^4\omega$ must disappear and they do. In the explicit formula for $p = 2$, multiplication produces

$$P(\omega) = 1 + \frac{3}{2}\cos\omega - \frac{1}{2}\cos^3\omega.$$

The halfband property is $P(\omega) + P(\omega + \pi) \equiv 2$. This addition cancels the odd powers of $\cos\omega$, and the even powers are not present (except the constant term 1). This identity follows immediately from (2.64) because $1 - y = \frac{1+\cos\omega}{2} = \frac{1-\cos(\omega+\pi)}{2}$:

$$\widetilde{P}(y) + \widetilde{P}(1-y) \equiv 2 \ \text{ becomes } \ P(\omega) + P(\omega+\pi) \equiv 2.$$

The reader recognizes this "Condition O" as $|C(\omega)|^2 + |C(\omega+\pi)|^2 = 2$.

The halfband property is immediate in the Meyer form, with absolutely no calculations. Replace y by $(1 - \cos\omega)/2$ in (2.63) to find $P'(\omega)d\omega$:

$$-Cy^{p-1}(1-y)^{p-1}dy = -C\left(\frac{1-\cos\omega}{2}\right)^{p-1}\left(\frac{1+\cos\omega}{2}\right)^{p-1}\frac{\sin\omega}{2}d\omega.$$

This is $-c(1-\cos^2\omega)^{p-1}\sin\omega\, d\omega$, which is also $-c(\sin\omega)^{2p-1}d\omega$. Its integral is

$$-c\int\left(1-\cos^2\omega\right)^{p-1}\sin\omega\ d\omega = \text{odd powers of } \cos\omega.$$

The only even frequency is a constant of integration. The filter is halfband.

The flatness condition requires first of all that $P(\pi) = 0$. The constant c makes this true. The derivative $P'(\omega) = -c(\sin\,\omega)^{2p-1}$ has a zero of order $2p-1$ at $\omega = \pi$. Then P itself has a zero of order $2p$. Its factor C has a zero of order p. Condition A_p is satisfied and Meyer's formula is confirmed.

Note that $P(\omega)$ *decreases monotonically* from $P(0) = 2$ to $P(\pi) = 0$. Then $P' = -c(\sin\,\omega)^{2p-1}$ is everywhere negative between 0 and π. There are no ripples. Therefore $P(\omega) \geq 0$ for all ω, and a factorization into $|C(\omega)|^2$ is assured.

The transition from passband (low frequencies) to stopband (high frequencies) becomes steeper and sharper as p increases. The slope at the midpoint $\omega = \frac{\pi}{2}$ is $-c\left(\sin\frac{\pi}{2}\right)^{2p-1}$, which is $-c$. We will show that c increases asymptotically like $\sqrt{p}$ as $p \to \infty$. Thus the transition band has width of order $1/\sqrt{p}$.

The halfband filter $P(z)$

Now we change from y and ω to the complex variable z. This will produce the filter coefficients in $P(z)$. That polynomial will be halfband and centered. The shifted polynomial $P_0(z) = z^{-N}P(z) = z^{1-2p}P(z)$ will be halfband and causal. The change of variables comes from $z = e^{i\omega}$:

$$\frac{z+z^{-1}}{2} = \cos\omega = 1-2y.$$

Thus $y=0$ and $\omega=0$ give $z=1$. Similarly $y=1$ and $\omega=\pi$ give $z=-1$.

Notice that the midpoints $y=\frac{1}{2}$ and $\omega=\frac{\pi}{2}$ give $z=\pm i$. **There are two** z's **for each** y, from $z+z^{-1}=2-4y$. (This is a quadratic equation for z.) One z is inside the unit circle, while z^{-1} is outside. This "Joukowski transformation" is also central in fluid flow. The endpoints $z=1$ and $z=-1$ are really double roots of $z+z^{-1}=2$ and $z+z^{-1}=-2$.

The change of variable gives $1-y$ and y in factored form:

$$1-y=\frac{1+\cos\omega}{2}=\left(\frac{1+z}{2}\right)\left(\frac{1+z^{-1}}{2}\right)$$

and

$$y=\frac{1-\cos\omega}{2}=\left(\frac{1-z}{2}\right)\left(\frac{1-z^{-1}}{2}\right).$$

Substituting in $\widetilde{P}(y)$, the maxflat filter in the z-domain becomes $P(z)$:

$$P(z)=2\left(\frac{1+z}{2}\right)^p\left(\frac{1+z^{-1}}{2}\right)^p\sum_{k=0}^{p-1}\binom{p+k-1}{k}\left(\frac{1-z}{2}\right)^k\left(\frac{1-z^{-1}}{2}\right)^k.$$

This factors into $P(z)=C(z)C(z^{-1})$ when $P(\omega)$ factors into $|C(\omega)|^2$. The p zeros at $y=1$ and $\omega=\pi$ are now $2p$ zeros at $z=-1$. Half of them go into $C(z)$. The $p-1$ complex zeros of the other factor $B_p(y)$ become $2p-2$ zeros of $P(z)$. Half of those (the $p-1$ zeros inside the circle $|z|=1$, if we want minimum phase) also go into $C(z)$. So the spectral factor $C(z)$ can be computed in two steps:

1. find the $p-1$ zeros of $B_p(y)$ and the $p-1$ corresponding z's with $|z|<1$;
2. include p zeros at $z=-1$: then $C(z)$ has these $2p-1$ zeros.

Example 2.12 $p=2$ gives Daubechies D_4 from $B_2(y)=1+2y=\frac{1}{2}(-z+4-z^{-1})$. The zero is at $y=-\frac{1}{2}$. Therefore $z+z^{-1}=4$. This quadratic equation has roots $z=2\pm\sqrt{3}$. Then the $2p-1$ roots of $C(z)$ are $-1,-1,2-\sqrt{3}$. The coefficients of D_4 are approximately .4830, .8365, .2241, and $-.1294$:

$$C(z)=\left[(1+\sqrt{3})+(3+\sqrt{3})z^{-1}+(3-\sqrt{3})z^{-2}+(1-\sqrt{3})z^{-3}\right]/4\sqrt{2}.$$

3 Eigenvalues of $(\downarrow 2)H$ and convergence of the cascade algorithm

The key operator in multirate filtering is $(\downarrow 2)H$. The input signal x is filtered by H and then downsampled. We keep the even-numbered components of Hx. This is the lowpass channel in the analysis half of a filter bank. When $(\downarrow 2)H$ is iterated, either finitely often in practice or infinitely often in the passage to scaling functions and wavelets, its eigenvalues become all-important. These eigenvalues are intimately related to the number of "zeros at π" in the frequency response, and to the equal number of "vanishing moments" in the wavelets. This section studies the eigenvalues and eigenvectors (left as well as right). We also determine when the cascade algorithm converges to the scaling function.

The cascade algorithm is the iteration $\varphi^{(i+1)} = M\varphi^{(i)} = (\downarrow 2)2H\varphi^{(i)}$. The extra factor 2 maintains constant area for the sequence of functions $\varphi^{(i+1)}(t) = \sum 2h(k)\varphi^{(i)}(2t-k)$, when the filter coefficients are normalized by $\sum h(k) = 1$. With double-shift orthogonality, $\sum 2h(k)h(k+2l) = \delta(l)$, convergence to $\varphi(t)$ is almost (but not quite) certain. In a *biorthogonal* filter bank, with a more general lowpass filter H, this convergence is not at all assured. Nevertheless these non-orthogonal filters are giving the best results in compression, and their condition numbers are often quite moderate. We determine when they lead to wavelets.

This section presents a simple proof of the necessary and sufficient condition for convergence to $\varphi(t)$. Since we work in the L^2 norm (convergence in energy), inner products play a decisive part. They lead to the "transition operator" $T = (\downarrow 2)2HH^T$, whose eigenvalues control the convergence. Those eigenvalues also determine the smoothness of the scaling function and wavelets.

We summarize now the main conclusions. Some are already in the literature, with different proofs, and some are new. The (real) filter coefficients $h(0), ..., h(N)$ yield the transfer function $H(z) = \sum h(n)z^{-n}$. The input signal transforms to $X(z) = \sum x(n)z^{-n}$, and the filtered signal is $H(z)X(z)$. In the z-domain, the key operators $M = (\downarrow 2)2H$ and $T = (\downarrow 2)2HH^T$ involve multiplication by $H(z)$ from the filter and an aliasing term (identified by $-z$) from the downsampling:

$$(MX)(z^2) = H(z)X(z) + H(-z)X(-z) \tag{3.1}$$
$$(TX)(z^2) = H(z)X(z)H(z^{-1}) + H(-z)X(-z)H(-z^{-1}). \tag{3.2}$$

Note the argument z^2. We are dealing with the even part, because $(\downarrow 2)$ removes the odd terms.

In the time domain, the i, k entry of M is $2h(2i-k)$. It is $2i$ that reflects the double shift from $(\downarrow 2)$. The entries of T are $2p(2i-k)$, where $P(z) = H(z)H(z^{-1})$ corresponds to HH^T. The calculations involve finite matrices, in which i and k range from 0 to $N-1$ for M and from $1-N$ to $N-1$ for T. The frequency responses of interest have p zeros at π. Thus

$H(z)$ has p zeros at $z = e^{j\pi} = -1$:

$$H(z) = \left(\frac{1+z^{-1}}{2}\right)^p Q(z) \qquad \text{with} \qquad Q(-1) \neq 0.$$

We state the conclusions in the time domain (for matrix eigenvalues), where they are easiest to check. We establish those conclusions in the z-domain, where they are easiest to prove.

Theorem 3.1 *Each time $H(z)$ is multiplied by $\frac{1+z^{-1}}{2}$, all the eigenvalues of M are multiplied by $\frac{1}{2}$ and a new eigenvalue $\lambda = 1$ is introduced. Thus the eigenvalues of M are*

$$1, \frac{1}{2}, \ldots, \left(\frac{1}{2}\right)^{p-1} \quad \text{together with } \tfrac{1}{2^p} \text{ times the eigenvalues for } (\downarrow 2)2Q. \tag{3.3}$$

Theorem 3.2 *When $H(z)$ is multiplied by $\left(\frac{1+z^{-1}}{2}\right)$, the new eigenvectors $\tilde{x}$ are the differences of the previous eigenvectors and the new left eigenvectors $\tilde{y}$ are the sums of the previous left eigenvectors:*

$$\tilde{X}(z) = \left(1 - z^{-1}\right) X(z) \quad \text{and} \quad \tilde{Y}(z) = \frac{Y(z)}{(1-z)}. \tag{3.4}$$

The extra eigenvalue $\lambda = 1$ has left eigenvector $e = [1 \ 1 \ \cdots \ 1]$. The right eigenvector gives the new values of the scaling function at the integers.

Theorem 3.3 *If $H(-1) = \sum (-1)^k h(k) = 0$, the periodized functions $P^{(i)}(t) = \sum \varphi^{(i)}(t-n)$ satisfy the identity*

$$P^{(i+1)}(t) = P^{(i)}(2t) \quad \text{and thus} \quad P^{(i)}(t) = P^{(0)}(2^i t).$$

The cascade algorithm cannot converge to $\varphi(t)$ unless $P^{(0)}(t) \equiv 1$. Otherwise $P^{(0)}(2^i t)$ will oscillate faster and faster. Thus Theorem 3.3 defines the acceptable class I of initial functions; they must satisfy $\sum \varphi^{(0)}(t-n) \equiv 1$. This is equivalent to the so-called Strang–Fix condition on the Fourier transform: $\widehat{\varphi}^{(0)}(2\pi n) = \delta(n)$. In that form, the requirement on $\varphi^{(0)}(t)$ was discovered and proved necessary by Durand and by Meyer and Paiva. Our proof uses the identity $P^{(1)}(t) = P^{(0)}(2t)$.

The central question is convergence from these acceptable $\varphi^{(0)}(t)$, and this is governed by the eigenvalues of T.

Theorem 3.4 *The cascade algorithm $\varphi^{(i+1)}(t) = \sum 2h(k)\varphi^{(i)}(2t-k)$ converges in L^2 for all $\varphi^{(0)}$ in I if and only if the eigenvalues of T satisfy Condition* **E**:

$$\lambda = 1 \text{ is a simple eigenvalue and all other eigenvalues have } |\lambda| < 1. \tag{3.5}$$

Condition **E** is also the Cohen–Daubechies requirement for the translates $\varphi(t-k)$ to be strongly independent. Jia has studied convergence and independence very carefully also in L^p.

The scaling function and wavelets are smoother by one more derivative, pointwise and in L^2, for every additional factor $(1+z^{-1})$ in $H(z)$. Splines come from the special choice $H(z) = \left(\frac{1+z^{-1}}{2}\right)^p$. They have no orthogonality, except in Haar's piecewise constant case $p = 1$, but they have maximum smoothness. $H(z)$ has binomial coefficients $h(k)$ divided by 2^p. For $p = 2$, 3, 4 we indicate the matrix $M = (\downarrow 2)2H$ and the eigenvalues predicted by Theorem 3.1:

$$\frac{1}{2}\begin{bmatrix} 1 & 0 \\ 1 & 2 \end{bmatrix} \qquad \frac{1}{4}\begin{bmatrix} 1 & 0 & 0 \\ 3 & 3 & 1 \\ 0 & 1 & 3 \end{bmatrix} \qquad \frac{1}{8}\begin{bmatrix} 1 & 0 & 0 & 0 \\ 6 & 4 & 1 & 0 \\ 1 & 4 & 6 & 4 \\ 0 & 0 & 1 & 4 \end{bmatrix}$$

$$\lambda = 1, \tfrac{1}{2} \qquad \lambda = 1, \tfrac{1}{2}, \tfrac{1}{4} \qquad \lambda = 1, \tfrac{1}{2}, \tfrac{1}{4}, \tfrac{1}{8}$$

The operator $(\downarrow 2)$ produces the double-shift between rows. The matrix on the right comes from the coefficients 1, 4, 6, 4, 1. This also illustrates the matrix T for the sequence $h(k) = 1, 2, 1$ because $h * h = 1, 4, 6, 4, 1$. (Actually the first row and column are dropped in T, so the eigenvalues are $1, \frac{1}{2}, \frac{1}{4}$.) Condition **E** in Theorem 3.4 is fully satisfied, and the cascade algorithm for the $\frac{1}{4}(1,2,1)$ filter converges quickly to the hat function—the linear spline.

In summary, the factor $(1+z^{-1})^p$ gives the zeros at π that produce flatness of $H(z)$ and smoothness of $\varphi(t)$. This pth order zero has important effects:

- p vanishing moments for the wavelets;
- p sum rules for the coefficients $h(k)$;
- pth order accuracy in approximation $f(t) \approx \sum_k a_k \varphi(t-k)$;
- pth order decay of wavelet coefficients for smooth $f(t) = \sum b_{jk} w_{jk}(t)$;
- all polynomials of degree $< p$ are combinations of the translates $\varphi(t-k)$.

The smoothness of $\varphi(t)$ is measured in the L^2 norm by using Parseval's equality. The scaling function has s derivatives when $|\omega|^s \hat{\varphi}(\omega)$ has finite energy. The supremum $s_{\max}$ depends on p and the largest eigenvalue $|\lambda_{\max}|$ of $(\downarrow 2)2QQ^T$:

$$s_{\max} = p - \log_4 |\lambda_{\max}| \,. \tag{3.6}$$

Villemoes [8, 9] has given a particularly neat analysis of this smoothness formula. It is the eigenvalue $\lambda_{\max}$ from $Q(z)$ that has no simple expression (but is easily computed). Then Theorem 3.1 shows how the factor $(\frac{1+z^{-1}}{2})^{2p}$ in HH^T divides it by $2^{2p} = 4^p$. Smaller eigenvalues of T mean more smoothness of $\varphi(t)$ and the wavelets.

Eigenvalues and eigenvectors of M

Theorems 3.1 and 3.2 will be proved together. By identifying the change in eigenvectors when $H(z)$ is multiplied by $\left(\frac{1+z^{-1}}{2}\right)$, we also confirm that the eigenvalues are cut in half. Starting from $Q(z)$ with no zeros at π, this multiplication occurs p times to reach the final $H(z)$ with p zeros at π. We go one step at a time, monitoring the eigenvectors. By eqn. (3.1), $(\downarrow 2)2Hx = \lambda x$ means

$$H(z)X(z) + H(-z)X(-z) = \lambda X(z^2). \tag{3.7}$$

Theorem 3.2 states that the step to $\left(\frac{1+z^{-1}}{2}\right) H(z)$ produces the new eigenfunction $\tilde{X}(z) = \left(1 - z^{-1}\right) X(z)$ with eigenvalue $\tilde{\lambda} = \frac{1}{2}\lambda$. If this is true, then eqn. (3.7) will hold for $\tilde{H}(z)$ and $\tilde{X}(z)$ and $\tilde{\lambda}$:

$$\begin{aligned}\left(\frac{1+z^{-1}}{2}\right) H(z)\left(1-z^{-1}\right) X(z) \quad + \quad & \left(\frac{1-z^{-1}}{2}\right) H(-z)\left(1+z^{-1}\right) X(-z) \\ = \quad & \frac{\lambda}{2}\left(1-z^{-2}\right) X\left(z^2\right). \end{aligned} \tag{3.8}$$

To verify (3.8), multiply (3.7) by $\frac{1}{2}\left(1-z^{-2}\right)$. That is the only step in the proof. Daubechies proved in a different way that $M = (\downarrow 2)2H$ has eigenvalues $1, \frac{1}{2}, \ldots, (\frac{1}{2})^{p-1}$.

Now consider the left eigenvectors. These are right eigenvectors of $M^T = 2H^T(\downarrow 2)^T$. In the z-domain this transposed operator takes $Y(z)$ into $2H(z^{-1})Y(z^2)$. Thus $yM = \lambda y$ means

$$2H(z^{-1})Y(z^2) = \lambda Y(z). \tag{3.9}$$

Theorem 3.2 says that eqn. (3.9) remains correct for

$$\tilde{H}(z) = \left(\frac{1+z^{-1}}{2}\right) H(z), \quad \tilde{\lambda} = \frac{1}{2}\lambda$$

when the eigenvector transforms to $\tilde{Y}(z) = \frac{Y(z)}{1-z}$. The left side is multiplied by $\frac{1+z}{2}$ and divided by $1 - z^2$. The right side is divided by $2(1-z)$. This agreement completes the proof of Theorem 3.2.

For Theorems 3.3 and 3.4 we refer to the full paper and to the textbook [7]. The convergence of the cascade algorithm becomes convergence of the power method for the matrix T. This requires Condition **E** and it yields scaling functions and wavelets in L^2. The theory goes onward from there.

4 Zeros of the Daubechies polynomials (*with J. Shen*)

We study the asymptotics of the maximally flat Daubechies filters, as the number of zeros at π is steadily increased. The product filter of degree

$4p-2$ with $2p$ zeros at π is factored into analysis times synthesis, $P(z) = F(z)H(z)$. When F is a time–reversal of H, the filter bank is orthogonal. Other factorizations yield linear phase filters by a different assignment of the zeros of $P(z)$. The zeros of this special $P(z)$ play an important role in the design of filters.

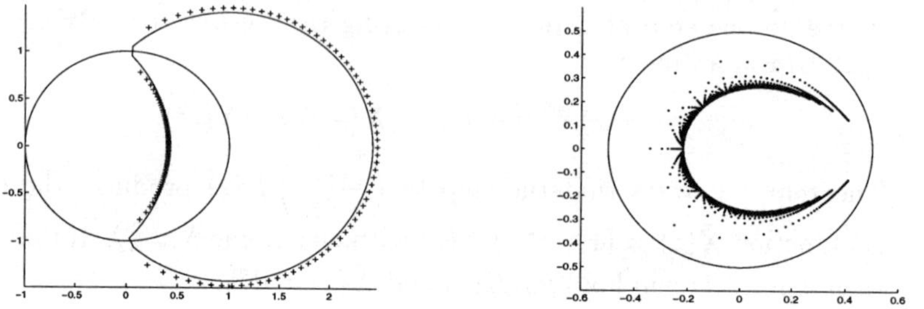

Left: The zeros are close to the limiting curve in the z–plane. Right: All zeros lie inside the circle of radius $1/2$ in the y–plane, $p = 1:1:60$.

Matlab shows a remarkable plot for $p = 70$. The zeros are close to a limiting curve $|z - z^{-1}| = 2$. This is a union of two circular arcs of radius $\sqrt{2}$, around $z = 1$ and $z = -1$. We prove that the zeros do approach this limit and we find their distribution along the curve. For finite p, a simple approximation to all the zeros is based on their asymptotic equidistribution on the circle of radius $1+\log(4\pi p)/2p$ in the w–plane. The final figure shows asymptotic vs. actual zeros for $p = 70$, with $w = 4y(1-y) = ((z-z^{-1})/2)^2$.

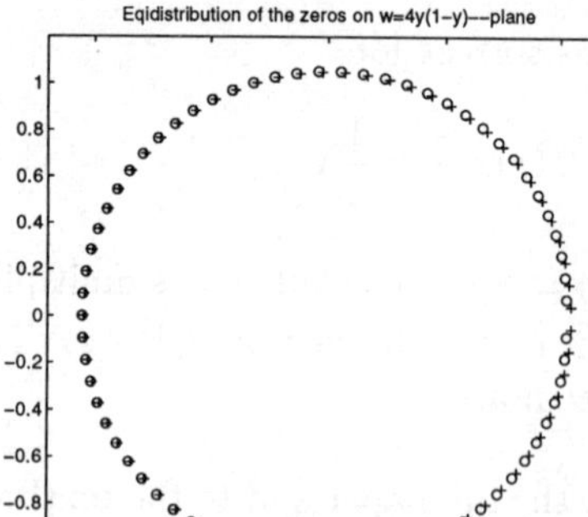

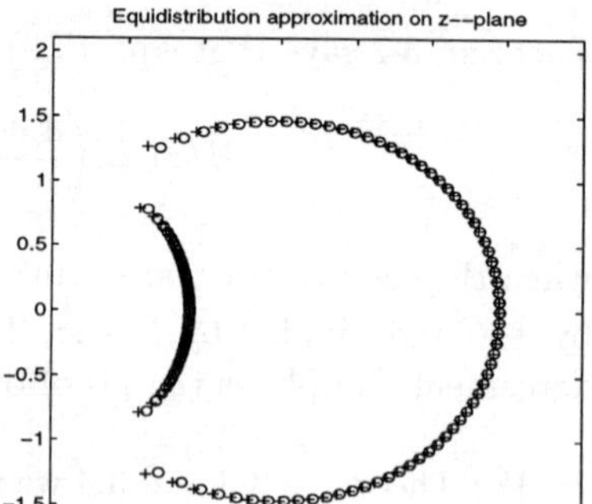

The wide dynamic range in the coefficients makes the zeros difficult to compute for large p. Rescaling y by 4 allows us to reach $p = 80$ by standard codes. This is "spectral factorization" of high order. The zeros at $z = -1$ stabilize the iteration of the lowpass filter in a wavelet filter bank.

Our starting point is the binomial series for $(1-y)^{-p}$, truncated after p terms. Its $p-1$ zeros give the $2p-2$ zeros of $P(z)$ inside and outside the unit circle, by the quadratic relation $z+z^{-1}=2-4y$. The zeros in the complex y-plane approach the curve $|4y(1-y)|=1$. All zeros have $|y|\leq 1/2$ and Re $z\geq 0$. The extreme zeros approach the singular point $y=1/2$ (corresponding to $z=\pm i$) with speed $p^{-1/2}$. Then the $2p$ additional zeros at $z=-1$ yield the maximally flat $P(z)$.

Other important questions about the asymptotics of the Daubechies filters and scaling functions remain open. An important paper by Lemarié and Kateb will appear in Revista Matematica Iberoamericana.

Bibliography

1. I. Daubechies, *Orthonormal bases of compactly supported wavelets.* Comm. Pure Appl. Math. **41**, 909–996 (1988).
2. D. Estaban and C. Galand, *Application of quadrature mirror filters to split-band voice coding schemes.* Proc. IEEE Int. Conf. ASSP, 191–195 (1977)
3. D. Kateb and P.G. Lemarie, *Asymptotic behaviour of the Daubechies filters.* Appl. Comp. Harmonic Analysis **2**, 398–399 (1995).
4. J. Shen and G. Strang, *Asymptotic analysis of Daubechies polynomials.* Proc. Amer. Math. Soc. **124**, 3819–3833 (1996).
5. M.J.T. Smith and T.P. Barnwell, *Exact reconstruction techniques for tree-structured subband coders.* IEEE Trans. ASSP **34**, 434–441 (1986).
6. G. Strang, *Eigenvalues of* $(\downarrow 2)$ *and convergence of the cascade algorithm.* IEEE Trans. Signal Processing **44**, 233–238 (1996).
7. G. Strang and T. Nguyen. Wavelets and Filter Banks. Wellesley-Cambridge Press (1996).
8. L. F. Villemoes, *Energy moments in time and frequency for two-scale difference equation solutions and wavelets.* SIAM J. Math. Anal. **23**, 1519–1543 (1992).
9. L. F. Villemoes, *Wavelet analysis of refinement equations.* SIAM J. Math. Anal. **25**, 1433–1460 (1994).

5 Further reading

5.1 Selected papers on wavelets

- A. S. Cavaretta, W. Dahmen, and C. A. Micchelli, Stationary Subdivision, *Amer. Math. Soc. Memoirs*, vol. 453, Providence RI (1991).
- A. Cohen and I. Daubechies, *A stability criterion for biorthogonal wavelet bases and their related subband coding scheme.* Duke Math. J. **68**, 313–335 (1992).

- A. Cohen, I. Daubechies, and J. C. Feauveau, *Biorthogonal bases of compactly supported wavelets.* Comm. Pure Appl. Math. **45**, 485-560 (1992).
- I. Daubechies, Ten Lectures on Wavelets. SIAM, Philadelphia (1992).
- I. Daubechies and J. Lagarias, *Two-scale difference equations: I. Existence and global regularity of solutions.* SIAM J. Math. Anal. **22**, 1388–1410 (1991). Also *Two-scale difference equations: II. Local regularity, infinite products of matrices and fractals.* SIAM J. Math. Anal. **23**, 1031–1079 (1992).
- T. Eirola, *Sobolev characterization of solutions of dilation equations.* SIAM J. Math. Anal. **23**, 1015–1030 (1992).
- C. Heil and D. Colella, *Dilation equations and the smoothness of compactly supported wavelets.* Wavelets: Mathematics and Applications, J. Benedetto and M. Frazier, eds., CRC Press (1993).
- R.–Q. Jia, *Subdivision schemes in L^p spaces.* Adv. in Comp. Math. **3**, 309-341 (1995).
- W. Lawton, *Necessary and sufficient conditions for constructing orthogonal wavelets.* J. Math. Phys. **32**, 52–61 (1992).
- W. Lawton, S. L. Lee, and Z. Shen, *Convergence of multidimensional cascade algorithm.* Preprint.

5.2 Books on wavelets and filter banks

- A.N. Akansu and R.A. Haddad. Multiresolution Signal Decomposition. Academic Press (1992).
- A.N. Akansu and M.J.T.Smith, eds. Subband and Wavelet Transforms. Kluwer (1995).
- J. Benedetto and M. Frazier, eds. Wavelets: Mathematics and Applications. CRC Press (1993).
- A.S. Cavaretta, W. Dahmen, and C.A. Micchelli, *Stationary Subdivision.* Amer. Math. Soc. Memoirs **453** (1991).
- W.K. Chen, ed. The Circuits and Filters Handbook. Chapters on filter banks and wavelets by T. Nguyen, I Djokovic, and P.P. Vaidyanathan. IEEE Press (1995).
- C.K. Chui. An Introduction to Wavelets. Academic Press (1992).
- C.K. Chui, ed. Wavelets: A Tutorial in Theory and Applications. Academic Press (1992).
- A. Cohen. Wavelets and Multiscale Signal Processing. Chapman and Hall (1995).
- J.M. Combes, A. Grossmann, and Ph. Tchamitchian, eds. Wavelets, Time-Frequency Methods, and Phase Space. Springer (1990).

- R.E. Crochiere and L.R. Rabiner. Multirate Digital Signal Processing. Prentice-Hall (1983).
- I. Daubechies. Ten Lectures on Wavelets. SIAM (1992).
- N.J. Fliege. Multirate Digital Signal Processing. John Wiley (1994).
- A. Gersho and R.M. Gray. Vector Quantization and Signal Compression. Kluwer (1992).
- R.M. Gray. Source Coding Theory. Kluwer (1990).
- N.J. Jayant and P. Noll. Digital Coding of Waveforms. Prentice-Hall (1984).
- G. Kaiser. A Friendly Guide to Wavelets. Birkhäuser (1994).
- S. Mallat. Wavelet Signal Processing. Academic Press (1996).
- H.S. Malvar. Signal Processing with Lapped Transforms. Artech House (1992).
- Y. Meyer. Wavelets and Operators. Translation of *Ondelettes et Opérateurs* (Hermann, 1990), Cambridge University Press (1993).
- Y. Meyer. Wavelets: Algorithms and Applications. SIAM (1993).
- Y. Meyer, ed. Proceedings of the Marseille Conference on Wavelets. Masson (1993).
- V. Oppenheim and R. W. Schafer. Discrete-Time Signal Processing. Prentice-Hall (1989).
- M.B. Ruskai et al., eds. Wavelets and Their Applications. Jones and Bartlett (1992).
- L.L. Schumaker and G. Webb, eds. Recent Advances in Wavelet Analysis. Academic Press (1994).
- P.P. Vaidyanathan. Multirate Systems and Filter Banks. Prentice-Hall (1992).
- M. Vetterli and J. Kovacevic. Wavelets and Subband Coding. Prentice-Hall (1995).
- G.G. Walter. Wavelets and Other Orthogonal Systems with Applications. CRC Press (1994).
- M.V. Wickerhauser. Adapted Wavelet Analysis from Theory to Software. AK Peters (1994).
- J.W. Woods, ed. Subband Image Coding. Kluwer (1991).
- Special Issue on Wavelets. IEEE Transactions on Signal Processing, **41** (12/93).
- Special Issue on Wavelet Transforms. IEEE Transactions on Information Theory, **38** (3/92).

An introduction to multilevel methods

Jinchao Xu

Pennsylvania State University[3]

Abstract

An introduction is given in this paper to the basic idea and the convergence theory of multilevel methods including overlapping domain decomposition methods and multigrid methods. Brief discussions are first given of some basic properties of some elementary linear iterative methods such as Jacobi and Gauss–Seidel iterations and preconditioned conjugate gradient methods, and then more detailed discussions are devoted to a general framework of subspace correction methods that can be applied to, among many other things, multilevel methods. A framework of auxiliary space methods is also briefly presented for the construction of preconditioners. The overlapping domain decomposition method and multigrid method are introduced with a model elliptic boundary value problem of second order. Convergence estimates are obtained for an overlapping domain decomposition method and especially for basic multigrid methods such as backslash (\) cycle, V-cycle and W-cycle. Two different approaches are used in the convergence analysis for multigrid methods. The first approach is the more traditional one that makes crucial use of elliptic regularity, while the second approach is based on the subspace correction framework that very weakly depends on the elliptic regularity. The first approach gives more precise estimates for simpler problems, while the second approach can be applied to more complex problems such as locally refined meshes and interface problems with large discontinuous jumps. As some more advanced topics, a general framework is briefly described on multigrid methods for nonnested multilevel subspaces and varying bilinear forms, and an optimal multigrid preconditioning technique is given for general unstructured grids using the auxiliary space framework. For nonsymmetric and/or indefinite problems, some two grid techniques are discussed. In addition to the aforementioned theoretical analysis, some discussions are also devoted to the implementation of some basic multigrid algorithms.

1 Introduction

Multilevel methods are among the most efficient modern techniques for solving large scale algebraic systems arising from the discretization of partial differential equations. In this paper, we shall give an introduction to

[3]The author was partially supported by NSF DMS94-03915-1 through Penn State.

these methods and their convergence properties by considering their applications to a model elliptic boundary value problem of second order.

Multilevel methods have been most efficiently used in solving the linear algebraic systems arising from the finite element discretizations of partial differential equations. The theory of the methods is an elegant combination of linear algebra, theory of finite element approximation and of partial differential equations. In this paper, we shall explore all these three aspects of the multilevel theory. We shall devote §2, §3 and §4 to the technical materials for the theory of multilevel methods.

§2 is on the basic linear iterative methods and preconditioning concepts. Many elementary iterative methods, such as Jacobi and Gauss–Seidel iterations, are often the major components in a multilevel procedure, and also a multilevel method is often used in conjunction with a preconditioned conjugate gradient method. Therefore the materials in §2 are fundamental to our multilevel algorithms and theory.

§3 is on an algebraic framework of subspace correction methods (following Xu [33]) that can be used in general for construction and analysis of linear iterative methods. This framework will be a main technical tool in the analysis of domain decomposition methods in §5 and multigrid methods in §6.

The most technical materials in this paper are perhaps those in §4 for finite element approximation theory. In this section, some basic materials in finite elements are reviewed and some approximation results concerning multiple level of finite element spaces are presented. Some of these results depend crucially on the regularity theory for elliptic boundary value problems.

Overlapping domain decomposition methods are presented and analysed in §5. In particular, it is demonstrated that a recursive application of nonoverlapping domain decomposition (corresponding to smallest possible subdomains) will naturally lead to a typical multigrid method.

The core of this paper is §6 in which many major multigrid algorithms are introduced and analysed. An attempt is made to explain the basic ideas behind multigrid methods and also to describe the implementation issues. But the major concern here is to present the multigrid convergence theory. The multigrid methods are analysed with two different approaches. The first approach is the more traditional one which makes crucial use of regularity theory of partial differential equations. The second approach is the subspace correction framework in §3.

The multigrid algorithms and their convergence analysis presented in §6 are only for the case that the underlying multilevel spaces are nested in the sense that the coarse spaces are subspaces of finer spaces. §7 is devoted to giving readers an idea of how multigrid methods can be applied to more complicated situations, where a general framework of nonnested multigrid methods which can be applied to cases like unstructured grids and nonconforming elements is presented, and a special technique is given for constructing optimal multigrid preconditioning techniques for unstructured grids using the framework of auxiliary space methods.

Multigrid methods have been extensively studied in a vast literature by researchers in many different areas; a short article like this can only give a glimpse of a small part of the whole subject. For further details, we refer to the research monographs of Hackbusch [20, 21], McCormick [23], Wesseling [31] and Bramble [5], and to the review articles of Xu [33] and Yserentant [44].

For convenience, following [33], the symbols $\lesssim, \gtrsim$ and $\eqsim$ will be used in this paper. That $x_1 \lesssim y_1, x_2 \gtrsim y_2$ and $x_3 \eqsim y_3$, mean that $x_1 \leq C_1 y_1$, $x_2 \geq c_2 y_2$ and $c_3 x_3 \leq y_3 \leq C_3 x_3$ for some constants C_1, c_2, c_3 and C_3 that are independent of mesh parameters.

2 Iterative and preconditioning methods

Assume $\mathcal{V}$ is a finite dimensional vector space. The goal of this section is to study iterative methods and preconditioning techniques for solving the following kind of equation:

$$Au = f. \tag{2.1}$$

Here $A : \mathcal{V} \mapsto \mathcal{V}$ is an SPD (symmetric positive definite) linear operator over $\mathcal{V}$ and $f \in \mathcal{V}$ is given.

2.1 Elementary linear iterative methods

A single step linear iterative method which uses an old approximation, u^{old}, of the solution u of (2.1), to produce a new approximation, u^{new}, usually consists of three steps:

(1) form $r^{\text{old}} = f - Au^{\text{old}}$;
(2) solve $Ae = r^{\text{old}}$ approximately: $\hat{e} = Br^{\text{old}}$;
(3) update $u^{new} = u^{\text{old}} + \hat{e}$;

where B is a linear operator on $\mathcal{V}$ and can be thought of as an approximate inverse of A.

As a result, we have the following iterative algorithm.

Algorithm 2.1 *Given $u^0 \in \mathcal{V}$,*

$$u^{k+1} = u^k + B(f - Au^k), \quad k = 0, 1, 2, \cdots. \tag{2.2}$$

The core of the above iterative scheme is the operator B. Notice that if $B = A^{-1}$, after one iteration, u^1 is then the exact solution. B will be called an iterator of A.

We say that an iterative scheme like (2.2) converges if $\lim_{k\to\infty} u_k = u$ for any $u_0 \in \mathcal{V}$. Assume that u and u^k are solutions of (2.1) and (2.2) respectively. Then

$$u - u^k = (I - BA)^k (u - u_0).$$

Therefore the iterative scheme (2.2) converges iff $\rho(I - BA) < 1$.
Symmetrization. Sometimes it is more desirable that the iterator B is symmetric. If B is not symmetric, there is a natural way to symmetrize it. Consider the following iteration

$$\begin{aligned} u^{k+1/2} &= u^k + B(f - Au^k) \\ u^{k+1} &= u^{k+1/2} + B^t(f - Au^{k+1/2}) \end{aligned}$$

where 't' denotes the adjoint operator with respect to $(\cdot,\cdot)$. Eliminating the intermediate $u^{k+1/2}$ gives

$$u - u^{k+1} = (I - B^tA)(I - BA)(u - u^k)$$

or

$$u^{k+1} = u^k + \bar{B}(f - Au^k) \tag{2.3}$$

where, with '$*$' denoting the adjoint operator with respect to $(\cdot,\cdot)_A$,

$$\bar{B} = (I - (I - BA)^*(I - BA))A^{-1} = B^t + B - B^tAB \tag{2.4}$$

or

$$I - \bar{B}A = (I - BA)^*(I - BA). \tag{2.5}$$

Obviously $\bar{B}$ is symmetric and will be called the *symmetrization* of the iterator B. The following identities obviously hold:

$$(\bar{B}Av, v)_A = ((2I - BA)v, BAv)_A \quad \forall\, v \in \mathcal{V}, \tag{2.6}$$

and

$$\|v\|_A^2 - \|(I - BA)v\|_A^2 = (\bar{B}Av, v)_A \quad \forall\, v \in \mathcal{V}. \tag{2.7}$$

A simple consequence of (2.7) is that

$$\lambda_{\max}(\bar{B}A) \le 1.$$

Theorem 2.1 *The following are equivalent.*

1. *The symmetrized scheme (2.3) is convergent.*
2. *The operator $\bar{B}$ given by (2.4) is SPD.*
3. *The matrix $B^{-t} + B^{-1} - A$ is SPD*
4. *There exists a constant $\omega_1 \in (0,2)$ such that any one of the following is satisfied for any $v \in \mathcal{V}$:*

$$(BAv, BAv)_A \le \omega_1 (BAv, v)_A; \tag{2.8}$$

$$(Av, v) \le \omega_1 (B^{-1}v, v); \tag{2.9}$$

$$\left(\frac{2}{\omega_1} - 1\right)(Av, v) \le ((B^{-1} + B^{-t} - A)v, v); \tag{2.10}$$

$$(2 - \omega_1)(Bv, v) \le (\bar{B}v, v). \tag{2.11}$$

Furthermore, the scheme (2.2) converges if (and only if, when B is symmetric) its symmetrized scheme (2.3) converges.

The above results can be proved easily by definition. We further notice that

$$(2 - \omega_1)B \le \bar{B} \le 2B. \tag{2.12}$$

Richardson iterative methods. Richardson iteration is perhaps the simplest iterative method which corresponds to (2.2) with $B = \frac{\omega}{\rho(A)}I$. Namely,

$$u^{k+1} = u^k + \frac{\omega}{\rho(A)}(f - Au^k), \quad k = 0, 1, 2, \cdots, \tag{2.13}$$

One can imagine that the Richardson method is not very efficient, but it is theoretically very important. One of the most important properties of this method is its 'smoothing property' that will be discussed now.

Let $A\varphi_i = \lambda_i \varphi_i$ with $\lambda_1 < \lambda_2 \le \ldots \lambda_n$, $(\varphi_i, \varphi_j) = \delta_{ij}$, and $u - u^0 = \sum \alpha_i \varphi_i$. Then

$$u - u^k = \sum_i \alpha_i (1 - \omega\lambda_i/\lambda_n)^k \varphi_i.$$

For a fixed $\omega \in (0, 2)$, it is clear that $(1 - \omega\lambda_i/\lambda_n)^k$ converges to zero very fast as $k \to \infty$ if λ_i is close to λ_n. This exactly means that the high frequency modes in the error get damped out very quickly.

An iterative method (2.2) is said to be Richardson-like if there exists an $\omega \in (0, 2)$ such that

$$\|(I - BA)v\|_A \le \|(I - \frac{\omega}{\rho(A)}A)v\|_A \quad \forall\, v \in \mathcal{V}. \tag{2.14}$$

Lemma 2.2 *For the iterative method (2.2), the following are equivalent.*

1. *The inequality (2.14) is satisfied with* $\omega = C_0^{-1}$.
2. $(C_0\rho(A))^{-1}\|v\|^2 \le (\bar{B}v, v) \quad \forall\, v \in \mathcal{V}$.
3. $(C_0\rho(A))^{-1}\|Av\|^2 \le \|v\|_A^2 - \|(I - BA)v\|_A^2 \quad \forall\, v \in \mathcal{V}$.

2.2 Jacobi and Gauss–Seidel methods

Assume $\mathcal{V} = \mathbb{R}^n$ and $A = (a_{ij}) \in \mathbb{R}^{n \times n}$ is the usual SPD matrix. We write $A = D - L - U$ with D being the diagonal of A and $-L$ and $-U$ the lower and upper triangular parts of A respectively. The easiest approximate inverses of A are perhaps

$$B = D^{-1} \quad \text{or} \quad B = (D - L)^{-1}.$$

As we shall see these two choices of B result in the well known Jacobi and Gauss–Seidel methods. More generally, we have the following choices of B that result in various different iterative methods:

$$B = \begin{cases} \omega & \text{Richardson;} \\ D^{-1} & \text{Jacobi;} \\ \omega D^{-1} & \text{Damped Jacobi;} \\ (D-L)^{-1} & \text{Gauss–Seidel;} \\ \omega(D-\omega L)^{-1} & \text{SOR.} \end{cases} \tag{2.15}$$

The symmetrization of the aforementioned Gauss–Seidel method is called the symmetric Gauss–Seidel method.

Theorem 2.3 *Assume A is SPD. Then*

- *the Richardson method converges iff $0 < \omega < 2/\rho(A)$;*
- *the Jacobi method converges iff $2D - A$ is SPD;*
- *the Damped Jacobi method converges iff $0 < \omega < 2/\rho(D^{-1}A)$;*
- *the Gauss–Seidel method always converges;*
- *the SOR method converges iff $0 < \omega < 2$.*

The proof of the above results follow directly from Theorem 2.1 by (2.15) to compute $B^{-t}+B^{-1}-A$. For example, for the SOR method, $B^{-t}+B^{-1}-A = \frac{2-\omega}{\omega}D$.

2.3 Alternative formulations of iterative schemes

Assume that $\mathcal{V}$ and $\mathcal{W}$ are two vector spaces and $A \in L(\mathcal{V},\mathcal{W})$. By convention, the matrix representation of A with respect to a basis $(\varphi_1,\cdots,\varphi_n)$ of $\mathcal{V}$ and a basis $(\psi_1,\cdots,\psi_m)$ of $\mathcal{W}$ is the matrix $\widetilde{A} \in \mathbb{R}^{m\times n}$ satisfying

$$(A\varphi_1,\cdots,A\varphi_n) = (\psi_1\cdots,\psi_m)\widetilde{A}.$$

Given any $v \in \mathcal{V}$, there exists a unique $\nu = (\nu_i) \in \mathbb{R}^n$ such that $v = \sum_{i=1}^n \nu_i\varphi_i$. The vector ν can be regarded as the matrix representation of v, denoted by $\nu = \widetilde{v}$.

By definition, we have, for any two operators A, B and a vector v

$$\widetilde{AB} = \widetilde{A}\widetilde{B} \quad \text{and} \quad \widetilde{Av} = \widetilde{A}\widetilde{v}. \tag{2.16}$$

Under the basis (φ_k), we define the so-called mass matrix and stiffness matrix as follows

$$\mathcal{M} = ((\varphi_i,\varphi_j))_{n\times n} \quad \text{and} \quad \mathcal{A} = ((A\varphi_i,\varphi_j))_{n\times n},$$

respectively. It can easily be shown that

$$\mathcal{A} = \mathcal{M}\widetilde{A},$$

and that $\mathcal{M}$ is the matrix representation of the operator defined by

$$Rv = \sum_{i=1}^{n}(v, \varphi_i)\varphi_i, \quad \forall\, v \in \mathcal{V}. \tag{2.17}$$

Under a given basis (φ_k), the equation (2.1) can be transformed to an algebraic system

$$\mathcal{A}\mu = \eta. \tag{2.18}$$

Similarly to (2.2), a linear iterative method for (2.18) can be written as

$$\mu^{k+1} = \mu^k + \mathcal{B}(\eta - \mathcal{A}\mu^k), \quad k = 0, 1, 2, \cdots, \tag{2.19}$$

where $\mathcal{B} \in \mathbb{R}^{n\times n}$ is an iterator of the matrix $\mathcal{A}$.

Proposition 2.4 *Assume that $\tilde{u} = \mu, \tilde{f} = \beta$ and $\eta = \mathcal{M}\beta$. Then u is the solution of (2.1) if and only if μ is the solution of (2.18). The linear iterations (2.2) and (2.19) are equivalent if and only if $\widetilde{B} = \mathcal{B}\mathcal{M}$. In this case the condition numbers are related by $\kappa(\mathcal{B}\mathcal{A}) = \kappa(BA)$.*

In the following, we shall call $\mathcal{B}$ the algebraic representation of B.

Using the properties of the operator defined by (2.17), we can prove the following simple result.

Proposition 2.5 *The scheme (2.2) represents the Richardson iteration for the equation (2.18) if B is given by*

$$Bv = \omega\rho(\mathcal{A})^{-1}\sum_{i=1}^{n}(v, \varphi_i)\varphi_i, \quad \forall\, v \in \mathcal{V},$$

and it represents the damped Jacobi iteration if B is given by

$$Bv = \omega\sum_{i=1}^{n}(A\varphi_i, \varphi_i)^{-1}(v, \varphi_i)\varphi_i, \quad \forall\, v \in \mathcal{V}.$$

2.4 Preconditioned conjugate gradient method

The well known conjugate gradient method is the basis of all the preconditioning techniques to be studied in this paper. The preconditioned conjugate gradient (PCG) method can be viewed as a conjugate gradient method applied to the preconditioned system:

$$BAu = Bf. \tag{2.20}$$

Here $B : V \mapsto V$ is another SPD operator and known as a preconditioner for A. Note that BA is symmetric with respect to the inner product $(B^{-1}\cdot, \cdot)$. One version of this algorithm is as follows: *given* u_0; $r_0 = f - Au_0$; $p_0 = Br_0$;

for $k = 1, 2, \ldots,$

$$u_k = u_{k-1} + \alpha_k p_{k-1},\ r_k = r_{k-1} - \alpha_k A p_{k-1},\ p_k = B r_k + \beta_k p_{k-1},$$
$$\alpha_k = (B r_{k-1}, r_{k-1})/(A p_{k-1}, p_{k-1}),\ \beta_k = (B r_k, r_k)/(B r_{k-1}, r_{k-1}).$$

It is well known that

$$\|u - u_k\|_A \le 2\Big(\frac{\sqrt{\kappa(BA)} - 1}{\sqrt{\kappa(BA)} + 1}\Big)^k \|u - u_0\|_A, \tag{2.21}$$

which implies that PCG converges faster with smaller condition number $\kappa(BA)$.

Observing the formulae in the PCG method and the convergence estimate (2.21), one sees that the efficiency of a PCG method depends on two main factors: the action of B and the size of $\kappa(BA)$. Hence, a good preconditioner should have the properties that the action of B is relatively easy to compute and that $\kappa(BA)$ is relatively small (at least smaller than $\kappa(A)$).

3 Iterative methods by subspace correction

Following Xu [33] (see also Bramble et al. [8, 7]), a general framework for constructing linear iterative methods and/or preconditioners can be obtained by the concept of *space decomposition* and *subspace correction.* This framework will be presented here from a purely algebraic point of view. Some simple examples are given for illustration and more important applications are given in the later sections for multigrid methods. This framework can also be applied directly to domain decomposition methods.

The presentation here more or less follows Xu [33]. The main modification is that the subspace solvers here may not be symmetric. For related topics, we refer to Bramble [5].

3.1 Preliminaries

A decomposition of a vector space $\mathcal{V}$ consists of a number of subspaces $\mathcal{V}_i \subset \mathcal{V}$ (for $0 \le i \le J$) such that

$$\mathcal{V} = \sum_{i=0}^{J} \mathcal{V}_i. \tag{3.1}$$

This means that, for each $v \in \mathcal{V}$, there exist $v_i \in \mathcal{V}_i$ $(0 \le i \le J)$ such that $v = \sum_{i=0}^{J} v_i$. This representation of v may not be unique in general, namely (3.1) is not necessarily a direct sum.

For each i, we define $Q_i, P_i : \mathcal{V} \mapsto \mathcal{V}_i$ and $A_i : \mathcal{V}_i \mapsto \mathcal{V}_i$ by

$$(Q_i u, v_i) = (u, v_i), \quad (P_i u, v_i)_A = (u, v_i)_A, \quad u \in \mathcal{V}, v_i \in \mathcal{V}_i, \tag{3.2}$$

and

$$(A_i u_i, v_i) = (A u_i, v_i), \quad u_i, v_i \in \mathcal{V}_i. \tag{3.3}$$

Q_i and P_i are both orthogonal projections and A_i is the restriction of A on $\mathcal{V}_i$ and is SPD. It follows from the definition that

$$A_i P_i = Q_i A. \tag{3.4}$$

This identity is of fundamental importance and will be used frequently in this chapter. A consequence of it is that, if u is the solution of (2.1), then

$$A_i u_i = f_i \tag{3.5}$$

with $u_i = P_i u$ and $f_i = Q_i f$. This equation may be regarded as the restriction of (2.1) to $\mathcal{V}_i$.

We note that the solution u_i of (3.5) is the best approximation of the solution u (2.1) in the subspace $\mathcal{V}_i$ in the sense that

$$J(u_i) = \min_{v \in \mathcal{V}_i} J(v), \quad \text{with } J(v) = \frac{1}{2}(Av, v) - (f, v)$$

and

$$\|u - u_i\|_A = \min_{v \in \mathcal{V}_i} \|u - v\|_A.$$

In general the subspace equation (3.5) will be solved approximately. To describe this, we introduce, for each i, another nonsingular operator $R_i : \mathcal{V}_i \mapsto \mathcal{V}_i$ that represents an approximate inverse of A_i in a certain sense. Thus an approximate solution of (3.5) may be given by $\hat{u}_i = R_i f_i$.

Example 3.1 *Consider the space $\mathcal{V} = R^n$ and the simplest decomposition:*

$$\mathbb{R}^n = \sum_{i=1}^{n} \operatorname{span}\{e^i\},$$

where e^i is the i-th column of the identity matrix. For an SPD matrix $A = (a_{ij}) \in \mathbb{R}^{n \times n}$

$$A_i = a_{ii}, \quad Q_i y = y_i e^i,$$

where y_i is the i-th component of $y \in \mathbb{R}^n$.

3.2 Basic algorithms

From the viewpoint of subspace correction, most linear iterative methods can be classified into two major algorithms, namely the *parallel subspace correction* (PSC) method and the *successive subspace correction* method (SSC).

PSC: Parallel subspace correction This type of algorithm is similar to the Jacobi method. The idea is to correct the residue equation on each subspace in parallel.

Let $u^{\rm old}$ be a given approximation of the solution u of (2.1). The accuracy of this approximation can be measured by the residual: $r^{\rm old} = f - Au^{\rm old}$. If $r^{\rm old} = 0$ or is very small, we are done. Otherwise, we consider the residual equation:

$$Ae = r^{\rm old}.$$

Obviously $u = u^{\rm old} + e$ is the solution of (2.1). Instead we solve the restricted equation on each subspace $\mathcal{V}_i$

$$A_i e_i = Q_i r^{\rm old}.$$

It should be helpful to note that the solution e_i is the best possible correction $u^{\rm old}$ in the subspace $\mathcal{V}_i$ in the sense that

$$J(u^{\rm old} + e_i) = \min_{e \in \mathcal{V}_i} J(u^{\rm old} + e), \quad \text{with } J(v) = \frac{1}{2}(Av, v) - (f, v)$$

and

$$\|u - (u^{\rm old} + e_i)\|_A = \min_{e \in \mathcal{V}_i} \|u - (u^{\rm old} + e)\|_A.$$

As we are only seeking a correction, we only need to solve this equation approximately using the subspace solver R_i described earlier

$$\hat{e}_i = R_i Q_i r^{\rm old}.$$

An update of the approximation of u is obtained by

$$u^{new} = u^{\rm old} + \sum_{i=0}^{J} \hat{e}_i$$

which can be written as

$$u^{new} = u^{\rm old} + B(f - Au^{\rm old}),$$

where

$$B = \sum_{i=0}^{J} R_i Q_i. \tag{3.6}$$

We have therefore

Algorithm 3.1 *Given $u_0 \in \mathcal{V}$, apply the iterative scheme (2.2) with B given in (3.6).*

Example 3.2 *With $\mathcal{V} = \mathbb{R}^n$ and the decomposition given by Example 3.1, the corresponding Algorithm 3.1 is just the Jacobi iterative method.*

It is well known that the Jacobi method is not convergent for all SPD problems (see Theorem 2.3) hence Algorithm 3.1 is not always convergent. However the preconditioner obtained from this algorithm is of great importance. We note that the operator B given by (3.6) is SPD if each $R_i : \mathcal{V}_i \to \mathcal{V}_i$ is SPD.

Algorithm 3.2 *Apply the CG method to equation (2.1), with B defined by (3.6) as a preconditioner.*

Example 3.3 *The preconditioner B corresponding to Example 3.1 is*

$$B = \mathrm{diag}(a_{11}^{-1}, a_{22}^{-1}, \cdots, a_{nn}^{-1})$$

which is the well known diagonal preconditioner for the SPD matrix A.

SSC: Successive subspace correction This type of algorithm is similar to the Gauss–Seidel method.

To improve the PSC method that makes simultaneous correction, we make the correction here in one subspace at a time by using the most updated approximation of u. More precisely, starting from $v^{-1} = u^{\mathrm{old}}$ and correcting its residue in $\mathcal{V}_0$ gives

$$v^0 = v^{-1} + R_0 Q_0 (f - Av^{-1}).$$

By correcting the new approximation v^1 in the next space $\mathcal{V}_1$, we get

$$v^1 = v^0 + R_1 Q_1 (f - Av^0).$$

Proceeding this way successively for all $\mathcal{V}_i$ leads to

Algorithm 3.3 *Given* $u^0 \in \mathcal{V}$,

for $k = 0, 1, \ldots$ till convergence
  $v \leftarrow u^k$
  for $i = 0 : J$ $v \leftarrow v + R_i Q_i (f - Av)$ endfor
  $u^{k+1} \leftarrow v.$
endfor

Example 3.4 *Corresponding to decomposition in Example 3.1, the Algorithm 3.3 is the Gauss–Seidel iteration.*

Example 3.5 *More generally, decompose* $\mathbb{R}^n$ *as*

$$\mathbb{R}^n = \sum_{i=0}^{J} \mathrm{span}\{e^{l_i}, e^{l_i+1}, \cdots, e^{l_{i+1}-1}\},$$

where $1 = l_0 < l_1 < \cdots < l_{J+1} = n + 1$. *Then Algorithms 3.1, 3.2 and 3.3 are the block Jacobi method, block diagonal preconditioner and block Gauss–Seidel methods respectively.*

Let $T_i = R_i Q_i A$. By (3.4), $T_i = R_i A_i P_i$. Note that $T_i : \mathcal{V} \mapsto \mathcal{V}_i$ is symmetric with respect to $(\cdot,\cdot)_A$ and nonnegative and that $T_i = P_i$ if $R_i = A_i^{-1}$.

If u is the exact solution of (2.1), then $f = Au$. Let v^i be the i-th iterate (with $v^0 = u^k$) from Algorithm 3.3. We have by definition

$$u - v^{i+1} = (I - T_i)(u - v^i), \quad i = 0, \cdots, J.$$

A successive application of this identity yields

$$u - u^{k+1} = E_J(u - u^k), \tag{3.7}$$

where

$$E_J = (I - T_J)(I - T_{J-1}) \cdots (I - T_1)(I - T_0). \tag{3.8}$$

Remark 3.1 It is interesting to look at the operator E_J in the special case that $R_i = \omega A_i^{-1}$ for all i. The corresponding SSC iteration is a generalization of the classic SOR method. In this case, we have

$$E_J = (I - \omega P_J)(I - \omega P_{J-1}) \cdots (I - \omega P_1)(I - \omega P_0).$$

One trivial fact is that E_J is invertible when $\omega \neq 1$. Following an argument by Nicolaides [27] for the SOR method, let us take a look at the special case $\omega = 2$. Since, obviously, $(I - 2P_i)^{-1} = I - 2P_i$ for each i, we conclude that $E_J^{-1} = E_J^*$ where $*$ is the adjoint with respect to the inner product $(\cdot,\cdot)_A$. This means that E_J is an orthogonal operator and, in particular, $\|E_J\|_A = 1$. As a consequence, the SSC iteration cannot converge when $\omega = 2$. In fact, as we shall see in Proposition 3.16 below, in this special case, the SSC method converges if and only if $0 < \omega < 2$.

The symmetrization of Algorithm 3.3 can also be implemented as follows.

Algorithm 3.4 *Given* $u^0 \in \mathcal{V}$, $v \leftarrow u^0$
for $k = 0, 1, \ldots$ till convergence
for $i = 0 : J$ and $i = J : -1 : 0$ $\quad v \leftarrow v + R_i Q_i (f - Av)$ endfor
endfor

The advantage of the symmetrized algorithm is that it can be used as a preconditioner. In fact, Algorithm 3.4 can be formulated in the form of (2.2) with operator B defined as follows: for $f \in \mathcal{V}$, let $Bf = u^1$ with u^1 obtained by Algorithm 3.4 applied to (2.1) with $u^0 = 0$.

(3.9) Colorization and parallelization of SSC iteration.

Definition 3.6 *Associated with a given partition (3.1), a coloring of the set* $\mathcal{J} = \{0, 1, 2, \ldots, J\}$ *is a disjoint decomposition:*

$$\mathcal{J} = \bigcup_{t=1}^{J_c} \mathcal{J}(t)$$

such that

$$P_i P_j = 0 \text{ for any } i, j \in \mathcal{J}(t), i \neq j \ (1 \leq t \leq J_c).$$

We say that i, j have the same color if they both belong to some $\mathcal{J}(t)$.

The important property of the coloring is that the SSC iteration can be carried out in parallel in each color.

Algorithm 3.5. (Colored SSC) *Given $u^0 \in \mathcal{V}$, $v \leftarrow u^0$*
for $k = 0, 1, \ldots$ till convergence
  for $t = 1 : J_c$ $\quad v \leftarrow v + \sum_{i \in \mathcal{J}(t)} R_i Q_i (f - Av)$ $\quad$ endfor
endfor

We note that the terms under the sum in the above algorithm can be evaluated in parallel (for each t, namely within the same color).

3.3 Convergence theory

The purpose of this section is to establish an abstract theory for algorithms described in previous sections.

In view of Theorem 2.1, it suffices to study Algorithms 3.2 and 3.3. Two fundamental theorems will be presented.

For the preconditioner of Algorithm 3.2, we need to estimate the condition number of

$$T = BA = \sum_{i=0}^{J} T_i,$$

where B is defined by (3.6) and $T_i = R_i A_i P_i$.

It is interesting to note the following special case:

$$BA = \sum_{i=0}^{J} P_i, \text{ if } R_i = A^{-1}.$$

For Algorithm 3.3, we need to establish the contraction property: there exists a constant $0 < \delta < 1$ such that

$$\|E_J\|_A \leq \delta \quad \text{with} \quad \|E_J\|_A = \sup_{v \in \mathcal{V}} \frac{\|E_J v\|_A}{\|v\|_A},$$

where E_J is given by (3.8). Applying this estimate to (3.7) yields

$$\|u - u^k\|_A \leq \delta^k \|u - u^0\|_A.$$

3.3.1 *Important parameters*

The convergence theory here is to be built upon several parameters associated with the space decomposition and subspace solvers.

Parameter ω_1 The first constant, named ω_1, is the smallest constant satisfying

$$(T_i v, T_i v)_A \leq \omega_1 (T_i v, v)_A \quad \forall\, v \in \mathcal{V},\ 0 \leq i \leq J, \tag{3.10}$$

or equivalently

$$(v_i, A_i v_i) \le \omega_1 (R_i^{-1} v_i, v_i) \quad \forall\, v \in \mathcal{V},\ 0 \le i \le J. \tag{3.11}$$

We assume that R_i is chosen in such a way that ω_1 is well defined. If all R_i are SPD, then ω_1 is obviously well defined and in fact

$$\omega_1 = \max_{0 \le i \le J} \rho(R_i A_i) = \max_{0 \le i \le J} \rho(T_i).$$

The constant ω_1 is, in most cases, very easy to estimate and its boundedness often comes as an assumption. For example, while all the subspace solvers are exact, namely $R_i = A_i^{-1}$, then $\omega_1 = 1$. As we shall see later, the convergence of an SSC method is assured if the following condition holds:

$$\omega_1 < 2.$$

This condition is equivalent to saying that the symmetrized schemes for all R_i are convergent schemes (see Theorem 2.1) and in particular the iterative schemes given by all R_i are convergent schemes.

Parameters K_0 and $\bar{K}_0$ The parameter K_0 to be introduced now plays the most crucial rôle in most applications and it is also most difficult to estimate in applications. It measures the correlation between space decomposition and the choice of subspace solvers. We define

$$K_0 = \sup_{\|v\|_A = 1} \inf_{\substack{v_i \in \mathcal{V}_i \\ \sum v_i = v}} \sum_i (R_i^{-1} v_i, v_i).$$

and

$$\bar{K}_0 = \sup_{\|v\|_A = 1} \inf_{\substack{v_i \in \mathcal{V}_i \\ \sum v_i = v}} \sum_i (\bar{R}_i^{-1} v_i, v_i).$$

In other words, for any $v \in \mathcal{V}$, there exists a decomposition $v = \sum_{i=0}^J v_i$ for $v_i \in \mathcal{V}_i$ such that

$$\sum_{i=0}^{J} (R_i^{-1} v_i, v_i) \le K_0 (Av, v). \tag{3.12}$$

Lemma 3.7 *Assume, for any $v \in \mathcal{V}$, there is a decomposition $v = \sum_{i=0}^J v_i$ with $v_i \in \mathcal{V}_i$ satisfying*

$$\sum_{i=0}^{J} (v_i, v_i)_A \le C_0 (v, v)_A; \tag{3.13}$$

then

$$\bar{K}_0 \le \frac{C_0}{\bar{\omega}_0}, \quad \text{with} \quad \bar{\omega}_0 = \min_{0 \le i \le J} \lambda_{\min}(\bar{R}_i A_i)$$

and, if all R_i are SPD,

$$K_0 \leq \frac{C_0}{\omega_0} \quad \text{with } \omega_0 = \min_{0 \leq i \leq J} \lambda_{\min}(R_i A_i).$$

The above lemma is most useful in domain decomposition applications. A good upper bound for K_0 relies on a good lower bound for ω_0, which means that each subspace solver R_i should resolve the whole range of the spectrum of A_i. In other words, the subspace problems should be very well solved or preconditioned.

The constant C_0 in (3.13) only depends on the partition (decomposition) of the space and it is sometimes called the *partition constant.*

Lemma 3.8 *Assume, for any $v \in \mathcal{V}$, that there is a decomposition $v = \sum_{i=0}^J v_i$ with $v_i \in \mathcal{V}_i$ satisfying*

$$\sum_{i=0}^{J} \rho(A_i)(v_i, v_i) \leq \hat{C}_0(v, v)_A;$$

then

$$K_0 \leq \frac{\hat{C}_0}{\breve{\omega}_0} \quad \text{with} \quad \breve{\omega}_0 = \min_{0 \leq i \leq J} (\lambda_{\min}(\bar{R}_i)\rho(A_i)),$$

and, if all R_i are SPD,

$$K_0 \leq \frac{\hat{C}_0}{\hat{\omega}_0} \quad \text{with} \quad \hat{\omega}_0 = \min_{0 \leq i \leq J} (\lambda_{\min}(R_i)\rho(A_i)).$$

The above lemma is most useful in multigrid applications. A good upper bound of K_0 relies on a good lower bound of $\hat{\omega}_0$, which means that each subspace solver R_i only needs to resolve the 'upper' range of the spectrum of A_i. In other words, each subspace solver R_i should be spectrally equivalent to $(\rho(A_i))^{-1}$.

Parameters K_1 and $\bar{K}_1$ These parameters measure the interaction among subspaces together with the subspace solvers.

If each R_i is SPD, we define $\epsilon_{ij} \in (0, 1]$, for $j < i$, by

$$\epsilon_{ij}^2 = \rho(P_j T_i P_j)/\omega_1 \quad \text{and} \quad \epsilon_{ji} = \epsilon_{ij}, \quad \epsilon_{ii} = 1. \tag{3.14}$$

We also define $\bar{\epsilon}_{ij} \in (0, 1]$, for $j < i$, by

$$\bar{\epsilon}_{ij}^2 = \rho(P_j \bar{T}_i P_j) \quad \text{and} \quad \epsilon_{ji} = \epsilon_{ij}, \quad \epsilon_{ii} = 1, \tag{3.15}$$

where with $\bar{R}_i$ being the symmetrization of R_i (see eqn. (2.4)),

$$\bar{T}_i = \bar{R}_i A_i P_i. \tag{3.16}$$

Note that, for each $i \geq j$, ϵ_{ij} and $\bar{\epsilon}_{ij}$ are the smallest numbers satisfying

$$(T_i v_j, v_j)_A \leq \omega_1 \epsilon_{ij}^2 (v_j, v_j)_A, \quad (\bar{T}_i v_j, v_j)_A \leq \bar{\epsilon}_{ij}^2 (v_j, v_j)_A \quad \forall\, v_j \in \mathcal{V}_j.$$

Lemma 3.9 *If each R_i is SPD, then*

$$(T_i u, T_j v)_A \leq \omega_1 \epsilon_{ij} (T_i u, u)_A^{\frac{1}{2}} (T_j v, v)_A^{\frac{1}{2}} \quad \forall\, u, v \in \mathcal{V}; \tag{3.17}$$

Proof Without loss of generality, we may assume that $i \geq j$. It follows from the Cauchy–Schwarz inequality that

$$\begin{aligned} (T_i u, T_j v)_A &\leq (T_i u, u)_A^{\frac{1}{2}} (T_i T_j v, T_j v)_A^{\frac{1}{2}} \\ &\leq \sqrt{\omega_1} \epsilon_{ij} (T_i u, u)_A^{\frac{1}{2}} (T_j v, T_j v)_A^{\frac{1}{2}} \\ &\leq \omega_1 \epsilon_{ij} (T_i u, u)_A^{\frac{1}{2}} (T_j v, v)_A^{\frac{1}{2}}. \end{aligned}$$

□

Remark 3.2 Clearly $\epsilon_{ij} \leq 1$ and $\epsilon_{ij} = 0$ if $P_i P_j = 0$. If $\epsilon_{ij} < 1$, the inequality (3.17) is often known as the *strengthened Cauchy–Schwarz inequality.*

Definition 3.10

$$K_1 = \min_{\mathcal{J}_0 \subset \{0:J\}} \left(|\mathcal{J}_0| + \max_{i \in \mathcal{J}_0^c} \sum_{j \in \mathcal{J}_0^c} \epsilon_{ij} \right)$$

and

$$\bar{K}_1 = \min_{\mathcal{J}_0 \subset \{0:J\}} \left(|\mathcal{J}_0| + \max_{i \in \mathcal{J}_0^c} \sum_{j \in \mathcal{J}_0^c} \bar{\epsilon}_{ij} \right).$$

Roughly speaking, K_1 is bounded if the matrix (ϵ_{ij}) is sparse except for a few rows and columns.

Lemma 3.11 *The parameter K_1 admits the following estimates:*

1. $K_1 \leq J + 1$;
2. $K_1 \leq 1 + \rho((\epsilon_{ij})_{i,j=1:J}) \leq 1 + \max_{1 \leq i \leq J} \sum_{j=1}^{n} \epsilon_{ij}$;
3. *if* $\epsilon_{ij} \lesssim \gamma^{|i-j|}$ *or* $\bar{\epsilon}_{ij} \lesssim \gamma^{|i-j|}$ *for some* $\gamma \in (0,1)$, *then*

$$\hat{K}_1 \lesssim \frac{1}{1-\gamma} \quad or \quad \bar{K}_1 \lesssim \frac{1}{1-\gamma}.$$

Lemma 3.12

$$\sum_{i>j} (\bar{T}_i u_i, T_j v_j)_A \leq (\bar{K}_1 - 1) \left(\sum_{i=0}^{J} (\bar{T}_i u_i, u_i)_A \right)^{1/2} \left(\sum_{j=0}^{J} (T_j v_j, T_j v_j)_A \right)^{1/2}.$$

If each R_i is SPD, then

$$\sum_{i>j}(T_iu_i, T_jv_j)_A \leq \omega_1(K_1-1)\Big(\sum_{i=0}^{J}(T_iu_i,u_i)_A\Big)^{1/2}\Big(\sum_{j=0}^{J}(T_jv_j,v_j)_A\Big)^{1/2}.$$

If each R_i is SPD, then for any $S \subset \{0:J\}\times\{0:J\}$,

$$\sum_{i,j\in S}(T_iu_i, T_jv_j)_A \leq \omega_1 K_1\Big(\sum_{i=0}^{J}(T_iu_i,u_i)_A\Big)^{1/2}\Big(\sum_{j=0}^{J}(T_jv_j,v_j)_A\Big)^{1/2}.$$

3.3.2 *Convergence theory*

With the parameters ω_1, K_0 and K_1 introduced above, the convergence estimates for the PSC and SSC methods can be neatly presented. The analysis for the PSC preconditioner is relatively easy whereas the analysis for SSC iteration is less straightforward.

We first give a lower bound for the spectrum of the PSC preconditioner.

Lemma 3.13 *Assume that all R_i are SPD. The PSC preconditioner B given by (3.6) satisfies*

$$\lambda_{\min}(BA) = K_0^{-1}.$$

Proof If $v=\sum_{i=0}^{J} v_i$ is a decomposition that satisfies (3.12), then

$$(v,v)_A = \sum_{i=0}^{J}(v_i,v)_A = \sum_{i=0}^{J}(v_i,P_iv)_A,$$

and by the Cauchy–Schwarz inequality

$$\begin{aligned}\sum_{i=0}^{J}(v_i,P_iv)_A &= \sum_{i=0}^{J}(v_i,A_iP_iv) \leq \sum_{i=0}^{J}(R_i^{-1}v_i,v_i)^{\frac12}(R_iA_iP_iv,v)_A^{\frac12}\\ &\leq \Big(\sum_{i=0}^{J}(R_i^{-1}v_i,v_i)\Big)^{\frac12}\Big(\sum_{i=0}^{J}(T_iv,v)_A\Big)^{\frac12} \leq \sqrt{K_0}\|v\|_A(Tv,v)_A^{\frac12}.\end{aligned}$$

Consequently

$$\|v\|_A^2 \leq K_0(Tv,v)_A.$$

This implies that $\lambda_{\min}(BA)\geq K_0^{-1}$.

Now for $v=\sum_{i=0}^{J} v_i$ with $v_i = T_iT^{-1}v$, we have

$$\begin{aligned}K_0 &\leq \max_{v\in\mathcal{V}}\frac{\sum_{i=0}^{J}(R_i^{-1}T_iT^{-1}v, T_iT^{-1}v)}{\|v\|_A^2}\\ &= \max_{v\in\mathcal{V}}\frac{(T^{-1}v,v)_A}{(v,v)_A} = (\lambda_{\min}(BA))^{-1}.\end{aligned}$$

The desired estimate then follows. □

Remark 3.3 It is easy to see from the above proof that the following slightly more general identity also holds:

$$(B^{-1}v, v) = \inf_{\substack{v_i \in \mathcal{V}_i \\ \sum v_i = v}} \sum_i (R_i^{-1} v_i, v_i) \tag{3.18}$$

which apparently implies Lemma 3.13.

Theorem 3.14 *Assume all R_i are SPDE. The PSC preconditioner B given by (3.6) satisfies*

$$\lambda_{\min}(BA) = K_0^{-1} \quad \textit{and} \quad \lambda_{\max}(BA) \le \omega_1 K_1,$$

and

$$\kappa(BA) \le \omega_1 K_0 K_1.$$

And in view of Lemmas 3.7 and 3.8,

$$\kappa(BA) \le \frac{\omega_1}{\omega_0} C_0 K_1, \quad \kappa(BA) \le \frac{\omega_1}{\hat{\omega}_0} \hat{C}_0 K_1,$$

Proof By Lemma 3.12,

$$\|Tv\|_A^2 = \sum_{i,j=0}^{J} (T_i v, T_j v)_A \le K_1 (Tv, v)_A \le K_1 \|Tv\|_A \|v\|_A,$$

which implies that $\lambda_{\max}(BA) \le K_1$. □

To present our next theorem, let us first prove a very simple but important lemma.

Lemma 3.15 *Denote $E_{-1} = I$ and for $0 \le i \le J$,*

$$E_i = (I - T_i)(I - T_{i-1}) \cdots (I - T_1)(I - T_0).$$

Then

$$I - E_i = \sum_{j=0}^{i} T_j E_{j-1}, \tag{3.19}$$

and for any $v \in \mathcal{V}$,

$$\|v\|_A^2 - \|E_J v\|_A^2 = \sum_{i=0}^{J} (\bar{T}_i E_{i-1} v, E_{i-1} v)_A \tag{3.20}$$

where $\bar{T}_i$ is given by (3.16). Furthermore if each R_i is symmetric then

$$\|v\|_A^2 - \|E_J v\|_A^2 \ge (2 - \omega_1) \sum_{i=0}^{J} (T_i E_{i-1} v, E_{i-1} v)_A. \tag{3.21}$$

Proof The identity (3.19) follows immediately from the trivial identity

$E_{i-1} - E_i = T_i E_{i-1}$. Similarly to (2.7) and (2.6), we have

$$\|E_{i-1}v\|_A^2 - \|E_i v\|_A^2 = ((2I - T_i)E_{i-1}v, T_i E_{i-1}v)_A = (\bar{T}_i E_{i-1}v, E_{i-1}v)_A.$$

Summing up these inequalities with respect to i gives (3.20). The estimate (3.21) follows by combining (3.20) and (2.12). □

Again let us take a look at the special case that $R_i = \omega A_i^{-1}$ for each i. In this case, we have

$$\|v\|_A^2 - \|E_J v\|_A^2 = \omega(2-\omega)\sum_{i=0}^{J}\|P_i E_{i-1}v\|_A^2.$$

This identity implies immediately that a necessary condition for the convergence of the related SSS method is that $0 < \omega < 2$. In fact, as for the SOR method, it is not hard to see that this condition is also sufficient for the convergence (see Corollary 3.19 below). Thus, we have the following simple generalization of a classic result for the SOR method (see also Remark 3.1).

Proposition 3.16 *The SSC method with $R_i = \omega A_i^{-1}$ for each i converges if and only if $0 < \omega_1 < 2$.*

Lemma 3.17 *Assume that $\omega_1 < 2$. If each R_i is SPD, then*

$$\sum_{i=0}^{J}(T_i v, v)_A \le (1+K_1)^2 \sum_{i=0}^{J}(T_i E_{i-1}v, E_{i-1}v)_A \quad \forall\, v \in \mathcal{V}, \tag{3.22}$$

and in general

$$\sum_{i=0}^{J}(\bar{T}_i v, v)_A \le \left(1 + \sqrt{\frac{\omega_1}{2-\omega_1}}(\bar{K}_1 - 1)\right)^2 \sum_{j=0}^{J}(\bar{T}_j E_{j-1}v, E_{j-1}v)_A. \tag{3.23}$$

Proof By (3.19)

$$\begin{aligned}(T_i v, v)_A &= (T_i v, E_{i-1}v)_A + (T_i v, (I - E_{i-1})v)_A \\ &= (T_i v, E_{i-1}v)_A + \sum_{j=0}^{i-1}(T_i v, T_j E_{j-1}v)_A.\end{aligned}$$

Applying the Cauchy–Schwarz inequality gives

$$\sum_{i=0}^{J}(T_i v, E_{i-1}v)_A \le \left(\sum_{i=0}^{J}(T_i v, v)_A\right)^{\frac{1}{2}} \left(\sum_{i=0}^{J}(T_i E_{i-1}v, E_{i-1}v)_A\right)^{\frac{1}{2}},$$

and, by Lemma 3.12,

$$\sum_{i=0}^{J}\sum_{j=0}^{i-1}(T_i v, T_j E_{j-1}v)_A$$

$$\leq \omega_1(K_1-1)\left(\sum_{i=0}^{J}(T_i v, v)_A\right)^{\frac{1}{2}}\left(\sum_{j=0}^{J}(T_j E_{j-1}v, E_{j-1}v)_A\right)^{\frac{1}{2}}.$$

Combining these three formulae then leads to (3.22) and hence completes the proof for (3.23).

With arguments similar to the above (essentially by replacing T_i by $\bar{T}_i$ in the above proof), it is easy to obtain that

$$\sum_{i=0}^{J}(\bar{T}_i v, v)_A \leq \left(\sum_{i=0}^{J}(\bar{T}_i v, v)_A\right)^{1/2}\left(\sum_{j=0}^{J}(\bar{T}_j E_{j-1}v, E_{j-1}v)_A\right)^{1/2}$$

$$+(\bar{K}_1-1)\left(\sum_{i=0}^{J}(\bar{T}_i v, v)_A\right)^{\frac{1}{2}}\left(\sum_{j=0}^{J}(T_j E_{j-1}v, T_j E_{j-1}v)_A\right)^{\frac{1}{2}}.$$

After cancelling the common factor and using the following inequalities (see (2.9) and (2.8)):

$$(T_j w, w) \leq (2-\omega_1)^{-1}(\bar{T}_j w, w), \quad (T_j w, T_j w) \leq \omega_1(2-\omega_1)^{-1}(\bar{T}_j w, w),$$

the estimate (3.23) then follows easily. □

Now we are in a position to present our second fundamental theorem.

Theorem 3.18 *Assume that $\omega_1 < 2$. If each R_i is SPD, then the iterator E_J (given by (3.8)) for the Algorithm 3.3 satisfies*

$$\|E_J\|_A^2 \leq 1 - \frac{2-\omega_1}{K_0(1+\omega_1(K_1-1))^2}; \qquad (3.24)$$

and, in general,

$$\|E_J\|_A^2 \leq 1 - \frac{2-\omega_1}{\bar{K}_0(\sqrt{2-\omega_1}+\sqrt{\omega_1}(\bar{K}_1-1))^2} \qquad (3.25)$$

Proof The estimate in (3.24) is obviously equivalent to

$$\|v\|_A^2 \leq \frac{K_0(1+K_1)^2}{2-\omega_1}(\|v\|_A^2 - \|E_J v\|_A^2) \quad \forall\, v \in \mathcal{V}.$$

Estimate (3.24) then follows by combining (3.22) with (3.20) and (2.9).

The second estimate (3.25) then follows by combining (3.23) with the fact that $\lambda_{\min}(\sum_i \bar{T}_i) = \bar{K}_0^{-1}$ (similar to Lemma 3.13). □

As a direct consequence of the above theorem, we have the following simple result.

Corollary 3.19 *A sufficient condition for the convergence of the SSC method is that*

$$\omega_1 < 2. \tag{3.26}$$

The condition (3.26) is also necessary in some sense; see Proposition 3.16.

Remark 3.4 Note that the convergence estimate in Theorem 3.18 is independent of the order in which Algorithm 3.3 is executed. Namely, if we shuffle the order in the decomposition (3.1), the corresponding estimate in Theorem 3.18 remains unchanged.

Theorem 3.20 *Under the assumptions in Lemma 3.7,*

$$\kappa(BA) \le \frac{\omega_1}{\omega_0} C_0 K_1$$

and

$$\|E_J\|_A^2 \le \begin{cases} 1 - \dfrac{(2-\omega_1)\omega_0}{C_0(1+\omega_1(K_1-1))^2} & \text{if each } R_i \text{ is SPD} \\[2ex] 1 - \dfrac{\bar{\omega}_0}{C_0(\sqrt{2-\omega_1}+\sqrt{\omega_1}(K_1-1))^2} & \text{otherwise.} \end{cases}$$

Theorem 3.21 *Under the assumptions in Lemma 3.8,*

$$\kappa(BA) \le \frac{\omega_1}{\hat{\omega}_0} \hat{C}_0 \; K_1$$

and

$$\|E_J\|_A^2 \le \begin{cases} 1 - \dfrac{\hat{\omega}_0}{\hat{C}_0(1+\omega_1(K_1-1))^2} & \text{if each } R_i \text{ is SPD} \\[2ex] 1 - \dfrac{(2-\omega_1)\breve{\omega}_0}{\hat{C}_0\sqrt{2-\omega_1}+\sqrt{\omega_1}(\bar{K}_1-1))^2} & \text{otherwise.} \end{cases}$$

3.4 Matrix representations of PSC and SSC methods

The PSC and SSC have been presented above in terms of projections and operators in abstract vector spaces. We shall now translate all these algorithms into explicit algebraic forms by using the simple techniques in §2.3.

For each k, let $\mathcal{I}_k \in \mathbb{R}^{n \times n_k}$ be the matrix representation of the natural inclusion $I_k : \mathcal{V}_k \mapsto \mathcal{V}$; to derive the algebraic representation of the preconditioner (3.6), we rewrite it in a slightly different form

$$B = \sum_{k=0}^{J} I_k R_k Q_k.$$

Applying (2.16) and the easily verifiable identity $\widetilde{Q}_k = \mathcal{M}_k^{-1}\mathcal{I}_k^t\mathcal{M}$ gives

$$\widetilde{B} = \sum_{k=0}^{J} \widetilde{I}_k \widetilde{R}_k \widetilde{Q}_k = \sum_{k=0}^{J} \mathcal{I}_k(\mathcal{R}_k\mathcal{M}_k)(\mathcal{M}_k^{-1}\mathcal{I}_k^t\mathcal{M}) = \mathcal{B}\mathcal{M}.$$

Here $\mathcal{R}_k$ is the algebraic representation of R_k and

$$\mathcal{B} = \sum_{k=0}^{J} \mathcal{I}_k \mathcal{R}_k \mathcal{I}_k^t. \tag{3.27}$$

Different choices of R_k yield the following three main different preconditioners:

$$\mathcal{B} = \begin{cases} \sum_{k=0}^{J} \rho(\mathcal{A}_k)^{-1}\mathcal{I}_k\mathcal{I}_k^t & \text{Richardson;} \\ \sum_{k=0}^{J} \mathcal{I}_k\mathcal{D}_k^{-1}\mathcal{I}_k^t & \text{Jacobi;} \\ \sum_{k=0}^{J} \mathcal{I}_k\mathcal{G}_k\mathcal{I}_k^t & \text{Gauss–Seidel.} \end{cases}$$

Here $\mathcal{G}_k = (\mathcal{D}_k - \mathcal{U}_k)^{-1}\mathcal{D}_k(\mathcal{D}_k - \mathcal{L}_k)^{-1}$, $\mathcal{A}_k = \mathcal{D}_k - \mathcal{L}_k - \mathcal{U}_k$, $\mathcal{D}_k$ is the diagonal of $\mathcal{A}_k$, $-\mathcal{L}_k$ and $-\mathcal{U}_k$ are, respectively, the lower and upper triangular parts of $\mathcal{A}_k$.

Following (2.4), we get

Proposition 3.22 *The PSC preconditioner for the stiffness matrix $\mathcal{A}$ is given by (3.27) and $\kappa(\mathcal{B}\mathcal{A}) = \kappa(BA)$.*

Similarly, we can derive the algebraic representation of Algorithm 3.3 for solving (2.18).

Algorithm 3.6 *$\mu^0 \in \mathbb{R}^n$ is given. Assume that $\mu^k \in \mathbb{R}^n$ is obtained. Then μ^{k+1} is defined by*

$$\mu^{k+i/J} = \mu^{k+(i-1)/J} + \mathcal{I}_i\mathcal{R}_i\mathcal{I}_i^t(\eta - \mathcal{A}\mu^{k+(i-1)/J})$$

for $i = 0 : J$.

4 Finite element approximations

In the following sections, we shall introduce the multigrid methods. Our presentations will be confined to a second order elliptic model problem with the linear finite element discretization.

This section is devoted to some basic properties of finite element spaces that will be used for the analysis of multigrid algorithms.

4.1 A model problem and finite element discretization

We consider the boundary value problem:

$$\begin{aligned} -\nabla \cdot a\nabla U &= F \quad \text{in } \Omega, \\ U &= 0 \quad \text{on } \partial\Omega, \end{aligned} \tag{4.1}$$

where $\Omega \subset \mathbb{R}^d$ is a polyhedral domain and a is a smooth function (or piecewise smooth) on $\bar{\Omega}$ with a positive lower bound.

Let $H^1(\Omega)$ be the standard Sobolev space consisting of square integrable functions with square integrable (weak) derivatives of first order, and $H_0^1(\Omega)$ the subspace of $H^1(\Omega)$ consisting of functions that vanish on $\partial\Omega$. Then $U \in H_0^1(\Omega)$ is the solution of (4.1) if and only if

$$A(U, \chi) = (F, \chi) \quad \forall \chi \in H_0^1(\Omega), \tag{4.2}$$

where

$$A(U, \chi) = \int_\Omega a\nabla U \cdot \nabla \chi dx, \quad (F, \chi) = \int_\Omega F\chi dx.$$

Introduce the fractional order Sobolev spaces

$$H^{m+\sigma}(\Omega) \ \ (m \geq 0, 0 < \sigma < 1)$$

defined by the completion of smooth functions in the following norm:

$$||v||_{H^{m+\sigma}(\Omega)} = \left(||v||^2_{H^m(\Omega)} + |v|^2_{H^{m+\sigma}(\Omega)}\right)^{\frac{1}{2}},$$

where

$$|v|^2_{H^{m+\sigma}(\Omega)} = \sum_{|\alpha|=m} \int_\Omega \int_\Omega \frac{|D^\alpha v(x) - D^\alpha v(y)|^2}{|x-y|^{d+2\sigma}} dx\, dy.$$

It is well known that there exists a constant $\alpha \in (0, 1]$ such that

$$||U||_{H^{1+\alpha}(\Omega)} \leq C||F||_{H^{\alpha-1}(\Omega)}, \tag{4.3}$$

for the solution U of (4.2), where C is a constant depending on the domain Ω and the coefficient $a(x)$.

Assume that Ω is triangulated with $\Omega = \cup_i \tau_i$, where the τ_i are nonoverlapping simplices of size $h \in (0, 1]$ and are quasi-uniform, i.e. there exist constants C_0 and C_1 not depending on h such that each simplex τ_i is contained in (contains) a ball of radius $C_1 h$ (respectively $C_0 h$). Define

$$\mathcal{V} = \{v \in H_0^1(\Omega) : v|_{\tau_i} \in \mathcal{P}_1(\tau_i), \quad \forall\, \tau_i\},$$

where $\mathcal{P}_1$ is the space of linear polynomials.

We shall now mention some properties of the finite element space. For any $v \in \mathcal{V}$, we have

$$\|v\|_{L^\infty(\Omega)} \lesssim h^{-d/p}\|v\|_{L^p(\Omega)}, p \geq 1, \tag{4.4}$$
$$\|v\|_{H^1(\Omega)} \lesssim h^{-1}\|v\|, \tag{4.5}$$
$$\|v\|_{H^{1+\sigma}(\Omega)} \lesssim h^{-\sigma}\|v\|_{H^{H^1(\Omega)}(\Omega)} \quad \sigma \in (0, \frac{1}{2}), \tag{4.6}$$
$$\|v\|_{H^s(\Omega)} \lesssim h^{t-s}\|v\|_{H^t(\Omega)} \quad s,t \in [0,1],\ t \leq s, \tag{4.7}$$
$$\|v\|_{L^\infty(\Omega)} \lesssim c_d(h)\|v\|_{H^1(\Omega)}, \tag{4.8}$$

where $c_1(h) = 1, c_2(h) = |\log h|^{\frac{1}{2}}$ and $c_d(h) = h^{\frac{2-d}{2}}$ for $d \geq 3$. The *inverse* inequalities (4.4) and (4.5) can be found, for example, in Ciarlet [18] and a proof of the discrete Sobolev inequality (4.8) can be found in Bramble and Xu [11]. A proof of (4.6) and (4.7) may be found in Bramble, Pasciak and Xu [10] and Xu [32].

Theorem 4.1 *Assume that $P_h : H_0^1(\Omega) \mapsto \mathcal{V}$ is the Galerkin projection with respect to $A(\cdot,\cdot)$. Then*

$$\|(I - P_h)u\|_{H^{1-\alpha}(\Omega)} \lesssim h^\alpha \|u\|_{H^1(\Omega)} \quad \forall\, u \in H_0^1(\Omega), \tag{4.9}$$

and

$$\|(I - P_h)u\|_{H^1(\Omega)} \lesssim h^s\|u\|_{H^{1+s}(\Omega)}, \quad \forall\, u \in H_0^1(\Omega) \cap H^{1+s}(\Omega),\ 0 \leq s \leq \alpha \tag{4.10}$$

where α is as in (4.3).

Defining the L^2 projection $Q_h : L^2(\Omega) \mapsto \mathcal{V}$ by

$$(Q_h v, \chi) = (v, \chi), \quad \forall\, v \in L^2(\Omega), \chi \in \mathcal{V},$$

we have

$$\|v - Q_h v\| + h\|c - Q_h v\|_{H^1(\Omega)} \lesssim Ch\|v\|_{H^1(\Omega)}. \tag{4.11}$$

This estimate is well known; we refer to [32, 11] for a rigorous proof and related results.

By interpolation, we have (for $\sigma \in (0, \frac{1}{2})$)

$$\|Q_h v\|_{H^\sigma(\Omega)} \lesssim \|v\|_{H^\sigma(\Omega)} \quad \forall\, v \in H_0^1(\Omega). \tag{4.12}$$

and

$$\|v - Q_h v\|_{H^{1-\alpha}(\Omega)} \lesssim h^\alpha\|v\|_{H^1(\Omega)} \quad \forall\, v \in H_0^1(\Omega). \tag{4.13}$$

The finite element approximation to the solution of (4.1) is the function $u \in \mathcal{V}$ satisfying

$$A(u, v) = (F, v) \quad \forall v \in \mathcal{V}. \tag{4.14}$$

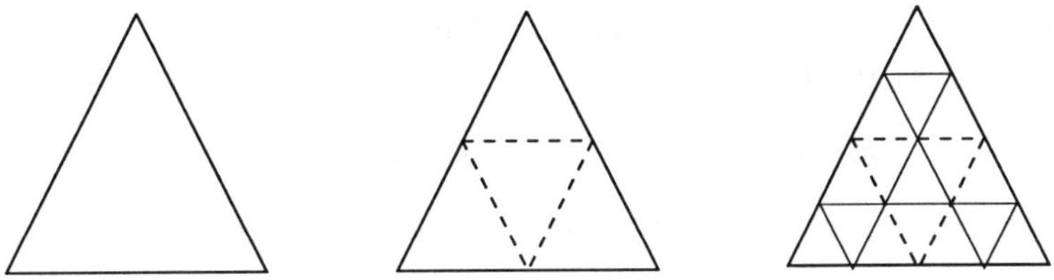

FIG. 1. Typical multilevel grids.

Define a linear operator $A : \mathcal{V} \mapsto \mathcal{V}$ by

$$(Au, v) = A(u, v), \quad u, v \in \mathcal{V}. \tag{4.15}$$

The equation (4.14) is then equivalent to (2.1) with $f = Q_h F$. The space $\mathcal{V}$ has a natural (nodal) basis $\{\varphi_i\}_{i=1}^n$ ($n = \dim\mathcal{V}$) satisfying

$$\varphi_i(x_l) = \delta_{il} \quad \forall\, i, l = 1, \ldots, n,$$

where $\{x_l : l = 1, \ldots, n\}$ is the set of all interior nodal points of $\mathcal{V}$. By means of these nodal basis functions, the solution of (4.14) is reduced to solving an algebraic system (2.18) with $\mathcal{A} = ((a\nabla\varphi_i, \nabla\varphi_l))_{n\times n}$ and $\eta = ((f, \varphi_i)_{n\times 1})$.

It is well known that, for all $\nu \in \mathbb{R}^n$,

$$h^d|\nu|^2 \lesssim \nu^t \mathcal{A}\nu \lesssim h^{d-2}|\nu|^2 \quad \text{and} \quad h^d|\nu|^2 \lesssim \nu^t \mathcal{M}\nu \lesssim h^d|\nu|^2. \tag{4.16}$$

Hence $\kappa(\mathcal{A}) \lesssim h^{-2}$ and $\kappa(\mathcal{M}) \eqsim 1$.

4.2 Finite element spaces on multiple levels

This section is to study the interaction between finite element spaces with different scales. We assume that Ω has been triangulated with a nested sequence of quasi-uniform triangulations $\mathcal{T}_k = \{\tau_k^i\}$ of size h for $k = 0, \ldots, j$ where the quasi-uniformity constants are independent of k. These triangulations should be nested in the sense that any triangle τ_{k-1}^l can be written as a union of triangles of $\{\tau_k^i\}$ (see Fig. 1). We further assume that there is a constant $\eta > 1$, independent of k, such that

$$h_k \eqsim \eta^{-k}.$$

Associated with each $\mathcal{T}_k$, a finite element space $\mathcal{M}_k \subset H_0^1(\Omega)$ can be defined. One has

$$\mathcal{M}_0 \subset \mathcal{M}_1 \subset \ldots \subset \mathcal{M}_k \subset \ldots \subset \mathcal{M}_J. \tag{4.17}$$

For each k, we define the interpolant $I_k : C(\bar{\Omega}) \mapsto \mathcal{M}_k$ by

$$(I_k u)(x) = u(x) \quad \forall\, x \in \mathcal{N}_k.$$

Here $\mathcal{N}_k$ is the set of all nodes in $\mathcal{T}_k$.

Let $Q_k, P_k : H_0^1(\Omega) \mapsto \mathcal{M}_k$ be the L^2 and H^1 projection defined, for all $u \in H_0^1(\Omega), v_k \in \mathcal{M}_k$, by

$$(Q_k u, v_k) = (u, v_k), \quad (\nabla P_k u, \nabla v_k) = (\nabla u, \nabla v_k). \tag{4.18}$$

Lemma 4.2 *Let R_k be any one of I_k, Q_k or P_k. Then*

(1) $R_i R_j = R_{i\wedge j}$, *where* $i \wedge j = \min(i, j)$.
(2) $(R_i - R_{i-1})(R_j - R_{j-1}) = 0$ *if* $i \neq j$.
(3) $(R_k - R_{k-1})^2 = R_k - R_{k-1} = (I - R_{k-1})R_k$.

Lemma 4.3

$$\|(I_k - I_{k-1})v\|^2 + h_k^2\|I_k v\|_A^2 \lesssim c_d(k) h_k^2 \|v\|_A^2, \quad v \in \mathcal{V}, \tag{4.19}$$

where $c_d(k) = 1, J - k$ *and* $2^{(d-2)(J-k)}$ *for* $d = 1, 2$ *and* $d \geq 3$, *respectively.*

4.3 Regularity and approximation property

Associated with each $\mathcal{M}_k$, we define, as in (3.3), $A_k : \mathcal{M}_k \mapsto \mathcal{M}_k$. The following result is instrumental in multigrid analysis.

Theorem 4.4 *Assume α is as in (4.3). Then*

$$A((I - P_{k-1})u, u) \lesssim (\lambda_k^{-1}\|A_k u\|^2)^\alpha A(u, u)^{1-\alpha} \quad \forall\, u \in \mathcal{M}_k. \tag{4.20}$$

Proof Let $u \in \mathcal{M}_k$. Applying the Cauchy–Schwarz inequality and the following norm equivalence (see Bank and Dupont [2])

$$\|A_k^{s/2} v\| \eqsim \|v\|_s \quad \forall\, H^s(\Omega) \cap H_0^1(\Omega) \quad s \in [0, 1].$$

we deduce that

$$\begin{aligned} A((I - P_{k-1})u, u) &\leq \|A_k^{\frac{1+\alpha}{2}} u\| \|A_k^{\frac{1-\alpha}{2}} (I - P_{k-1})u\| \\ &\leq \|A_k^{\frac{1+\alpha}{2}} u\| \|(I - P_{k-1})u\|_{H^{1-\alpha}(\Omega)}. \end{aligned}$$

By Hölder's inequality,

$$\|A_k^{\frac{1+\alpha}{2}} u\| \leq \left(A(u, u)^{1-\alpha} \|A_k u\|^{2\alpha}\right)^{1/2}, \tag{4.21}$$

and, since $\lambda_k \lesssim h_k^{-2}$, Theorem 4.1 gives

$$\begin{aligned} \|(I - P_{k-1})u\|_{H^{1-\alpha}(\Omega)} &\lesssim h_k^\alpha \|(I - P_{k-1})u\|_{H^1(\Omega)} \\ &\eqsim \lambda_k^{-\frac{\alpha}{2}} A((I - P_{k-1})u, (I - P_{k-1})u)^{\frac{1}{2}} \\ &\eqsim \lambda_k^{-\frac{\alpha}{2}} A((I - P_{k-1})u, u)^{\frac{1}{2}}. \end{aligned}$$

The theorem follows by combining these inequalities. □

4.4 Strengthened Cauchy–Schwarz inequalities

These types of inequalities were used as assumptions in §4 (see eqn. (3.17)). Here we shall establish them for multilevel spaces.

Lemma 4.5 *Let $i \geq j$; then*

$$A(u,v) \lesssim \gamma^{i-j} h_i^{-1} \|u\|_A \|v\| \quad \forall\, u \in \mathcal{M}_j, v \in \mathcal{M}_i.$$

Here we recall that $\gamma \in (0,1)$ is a constant such that $h_j \eqsim \gamma^{2j}$.

Proof Given $K \in \mathcal{T}_j$, it follows from Green's identity that

$$\begin{aligned}
\int_K a\nabla u \cdot \nabla v &= \int_K (\nabla a \cdot \nabla u) v + \int_{\partial K} a \frac{\partial u}{\partial n} v \\
&\lesssim \|u\|_{H^1(K)} \|v\| + \|\nabla u\|_{L^2(\partial K)} \|v\|_{L^2(\partial K)} \\
&\lesssim \|u\|_{H^1(K)} \|v\| + (h_j^{-1/2} \|\nabla u\|_{L^2(K)})(h_i^{-1/2} \|v\|_{L^2(K)}) \\
&\lesssim (h_j h_i)^{-1/2} \|\nabla u\|_{L^2(K)} \|v\|_{L^2(K)} \\
&\lesssim \gamma^{i-j} h_i^{-1} \|\nabla u\|_{L^2(K)} \|v\|_{L^2(K)}.
\end{aligned}$$

A repeated application of the Cauchy–Schwarz inequality yields

$$A(u,v) = \sum_{K \in \mathcal{T}_j} \int_K a\nabla u \cdot \nabla v \lesssim \gamma^{i-j} h_i^{-1} \sum_{K \in \mathcal{T}_j} \|u\|_{H^1(K)} \|v\|_{L^2(K)}$$

$$\begin{aligned}
&\lesssim \gamma^{i-j} h_i^{-1} \left(\sum_{K \in \mathcal{T}_j} \|u\|^2_{H^1(K)} \right)^{\frac{1}{2}} \left(\sum_{K \in \mathcal{T}_j} \|v\|^2_{L^2(K)} \right)^{\frac{1}{2}} \\
&= \gamma^{i-j} h_i^{-1} \|u\|_A \|v\|.
\end{aligned}$$

□

The inequality in the previous lemma is a generalization of the strengthened Cauchy inequality for hierarchical basis functions in Yserentant [43]. Our proof is similar in nature to that in [43], but appears to be a little shorter and more straightforward.

Lemma 4.6 *Let $\mathcal{V}_i = (I_i - I_{i-1})\mathcal{V}$ or $\mathcal{V}_i = (Q_i - Q_{i-1})\mathcal{V}$; then*

$$A(u,v) \lesssim \gamma^{|i-j|} \|u\|_A \|v\|_A \quad \forall\, u \in \mathcal{V}_i, v \in \mathcal{V}_j. \tag{4.22}$$

Proof By (4.11), we have

$$\|v\| \lesssim h_i \|v\|_A \quad \forall\, v \in \mathcal{V}_i.$$

The result then follows directly from Lemma 4.5. □

Lemma 4.7 *Assume that $T_k = R_k A_k P_k$ and that $R_k : \mathcal{M}_k \mapsto \mathcal{M}_k$ satisfies*

$$\|R_k A_k v\|^2 \lesssim \lambda_k^{-1}(A_k v, v) \quad \forall\, v \in \mathcal{M}_k,$$

where $\lambda_k = \rho(A_k)$. Then, for $0 \le i, j \le J$

$$(T_i u, T_j v)_A \lesssim \gamma^{\frac{|i-j|}{2}} (T_i u, u)_A^{\frac{1}{2}} (T_j v, v)_A^{\frac{1}{2}} \quad \forall\, u, v \in \mathcal{V}.$$

Proof If $i \le j$, an application of Lemma 4.5 yields

$$(u_i, T_j v)_A \lesssim \gamma^{j-i} h_j^{-1} \|u_i\|_A \|T_j v\|.$$

By the assumption on R_k,

$$\|T_j v\| = \|R_j A_j P_j v\| \lesssim h_j \|A_j^{\frac{1}{2}} P_j v\| \lesssim h_j \|v\|_A.$$

Consequently

$$(u_i, T_j v)_A \lesssim \gamma^{j-i} \|u_i\|_A \|v\|_A \quad \forall\, u_i \in \mathcal{V}_i, v \in \mathcal{V}.$$

The second inequality follows from the Cauchy–Schwarz inequality and the inequality just proved:

$$\begin{aligned}(T_i u, T_j v)_A &\le (T_j v, v)_A^{\frac{1}{2}} (T_j T_i u, T_i u)_A^{\frac{1}{2}} \\ &\lesssim \gamma^{\frac{j-i}{2}} (T_j v, v)_A^{\frac{1}{2}} \|T_i u\|_A \\ &\lesssim \gamma^{\frac{j-i}{2}} (T_i u, u)_A^{\frac{1}{2}} (T_j v, v)_A^{\frac{1}{2}}.\end{aligned}$$

□

4.5 An equivalent norm using multigrid splitting

If nested multilevel finite element spaces $\mathcal{M}_k$ are allowed to be refined in an infinite way, namely $k \to \infty$, then the Sobolev space H_0^1 can be characterized by these finite element spaces in a very elegant way. We shall give such a characterization.

Theorem 4.8 *For all $v \in H_0^1(\Omega)$,*

$$\|v\|_{H^1(\Omega)}^2 \eqsim \sum_{k=0}^{\infty} \|(Q_k - Q_{k-1})v\|_{H^1(\Omega)}^2 \eqsim \sum_{k=0}^{\infty} h_k^{-2} \|(Q_k - Q_{k-1})v\|^2.$$

Proof Let $\tilde{Q}_k = Q_k - Q_{k-1}$ and $v_i = (P_i - P_{i-1})v$. It follows that

$$\begin{aligned}\|\tilde{Q}_k v_i\|_{H^1(\Omega)}^2 &\lesssim h_k^{-2\alpha} \|\tilde{Q}_k v_i\|_{H^{1-\alpha}(\Omega)}^2 \quad \text{(by inverse inequality (4.7))} \\ &\lesssim h_k^{-2\alpha} \|v_i\|_{H^{1-\alpha}(\Omega)}^2 \quad \text{(by (4.12))} \\ &\lesssim h_k^{-2\alpha} h_i^{2\alpha} \|v_i\|_{H^1(\Omega)}^2 \quad \text{(by (4.13))}.\end{aligned}$$

Note that $v = \sum_i v_i$. Let $i \wedge j = \min(i,j)$: we have

$$\begin{aligned}
&\sum_{k=0}^{\infty} \|(Q_k - Q_{k-1})v\|^2_{H^1(\Omega)} \\
&= \sum_{k=0}^{\infty}\sum_{i,j=k}^{\infty} (\nabla \tilde{Q}_k v_i, \nabla \tilde{Q}_k v_j) \quad (\text{since } \tilde{Q}_k v_i = 0 \text{ if } i < k) \\
&= \sum_{i,j=1}^{\infty}\sum_{k=0}^{i\wedge j} (\nabla \tilde{Q}_k v_i, \nabla \tilde{Q}_k v_j) \quad (\text{change the order of sum: Fubini thm.}) \\
&\lesssim \sum_{i,j=1}^{\infty}\sum_{k=0}^{i\wedge j} h_k^{-2\alpha} h_i^{\alpha} h_j^{\alpha} \|v_i\|_{H^1(\Omega)} \|v_j\|_{H^1(\Omega)} \\
&\lesssim \sum_{i,j=1}^{\infty} h_{i\wedge j}^{-2\alpha} h_i^{\alpha} h_j^{\alpha} \|v_i\|_{H^1(\Omega)} \|v_j\|_{H^1(\Omega)} \\
&\lesssim \sum_{i,j=1}^{\infty} \eta^{\alpha|i-j|} \|v_i\|_{H^1(\Omega)} \|v_j\|_{H^1(\Omega)} \\
&\lesssim \sum_{i=1}^{\infty} \|v_i\|^2_{H^1(\Omega)} = \|v\|^2_{H^1(\Omega)}.
\end{aligned}$$

To prove the other inequality, we use the strengthened Cauchy–Schwarz inequality and obtain (Lemma 4.5)

$$\begin{aligned}
\|v\|^2_{H^1(\Omega)} &= \sum_{i,j=1}^{\infty} (\nabla \tilde{Q}_i v, \nabla \tilde{Q}_j v) \lesssim \sum_{i,j=1}^{\infty} \gamma^{|i-j|} \|\tilde{Q}_i v\|_{H^1(\Omega)} \|\tilde{Q}_j v\|_{H^1(\Omega)} \\
&\lesssim \sum_{i=1}^{\infty} \|\tilde{Q}_i v\|^2_{H^1(\Omega)}.
\end{aligned}$$

□

Theorem 4.9 *For all $v \in H_0^1(\Omega)$,*

$$\|v\|^2_{H^1(\Omega)} \eqsim \sum_{k=0}^{\infty} h_k^{-2} \|(I - Q_{k-1})v\|^2.$$

Proof By the previous theorem, we obviously have

$$\sum_{k=0}^{\infty} h_k^{-2} \|(I - Q_{k-1})v\|^2 \geq \sum_{k=0}^{\infty} h_k^{-2} \|(Q_k - Q_{k-1})v\|^2 \gtrsim \|v\|^2_{H^1(\Omega)}.$$

The proof for the other direction of inequality is identical to that of the previous theorem except using $\tilde{Q}_k = I - Q_{k-1}$ instead of $Q_k - Q_{k-1}$. □

Theorem 4.10 *For all $v \in \tilde{H}_0^s(\Omega)$ for $-1 \le s \le 1$, we have* [4]

$$\|v\|_{H^s(\Omega)}^2 \eqsim \sum_{k=0}^{\infty} \|(Q_k - Q_{k-1})v\|_{H^s(\Omega)}^2 \eqsim \sum_{k=0}^{\infty} h_k^{-2s} \|(Q_k - Q_{k-1})v\|^2.$$

Proof Set $B = \sum_{k=0}^{\infty} h_k^{-2}(Q_k - Q_{k-1})$. We then have $\|v\|^2 = (B^0 v, v)$ and, by the previous theorem, $\|v\|_{H^1(\Omega)}^2 \eqsim (Bv, v)$. An application of operator interpolation then gives that $\|v\|_{H^s(\Omega)}^2 = (B^s v, v)$ which implies the desired result. □

Remark 4.1 The above theorem is also valid for $-3/2 \le s \le 3/2$.

Remark 4.2 A relevant interesting identity is as follows:

$$\|v\|_{H^1(\Omega)}^2 \eqsim \sum_{k=0}^{\infty} h_k^2 \|A_k P_k\|^2 \quad \forall\, v \in H_0^1(\Omega).$$

5 Overlapping domain decomposition methods

We shall now discuss our first major algorithm which is based on a domain decomposition with overlappings. As the finite element space $\mathcal{V}$ is defined on a triangulation of the domain Ω, a finite element space restricted to a subdomain of Ω can naturally be regarded as a subspace of $\mathcal{V}$ and hence a decomposition of the domain naturally leads to a decomposition of the finite element space. This is the main viewpoint for the algorithms described in this section.

5.1 Preliminaries

We start by assuming that we are given a set of overlapping subdomains $\{\Omega_i\}_{i=1}^J$ of Ω whose boundaries align with the mesh triangulation defining $\mathcal{V}$. One way of defining the subdomains and the associated partition is by starting with disjoint open sets $\{\Omega_i^0\}_{i=1}^J$ with $\bar{\Omega} = \cup_{i=1}^J \bar{\Omega}_i^0$ and $\{\Omega_i^0\}_{i=1}^J$ quasi-uniform of size h_0. The subdomain Ω_i is defined to be a mesh subdomain containing Ω_i^0 with the distance from $\partial\Omega_i \cap \Omega$ to Ω_i^0 greater than or equal to ch_0 for some prescribed constant c.

Based on these subdomains, the subspaces $\mathcal{V}_i$ $(1 \le i \le J)$ are defined by

$$\mathcal{V}_i = \{v \in \mathcal{V} : v(x) = 0, \quad \forall\, x \in \Omega \setminus \Omega_i\}.$$

If the number of subdomains J is too large, the above subspaces are not sufficient to produce an optimal algorithm. To deal with this we introduce a coarse finite element subspace $\mathcal{V}_0$ defined from a quasi-uniform triangulation of Ω of size h_0.

[4] $\tilde{H}_0^s(\Omega)$ is the interpolation space $[L_2(\Omega), H_0^1(\Omega)]_s$.

Lemma 5.1 *For the subspaces $\mathcal{V}_i$ $(0 \le i \le J)$, we have*

$$\mathcal{V} = \sum_{i=0}^{J} \mathcal{V}_i. \tag{5.1}$$

Furthermore there is a constant C_0 that is independent of h, h_0 or J, such that for any $v \in \mathcal{V}$, there are $v_i \in \mathcal{V}_i$ that satisfy $v = \sum_{i=0}^{J} v_i$ and

$$\sum_{i=0}^{J} A(v_i, v_i) \le C_0 A(v, v). \tag{5.2}$$

Proof The main ingredient of the proof is a partition of unity, $\{\theta_i\}_{i=1}^{J}$, defined on Ω satisfying $\sum_{i=1}^{J} \theta_i = 1$ and, for $i = 1, \cdots, J$,

$$\text{supp}\theta_i \subset \Omega_i \cup \partial\Omega, \quad 0 \le \theta_i \le 1, \quad \|\nabla \theta_i\|_{\infty,\Omega_i} \le C h_0^{-1}.$$

Here $\|\cdot\|_{\infty,D}$ denotes the L^∞ norm of a function defined on a subdomain D.

The construction of such a partition of unity is standard. A partition $v = \sum_{i=0}^{J} v_i$ for $v_i \in \mathcal{V}_i$ can then be obtained with

$$v_0 = Q_0 v, \quad v_i = I_h(\theta_i(v - Q_0 v)), \quad i = 1, \cdots, J,$$

where I_h is the nodal value interpolant on $\mathcal{V}$.

For this decomposition, we prove that (5.2) holds. For any $\tau \in \mathcal{T}_h$, note that

$$\|\theta_i - \bar{\theta}_{i,\tau}\|_{L^\infty(\tau)} \lesssim h\|\nabla \theta_i\|_{L^\infty(\tau)} \lesssim \frac{h}{h_0},$$

where $\bar{\theta}_{i,\tau}$ denotes the average of θ_i over τ. Let $w = v - Q_0 v$; by the inverse inequality (4.5),

$$\begin{aligned} |v_i|_{H^1(\tau)} &\le |\bar{\theta}_{i,\tau} w|_{H^1(\tau)} + |I_h(\theta_i - \bar{\theta}_{i,\tau})w|_{H^1(\tau)} \\ &\lesssim |w|_{H^1(\tau)} + h^{-1}\|I_h(\theta_i - \bar{\theta}_{i,\tau})w\|_{L^2(\tau)}. \end{aligned}$$

It can easily be shown that

$$\|I_h(\theta_i - \bar{\theta}_{i,\tau})w\|_{L^2(\tau)} \lesssim \frac{h}{h_0}\|w\|_{L^2(\tau)}.$$

Consequently

$$|v_i|^2_{H^1(\tau)} \lesssim |w|^2_{H^1(\tau)} + \frac{1}{h_0^2}\|w\|^2_{L^2(\tau)}.$$

Summing over all $\tau \in \mathcal{T}_h \cap \Omega_i$ gives

$$|v_i|^2_{H^1(\Omega)} = |v_i|^2_{H^1(\Omega_i)} \lesssim |w|^2_{H^1(\Omega_i)} + \frac{1}{h_0^2}\|w\|^2_{L^2(\Omega_i)},$$

and

$$\begin{aligned}\sum_{i=1}^{J} A(v_i, v_i) &\lesssim \sum_{i=1}^{J} |v_i|_{H^1(\Omega_i)}^2 \lesssim \sum_{i=1}^{J} \left(|w|_{H^1(\Omega_i)}^2 + \frac{1}{h_0^2} \|w\|_{L^2(\Omega_i)}^2 \right) \\ &\lesssim \left(|v - Q_0 v|_{H^1(\Omega)}^2 + \|v - Q_0 v\|_{L^2(\Omega)}^2 \right) \lesssim \|v\|_{H^1(\Omega)}^2.\end{aligned}$$

For $i = 0$, we apply (4.11) and get

$$\|v_0\|_{H^1(\Omega)} \lesssim \|v\|_{H^1(\Omega)}.$$

The desired result then follows. □

Lemma 5.2 *The parameters K_0 and K_1 of subsection 3.3.1 satisfy*

$$K_0 \le C_0/\omega_0 \quad \textit{and} \quad K_1 \le C.$$

Proof The first estimate follows directly from Lemmas 3.8 and 5.1. To prove the second estimate, we define

$$Z_i = \{1 \le j \le J : \Omega_i \cap \Omega_j \ne \emptyset\}.$$

By the construction of the domain decomposition, there exists a fixed integer n_0 such that

$$|Z_i| \le n_0, \quad \forall\, 1 \le i \le J.$$

Note that if $P_i P_j \ne 0$ or $P_j P_i \ne 0$, then $j \in Z_i$. It therefore follows that $K_1 \le 1 + n_0$. □

Remark 5.1 A slightly more careful analysis would give that the constant in (5.2) depends on the overlapping size, say δ_0, in the following way

$$C_0 = O(\frac{h_0}{\delta_0}).$$

5.2 Domain decomposition methods with overlappings

By Theorem 3.14 and Lemma 5.2, we get

Theorem 5.3 *The SSC preconditioner B given by (3.6) associated with the decomposition (5.1) satisfies*

$$\kappa(BA) \lesssim \frac{\omega_1}{\omega_0}.$$

The proof of the above theorem follows from Theorem 3.14 and Lemma 5.2.

Combing Theorem 3.18 with Lemma 5.2, we also obtain

Theorem 5.4 *The Algorithm 3.3 associated with the decomposition (5.1) satisfies*

$$\|E_J\|_A^2 \le 1 - \frac{\omega_0(2 - \omega_1)}{C}, \tag{5.3}$$

where C is a constant independent of the number of subdomains J and the mesh size h.

We note that in our theory, the subdomain problems do not have to be solved exactly and only the spectrum of the inexact solvers matters. In the estimate (5.3), ω_0 should not be too small and ω_1 should stay away from 2. It is easy to see that one iteration of a V-cycle multigrid on each subdomain always satisfies this requirement.

As for the implementation of these domain decomposition methods, the algebraic formulations (3.27) and Algorithm 3.6 can be used. For example, if exact solvers are used in each subspace, the PSC preconditioner is

$$\mathcal{B} = \sum_{i=0}^{J} \mathcal{I}_i \mathcal{A}_i^{-1} \mathcal{I}_i^t.$$

Here $\mathcal{I}_i \in \mathbb{R}^{n \times n_i}$ is defined by

$$(\varphi_1^i, \cdots, \varphi_{n_i}^i) = (\varphi_1, \cdots, \varphi_n)\mathcal{I}_i,$$

where $(\varphi_1^i, \cdots, \varphi_{n_i}^i)$ is the nodal basis of $\mathcal{V}_i$ and $(\varphi_1, \cdots, \varphi_n)$ is the nodal basis of $\mathcal{V}$. Note that if $i \neq 0$, the entries of matrix $\mathcal{I}_i$ consist of 1 and 0, since $\{\varphi_1^i, \cdots, \varphi_{n_i}^i\}$ is a subset of $\{\varphi_1, \cdots, \varphi_n\}$.

5.3 A recursive application of domain decomposition

As we shall see later, multigrid and domain decomposition methods can be viewed in the same mathematical framework. In fact, as we shall now demonstrate, a typical multigrid method may be derived by a recursive application of the overlapping domain decomposition method discussed above.

Let $\mathcal{T}_J$ be the finest triangulation in the multilevel structure described earlier with nodes $\{x_i\}_{i=1}^{n_J}$. With such a triangulation, a natural domain decomposition is

$$\bar{\Omega} = \bar{\Omega}_0^h \bigcup \bigcup_{i=1}^{n_J} \operatorname{supp} \varphi_i,$$

where φ_i is the nodal basis function in $\mathcal{M}_J$ associated with the node x_i and Ω_0^h (maybe empty) is the region where all functions in $\mathcal{M}_J$ vanish.

In view of our earlier discussions, the corresponding decomposition method without the coarser space is exactly the Gauss–Seidel method, which as we know is not very efficient (its convergence rate is easily shown to be $1-O(h^2)$). The more interesting case is when a coarser space is introduced. The choice of such a coarse space is clear here, namely a subspace defined on a triangulation with characteristic mesh size being the same as the subdomains' size which is $2h$, therefore the coarse space $\mathcal{M}_{J-1}$ is the most natural candidate. But the space $\mathcal{M}_{J-1}$ is in general still very large and an efficient solver for this space is needed. The most natural way to obtain such a solver for $\mathcal{M}_{J-1}$ is simply to repeat the above process by using the space $\mathcal{M}_{J-2}$ as a 'coarser' space with the supports of the nodal basis function in $\mathcal{M}_{J-1}$ as a domain decomposition. We continue in this way until we reach a coarse space $\mathcal{M}_0$ where a direct solver can be used. As a result, a multilevel algorithm based on domain decomposition is obtained. In fact, this recursively defined domain decomposition method is just a multigrid algorithm with the Gauss–Seidel method as a smoother. Multigrid methods will be the main topic in the rest of this paper.

6 Multigrid methods

This section is devoted to multigrid methods and their convergence properties. The following topics will be studied: classic multigrid iterative methods, BPX preconditioners, hierarchical basis methods, methods for locally refined meshes and the full multigrid principle.

6.1 Analysis for smoothers

The most crucial step in developing a multigrid solver is the design of a relaxation scheme. A relaxation scheme is also the most problem-dependent part of a multigrid solver as most other parts (such as prolongation and restriction operators) are usually quite standard. The rôle of relaxation is not to reduce the overall error, but to smooth it out (namely damp out the non-smooth or high frequency components) so that it can be well approximated by functions on a coarser grid.

The smoother will be analysed by three approaches in this section. The first approach is through numerical experiments, which would give an intuitive idea of the numerical behavior of a smoother. The second approach is Brandt's local mode analysis. This approach, using local Fourier analysis, can give a good insight on the rôle of a smoother. The third approach is to build technical machinery for the convergence analysis of multigrid methods.

6.1.1 *A model problem and some numerical examples*

Consider the Poisson equation with homogeneous Dirichlet condition on the unit square discretized with a uniform triangulation. The discretized equation can be expressed as

$$4u_{ij} - (u_{i+1,j} + u_{i-1,j} + u_{i,j+1} + u_{i,j-1}) = b_{i,j}, \quad 1 \le i, j \le n. \tag{6.1}$$

The damped Jacobi and Gauss–Seidel methods are among the most popular relaxation schemes for this problem. The damped Jacobi (or Richardson) iteration can be written as

$$4\tilde{u}_{ij} = \omega(\bar{u}_{i+1,j} + \bar{u}_{i-1,j} + \bar{u}_{i,j+1} + \bar{u}_{i,j-1}) + b_{i,j} \tag{6.2}$$

where $\tilde{u}_{ij}$ denotes the new value of u while $\bar{u}_{ij}$ denotes the old value of u, and the (point) Gauss–Seidel iteration (with lexicographical order on nodal

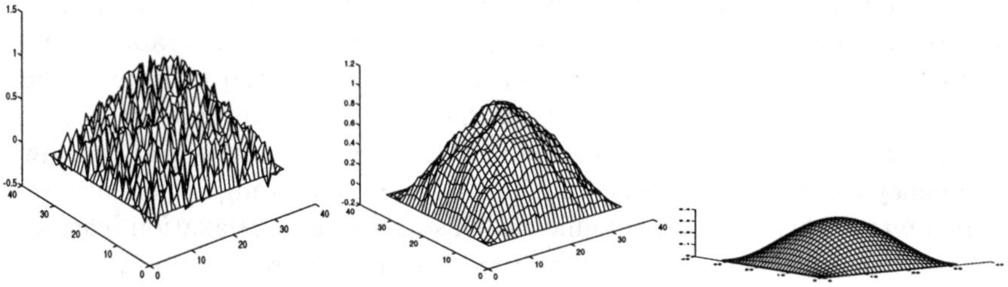

FIG. 2. Residual after 0, 2 and 100 iterations, respectively, with 961 unknowns.

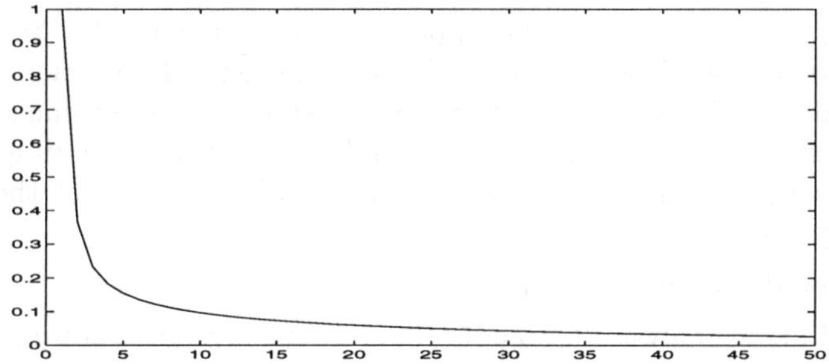

FIG. 3. Convergence history of Gauss–Seidel method within 50 iterations.

points, from left to right and bottom to top):

$$4\tilde{u}_{ij} = (\bar{u}_{i+1,j} + \tilde{u}_{i-1,j} + \bar{u}_{i,j+1} + \tilde{u}_{i,j-1}) + b_{i,j}. \tag{6.3}$$

The Gauss–Seidel method has a good smoothing property. Let us illustrate this by a simple numerical example. Consider eqn. (6.1) with an initial residual $u - u^0$ shown on the left plot of Fig. 2. The initial residual apparently contains a lot of oscillations. The middle plot in Fig. 2 is the residual after 2 Gauss–Seidel iterations. As we see the error components are smoothed out very quickly with only two Gauss–Seidel iterations although the global errors are still very large. The right plot in Fig. 2 is the residual after one hundred iterations and as we see the error is still quite big.

Basic ideas in a multigrid strategy The above numerical examples show that high frequency errors, which involve local variations in the solution, are well annihilated by simple relaxation methods such as Gauss–Seidel iterations. Low frequency or more global errors are much more insensitive to the application of simple relaxation methods. In fact, as shown in Fig. 3, the convergence rate of the Gauss–Seidel iteration consists of a rather rapid initial residual reduction phase, which gradually develops into a much slower residual reduction phase, corresponding to a situation where all high frequency errors have been damped down and low frequency errors dominate. A multigrid methodology capitalizes on this rapid initial reduction of high frequency errors associated with an initial solution on the fine grid, using a simple relaxation scheme such as Gauss–Seidel iteration. Therefore, the solution is transferred to a coarse grid. On this grid, the low frequency errors of the fine grid manifest themselves as high frequency errors, and are thus damped out efficiently using the same relaxation scheme. The coarse grid corrections computed in this manner are interpolated back to the fine grid in order to update the solution. This procedure can be applied recursively on a sequence of coarser and coarser grids, where each grid-level is responsible for eliminating a particular frequency bandwidth of errors.

Multigrid strategies may be applied to any existing relaxation technique. The success of the overall solution strategy depends on a close matching between the bandwidth of errors which can be efficiently smoothed on a given grid using the particular chosen relaxation strategy, with a careful construction of a sequence of coarse grids, in order to represent the entire error frequency range.

6.1.2 *Brandt's local mode analysis*

The local mode analysis of Brandt [13] is an effective general tool to analyze and predict the performance of a multigrid solver and in particular the performance of a smoother. This method is based on the fact that a relaxation process is often a local process in which the information propagates just a few mesh-sizes per sweep. Therefore, one can assume the problem to be in an unbounded domain, with constant (frozen) coefficients, in which case the algebraic error can be expanded in terms of Fourier series.

The local mode analysis for a smoother can in fact be applied in a rigorous fashion to the model problem discussed in the previous section. Let us first recall the discrete Fourier theory. For clarity, we confine our discussion to the two dimensional case. The discrete Fourier transform theory says that every discrete function $u : I_n \mapsto \mathbb{R}$, with $I_n = \{(i,j) : 0 \le i,j \le n\}$, can be written as

$$u_{i,j} = \sum_{\theta \in \Theta_n} c_\theta \psi_{i,j}(\theta), \quad \psi_{i,j}(\theta) = e^{\mathbf{i}(i\theta_1 + j\theta_2)}, \quad \mathbf{i} = \sqrt{-1}, \quad \theta = (\theta_1, \theta_2).$$

where

$$c_\theta = \frac{1}{(n+1)^2} \sum_{(k,l)\in I_n} u_{k,l}\psi_{k,l}(-\theta).$$

and

$$\Theta_n = \left\{ \frac{2\pi}{n+1}(k,l) \quad -m \le k,l \le m+p \right\},$$

where $p = 1, m = (n+1)/2$ for odd n and $p = 0, m = n/2 + 1$ for even n.

We now first use the discrete Fourier transform to analyze the damped Jacobi method. Let $\tilde{\epsilon}_{i,j} = u_{i,j} - \tilde{u}_{i,j}$ and $\bar{\epsilon}_{i,j} = u_{i,j} - \bar{u}_{i,j}$. It is easy to see that

$$\tilde{\epsilon}_{ij} = \bar{\epsilon}_{i,j} - \frac{\omega}{4}(4\bar{\epsilon}_{ij} - (\bar{u}_{i+1,j} + \bar{u}_{i-1,j} + \bar{u}_{i,j+1} + \bar{u}_{i,j-1})). \tag{6.4}$$

We write

$$\tilde{\epsilon}_{i,j} = \sum_{\theta \in \Theta_n} \tilde{c}_\theta \psi_{i,j}(\theta) \tag{6.5}$$

and

$$\bar{\epsilon}_{i,j} = \sum_{\theta \in \Theta_n} \bar{c}_\theta \psi_{i,j}(\theta). \tag{6.6}$$

Substituting the above expressions into (6.4) and comparing the coefficients of each $\psi_{ij}(\theta)$, we obtain that

$$\lambda(\theta) = 1 - \omega\left(1 - \frac{\cos\theta_1 + \cos\theta_2}{2}\right). \tag{6.7}$$

where

$$\lambda(\theta) \equiv \frac{\tilde{c}(\theta)}{\bar{c}(\theta)} \tag{6.8}$$

is called the amplification factor of the local mode $\psi_{i,j}(\theta)$.

The smoothing factor introduced by Brandt is the quantity

$$\bar{\rho} = \sup\{|\lambda(\theta)|, \pi/2 \le |\theta_k| \le \pi, k = 1, 2\}. \tag{6.9}$$

Roughly speaking, the smoothing factor $\bar{\rho}$ is the maximal amplification factor corresponding to those high frequency local modes that oscillate within range $2h$ (and hence cannot be resolved by a coarse grid of size $2h$).

For the damped Jacobi method, it is easy to see that

$$\bar{\rho} = \max\{|1 - 2\omega|, |1 - \omega/2|, |1 - 3\omega/2|\}.$$

The optimal ω that minimizes the smoothing factor is

$$\omega = 4/5, \quad \bar{\rho} = 3/5.$$

For $\omega = 1$ we have $\bar{\rho} = 1$. This means that the undamped Jacobi method for this model problem, although convergent as an iterative method by itself, should not be used as a smoother.

We next examine the smoothing property of the Gauss–Seidel iteration. Unlike the Jacobi method, the Gauss–Seidel method depends on the ordering of the unknowns. The most natural ordering is perhaps the lexicographic order which was used in the numerical examples given earlier and the corresponding Gauss–Seidel method reads

$$\tilde{\epsilon}_{ij} = \frac{1}{4}(\bar{\epsilon}_{i+1,j} + \tilde{\epsilon}_{i-1,j} + \bar{\epsilon}_{i,j+1} + \tilde{\epsilon}_{i,j-1}). \tag{6.10}$$

Again using the Fourier transform (6.5) and (6.6), we obtain the local amplification factor as follows:

$$\lambda(\theta) = \frac{e^{\mathbf{i}\theta_1} + e^{\mathbf{i}\theta_1}}{4 - e^{-\mathbf{i}\theta_1} - e^{-\mathbf{i}\theta_2}}.$$

It is elementary to see that

$$\bar{\rho} = |\lambda(\pi/2, \cos^{-1}(4/5))| = 1/2.$$

This means that the Gauss–Seidel method is a better smoother than the damped Jacobi method.

A more interesting ordering for the Gauss–Seidel method is the so-called red-black ordering. In this particular example, we say two grid points belong to the same color (see Definition 3.6) if and only they are not neighbors (in either the horizontal or vertical directions). It is easy to see that the uniform grid in our example can be grouped into two colors, often called red and black. The red-black ordering is to first order all the nodes in one color and then order the other points in another color. (The actual ordering within the same color is not crucial.)

The smoothing factor for the Gauss–Seidel method with red-black ordering cannot be obtained as easily as the lexicographical ordering, but it can indeed be proved that

$$\bar{\rho} = 1/4.$$

This means that the Gauss–Seidel method with red-black ordering is a better smoother than the one with the lexicographical ordering. Furthermore red-black Gauss–Seidel has much better parallel features (see Definition 3.6).

6.1.3 *General smoother analysis*

We shall now develop some technical results concerning the smoothing property of the Gauss–Seidel method. We choose to study the Gauss–Seidel method since it is one of the better smoothers for our model problems and also it is less obvious to analyse. The analysis for other smoothers is relatively simple (see the analysis for Richardson in (2.13)).

Lemma 6.1 *For the stiffness matrix $\mathcal{A} = \mathcal{D} - \mathcal{L} - \mathcal{U}$,*

$$\|(\mathcal{D} - \mathcal{L})\xi\|_2 \eqsim h^{d-2}\|\xi\|_2 \quad \forall\, \xi \in \mathbb{R}^N.$$

Proof Because of the sparsity, it is trivial to prove that

$$\|(\mathcal{D} - \mathcal{L})\xi\|_2 \lesssim h^{d-2}\|\xi\|_2.$$

Now it follows that

$$\begin{aligned} h^{2-d}(\xi,\xi) &\lesssim \frac{1}{2}(\mathcal{D}\xi,\xi) \le \frac{1}{2}((\mathcal{A}+\mathcal{D})\xi,\xi) \\ &= ((\mathcal{D}-\mathcal{L})\xi,\xi) \le \|(\mathcal{D}-\mathcal{L})\xi\|_2\|\xi\|_2. \end{aligned}$$

This completes the proof. □

The following result is an operator interpretation of the algebraic result given in Lemma 6.1, which means that Gauss–Seidel is basically like a Richardson iteration.

Lemma 6.2 *Assume that $R : \mathcal{V} \mapsto \mathcal{V}$ represents the iterator for symmetric Gauss–Seidel iteration. Then*

$$R \eqsim h^2 \eqsim \lambda_h^{-1}.$$

Proof By definition, the matrix representation of R is

$$\tilde{R} = (\mathcal{D}-\mathcal{U})^{-1}\mathcal{D}(\mathcal{D}-\mathcal{L})^{-1}\mathcal{M}.$$

Given $v \in \mathcal{V}$, let $\nu = \tilde{v}$; then it is easy to see that

$$(Rv, v) = \|\mathcal{D}^{\frac{1}{2}}(\mathcal{D}-\mathcal{L})^{-1}\mathcal{M}\nu\|_2^2$$

Thus it is equivalent to showing that

$$\|\mathcal{D}^{\frac{1}{2}}(\mathcal{D}-\mathcal{L})^{-1}\mathcal{M}\nu\|_2^2 \eqsim h^2(\mathcal{M}\nu,\nu).$$

Making a change of variable $\xi = (\mathcal{D}-\mathcal{L})^{-1}\mathcal{M}\nu$ and using the fact that $\mathcal{M}^{-1} \eqsim h^{-d}$, the above relation can be reduced to

$$\|(\mathcal{D}-\mathcal{L})\xi\|_2^2 \eqsim h^{d-2}(\mathcal{D}\xi,\xi) \eqsim h^{2(d-2)}\|\xi\|_2^2,$$

which was given by Lemma 6.1 □

Lemma 6.3 *For the stiffness matrix $\mathcal{A} = \mathcal{D} - \mathcal{L} - \mathcal{U}$*

$$(\mathcal{A}\xi,\xi) \le \frac{2}{1+k_0}((\mathcal{D}-\mathcal{L})\xi,\xi) \quad \forall\, \xi \in \mathbb{R}^N,$$

where k_0 is the maximal number of nonzero row entries of $\mathcal{A}$.

If $R : \mathcal{V} \mapsto \mathcal{V}$ represents the iterator for Gauss–Seidel iteration, then

$$(Av, v) \le \frac{2}{1+k_0}(R^{-1}v, v) \quad \forall\, v \in V_h.$$

Proof It is easy to see that the desired estimate is equivalent to the following:

$$(\mathcal{A}\xi,\xi) \le k_0(\mathcal{D}\xi,\xi) \quad \forall\, \xi \in \mathbb{R}^N$$

which can be obtained by a simple application of the Cauchy–Schwarz inequality. □

6.2 A basic multigrid cycle: the backslash (\) cycle

Although, as we shall see, multigrid methods have many variants, there is one particular multigrid algorithm which can be viewed as a basic multigrid cycle. This algorithm is sometimes called the backslash (\) cycle (we shall explain below why this algorithm is given this name).

We shall first present this method from a more classical point of view. This more classical approach makes it easier to introduce many different kinds of classical multigrid methods and also make it possible to use more classical approaches to analyze the convergence of multigrid methods.

A multigrid process can be viewed as defining a sequence of operators $B_k : \mathcal{M}_k \mapsto \mathcal{M}_k$ which are approximate inverses of A_k in the sense that $\|I - B_k A_k\|_A$ is bounded away from 1. A typical way of defining such a sequence of operators is the following *backslash* cycle multigrid procedure.

Algorithm 6.1 *For $k = 0$, define $B_0 = A_0^{-1}$. Assume that $B_{k-1} : \mathcal{M}_{k-1} \mapsto \mathcal{M}_{k-1}$ is defined. We shall now define $B_k : \mathcal{M}_k \mapsto \mathcal{M}_k$ which is an iterator for the equation of the form*

$$A_k v = g.$$

(1) Fine grid smoothing: *for $v^0 = 0$ and $l = 1, 2, \cdots, m$,*

$$v^l = v^{l-1} + R_k(g - A_k v^{l-1}).$$

(2) Coarse grid correction: *$e_{k-1} \in \mathcal{M}_{k-1}$ is the approximate solution of the residual equation $A_{k-1}e = Q_{k-1}(g - Av^m)$ by the iterator B_{k-1}:*

$$e_{k-1} = B_{k-1}Q_{k-1}(g - Av^m).$$

Define

$$B_k g = v^m + e_{k-1}.$$

After the first step, the residual $v - v^m$ is small at high frequencies. In other words, $v - v^m$ is smoother (see the middle plot in Fig. 2) and hence it can be very well approximated by a coarse space $\mathcal{M}_{k-1}$. The second step in the above algorithm plays the rôle of correcting the low frequencies by the coarser space $\mathcal{M}_{k-1}$ and the coarse grid solver B_{k-1} given by induction.

With the above defined B_k, we may consider the following simple iteration:

$$u^{k+1} = u^k + B_J(f - Au^k). \tag{6.11}$$

There are many different ways to make use of B_k, which will be discussed later.

Before we study its convergence, we now discuss briefly the algebraic version of the above algorithm.

Let $\Phi^k = (\varphi_1^k, \cdots, \varphi_{n_k}^k)$ be the nodal basis vector for the space $\mathcal{M}_k$; we define the so-called prolongation matrix $\mathcal{I}_k^{k+1} \in \mathbb{R}^{n_{k+1}\times n_k}$ as follows

$$\Phi^k = \Phi^{k+1}\mathcal{I}_k^{k+1}. \tag{6.12}$$

Algorithm 6.2. (Matrix version) *Let $\mathcal{B}_0 = \mathcal{A}_0^{-1}$. Assume that $\mathcal{B}_{k-1} \in \mathbb{R}^{n_{k-1}\times n_{k-1}}$ is defined; then for $\eta \in \mathbb{R}^{n_k}$, $\mathcal{B}_k \in \mathbb{R}^{n_k\times n_k}$ is defined as follows.*

(1) Fine grid smoothing: *for $\nu^0 = 0$ and $l = 1, 2, \cdots, m$*

$$\nu^l = \nu^{l-1} + \mathcal{R}_k(\eta - \mathcal{A}_k\nu^{l-1}).$$

(2) Coarse grid correction: *$\varepsilon_{k-1} \in \mathbb{R}^{n_{k-1}}$ is the approximate solution of the residual equation $\mathcal{A}_{k-1}\varepsilon = (\mathcal{I}_{k-1}^k)^t(\eta - \mathcal{A}_k\nu^m)$ by using $\mathcal{B}_{k-1}$,*

$$\varepsilon_{k-1} = \mathcal{B}_{k-1}(\mathcal{I}_{k-1}^k)^t(\eta - \mathcal{A}_k\nu^m).$$

Define

$$\mathcal{B}_k\eta = \nu^m + \mathcal{I}_{k-1}^k\varepsilon_{k-1}.$$

The above algorithm is given in recurrence form, but it can also be easily implemented in a nonrecursive fashion. For such an implementation, we refer to Algorithm 6.4

6.3 A convergence analysis using full elliptic regularity

With the assumption of full elliptic regularity (namely $\alpha = 1$ in (4.3), see also Theorem 4.4), a very sharp convergence estimate can be obtained in a very simple and elegant fashion.

We shall assume that the smoothers R_k are SPD and satisfy

$$\frac{c_0}{\lambda_k}(v, v) \le (R_k v, v) \le (A_k^{-1}v, v) \quad \forall\, v \in \mathcal{M}_k. \tag{6.13}$$

We would like to remark that the above assumptions on R_k can be much weakened and, for example, the R_k do not need to be symmetric (in this case, assumptions need to be made on the symmetrizations of the R_k, see, e.g., §6.5).

If the regularity estimate (4.3) holds with $\alpha = 1$, then there exists a positive constant c_1 independent of mesh parameters such that (see Theorem 4.4)

$$\|(I - P_{k-1})v\|_A^2 \le c_1\lambda_k^{-1}\|A_k v\|^2 \quad \forall\, v \in \mathcal{M}_k. \tag{6.14}$$

The next technical result shows that any function smoothened by local relaxation can be well approximated by a coarser grid.

Lemma 6.4

$$\|(I - P_{k-1})K_k^m v\|_A^2 \le \frac{c_1}{2mc_0}(\|v\|_A^2 - \|K_k^m v\|_A^2).$$

Proof

$$\begin{aligned} \|(I-P_{k-1})K_k^m v\|_A^2 &\le c_1\lambda_k^{-1}\|A_kK_k^m v\|^2 \\ &= \frac{c_1}{c_0}(R_kA_kK_k^m v, A_kK_k^m v) = \frac{c_1}{c_0}((I-K_k)K_k^{2m}v, v)_A \\ &\le \frac{c_1}{2mc_0}(\|v\|_A^2 - \|K_k^m v\|_A^2). \end{aligned}$$

The proof is completed by using the following elementary inequality:

$$\begin{aligned} ((I-K_k)K_k^{2m}v, v)_A &\le \frac{1}{2m}\sum_{j=0}^{2m-1}((I-K_k)K_k^j v, v)_A \\ &\le \frac{1}{2m}(\|v\|_A^2 - \|K_k^m v\|_A^2). \end{aligned} \tag{6.15}$$

□

Theorem 6.5 *For the Algorithm 6.1, we have*

$$\|I - B_kA_k\|_A^2 \le \frac{c_1}{2mc_0 + c_1}, \quad 1 \le k \le J.$$

Proof By definition of Algorithm 6.1, we have

$$I - B_kA_k = (I - P_{k-1}B_{k-1}A_{k-1})(I - R_kA_k)^m$$

and, thus, for all $v \in \mathcal{M}_k$

$$\begin{aligned} &\|(I-B_kA_k)v\|_A^2 \\ = &\ \|(I-P_{k-1})K_k^m v\|_A^2 + \|(I-B_{k-1}A_{k-1})P_{k-1}K_k^m v)\|_A^2. \end{aligned}$$

Let $\delta = c_1/(2mc_0+c_1)$. We shall prove the above estimate by induction. First of all it is obviously true for $k=0$. Assume it holds for $k-1$. In the case of k, we have from the above identity that

$$\begin{aligned} \|(I-B_kA_k)v\|_A^2 &\le \|(I-P_{k-1})K_k^m v\|_A^2 + \delta\|P_{k-1}K_k^m v\|_A^2 \\ &\le (1-\delta)\|(I-P_{k-1})K_k^m v\|_A^2 + \delta\|K_k^m v\|_A^2 \\ &\le (1-\delta)\frac{c_1}{2mc_0}(\|v\|_A^2 - \|K_k^m v\|_A^2) + \delta\|K_k^m v\|_A^2 \\ &\le \delta\|v\|_A^2. \end{aligned}$$

□

6.4 V-cycle and W-cycle

Two important variants of the above backslash cycle are the so-called V-cycle and W-cycle.

A V-cycle algorithm is obtained from the backslash cycle by performing more smoothings after the coarse grid corrections. Such an algorithm, roughly speaking, is like a backslash (\) cycle plus a slash (/) (a reversed backslash) cycle. The detailed algorithm is given as follows.

Algorithm 6.3 *For $k = 0$, define $B_0 = A_0^{-1}$. Assume that $B_{k-1} : \mathcal{M}_{k-1} \mapsto \mathcal{M}_{k-1}$ is defined. We shall now define $B_k : \mathcal{M}_k \mapsto \mathcal{M}_k$ which is an iterator for the equation of the form*

$$A_k v = g.$$

(1) Pre-smoothing: *for $v^0 = 0$ and $l = 1, 2, \cdots, m$*

$$v^l = v^{l-1} + R_k(g - A_k v^{l-1}).$$

(2) Coarse grid correction: *$e_{k-1} \in \mathcal{M}_{k-1}$ is the approximate solution of the residual equation $A_{k-1}e = Q_{k-1}(g - Av^m)$ by the iterator B_{k-1}:*

$$e_{k-1} = B_{k-1}Q_{k-1}(g - Av^m).$$

(3) Post-smoothing: *for $v^{m+1} = v^m + e_{k-1}$ and $l = m+2, 2, \cdots, 2m$*

$$v^l = v^{l-1} + R_k(g - A_k v^{l-1}).$$

A nonrecursive implementation The recursive formulation of the above algorithm makes it a little less straightforward to code sometimes. A nonrecursive version of the algorithm is given below in terms of matrices and vectors.

Algorithm 6.4. (V-cycle computation of $\mathcal{B}\beta$)

$\beta_J = \beta$;
for $l = J : 1$,
  $\alpha_l = \mathcal{R}_l \beta_l, \beta_{l-1} = (\mathcal{I}_{l-1}^l)^t(\beta_l - \mathcal{A}_l \alpha_l)$;
endfor

for $l = 2 : J$,
  $\alpha_l = \mathcal{R}_l^t(\alpha_l + \mathcal{I}_{l-1}^l \alpha_{l-1})$;
endfor

$\mathcal{B}\beta = \alpha_J$.

The reason why this algorithm is called a V-cycle is quite clear with the above implementation. The algorithm starts on the finest level and traverses all the grids, one at a time, until it reaches the coarsest grid. Then it traverses all the grids until it reaches the finest level.

It is easy to see that the operators B_k defined by the above V-cycle algorithm satisfy

$$I - B_k A_k = (I - B_k^{(\backslash)} A_k)^*(I - B_k^{(\backslash)} A_k)$$

where the $B_k^{(\backslash)}$ correspond to the operators defined by the backslash cycle (Algorithm 6.1) and $*$ is the adjoint operator with respect to the A-inner product. Consequently,

$$\|I - B_k A_k\|_A = \|I - B_k^{(\backslash)} A_k\|_A^2.$$

This means that the convergence of the V-cycle is a consequence of the convergence of the backslash cycle.

The W-cycle, roughly speaking, is like a V-cycle plus another V-cycle. In a V-cycle iteration, the coarse grid correction is only performed once, while in a W-cycle, the coarse grid correction is performed twice.

Algorithm 6.5 *For $k = 0$, define $B_0 = A_0^{-1}$. Assume that $B_{k-1} : \mathcal{M}_{k-1} \mapsto \mathcal{M}_{k-1}$ is defined. We shall now define $B_k : \mathcal{M}_k \mapsto \mathcal{M}_k$ which is an iterator for the equation of the form*

$$A_k v = g.$$

(1) Pre-smoothing: *for $v^0 = 0$ and $l = 1, 2, \cdots, m$*

$$v^l = v^{l-1} + R_k(g - A_k v^{l-1}).$$

(2) Coarse grid correction: *$e_{k-1} \in \mathcal{M}_{k-1}$ is the approximate solution of the residual equation $A_{k-1}e = Q_{k-1}(g - Av^m)$ by applying iterator B_{k-1} twice, $e_{k-1} = w^2$, with $w^0 = 0$ and*

$$w^j = w^{j-1} + B_{k-1}(Q_{k-1}(g - Av^m) - A_{k-1}w^{j-1}), \quad j = 0, 1, 2.$$

(3) Post-smoothing: *for $v^{m+1} = v^m + e_{k-1}$ and $l = m + 2, 2, \cdots, 2m$*

$$v^l = v^{l-1} + R_k(g - A_k v^{l-1}).$$

Again it is also easy to see that the convergence of the W-cycle is an easy consequence of the convergence of the V-cycle. But it is often the case that the W-cycle is easier to analyze. We shall now give an optimal estimate for the convergence of the W-cycle based on the elliptic regularity assumption (4.3) (which implies that (4.20) holds).

Theorem 6.6 *Under the elliptic regularity assumption (4.3) for some $\alpha \in (0, 1]$, the W-cycle iteration admits the following estimate:*

$$\|I - B_k A_k\|_A^2 \leq \frac{M}{m^\alpha + M} \tag{6.16}$$

for some constant M that is independent of mesh parameters.

Proof Let $\delta = \frac{M}{m^\alpha + M}$. The estimate (6.16) will be proved by induction. As there is nothing to prove for $k = 0$, we assume that (6.16) is valid for $k-1$. By definition, the following recurrence relation holds for any $v \in \mathcal{M}_k$.

$$A((I - B_k A_k)v, v) = A((I - P_{k-1})\tilde{v}, \tilde{v}) + A((I - B_{k-1}A_{k-1})P_{k-1}\tilde{v}, P_{k-1}\tilde{v})$$

where $\tilde{v} = K_k^m v$.

$$\begin{aligned}
&\|(I - B_k A_k)v\|_A^2 \\
&\leq \; \|(I - P_{k-1})\tilde{v}\|_A^2 + \delta\|P_{k-1}\tilde{v}\|_A^2 \quad \text{(by induction)} \\
&\leq \; (1 - \delta^2)\|(I - P_{k-1})\tilde{v}\|_A^2 + \delta^2\|\tilde{v}\|_A^2
\end{aligned}$$

$$\leq (1-\delta^2)(c_1\lambda_k^{-1}\|A_k\tilde{v}\|^2)^\alpha \|\tilde{v}\|_A^{2(1-\alpha)} + \delta^2\|\tilde{v}\|_A^2 \quad \text{(by (4.20))}$$
$$\leq (1-\delta^2)\left[\frac{c_1}{2mc_0}(\|v\|_A^2 - \|\tilde{v}\|_A^2)\right]^\alpha \|\tilde{v}\|_A^{2(1-\alpha)} + \delta^2\|\tilde{v}\|_A^2 \quad \text{(by (6.15))}.$$

Note that $t \equiv \|\tilde{v}\|_A^2/\|v\|_A^2 \in [0,1]$; the desired estimate then easily follows if we can prove the following elementary inequality:

$$(1-\delta^2)\left[\frac{c_1}{2mc_0}(1-t)\right]^\alpha t^{1-\alpha} + \delta^2 t \leq \delta \quad t \in [0,1]. \tag{6.17}$$

By the Hölder inequality, for any $\eta > 0$ we have

$$(1-t)^\alpha t^{1-\alpha} \leq \alpha\eta(1-t) + (1-\alpha)\eta^{\frac{\alpha}{\alpha-1}} t.$$

Thus (6.17) is a consequence of the inequality

$$(1-\delta^2)\left[\frac{c_1}{2mc_0}\right]^\alpha \alpha\eta(1-t) + ((1-\alpha)\eta^{\frac{\alpha}{\alpha-1}} + \delta^2)t \leq \delta, \quad t \in [0,1].$$

The choice of η is made to minimize the above left hand side, namely to equalize the coefficients of $1-t$ and t. The proof may be completed with some more elementary manipulations by choosing sufficiently large M in the expression for δ. □

6.5 Subspace correction interpretation

We shall now discuss the multigrid method from the subspace correction point of view. Let $\mathcal{M}_k$ $(k = 0, 1, \cdots, J)$ be the multilevel finite element spaces defined as in the preceding section. Again let $\mathcal{V} = \mathcal{M}_J$, but set $\mathcal{V}_k = \mathcal{M}_{J-k}$. In this case, the decomposition (3.1) is trivial.

We observe that, with the above choice of subspaces $\mathcal{V}_k$, there are redundant overlappings in the decomposition (3.1). The point is that these overlappings can be used advantageously in the choice of the subspace solvers in a simple fashion. Roughly speaking, the subspace solvers need only take care of those 'nonoverlapped parts' (which correspond to the so-called *high frequencies*). As we know, methods like the Gauss–Seidel method discussed earlier just satisfy such requirements.

With the above ingredients, the successive subspace correction method Algorithm 3.3 can be stated as follows.

Algorithm 6.6 *Given* $u^0 \in \mathcal{V}$.

```
for k = 0, 1, ... till convergence
    v ← u^k
    for i = J : -1 : 0    v ← v + R_i Q_i (f - Av)    endfor
    u^{k+1} ← v.
endfor
```

Lemma 6.7 *For the Algorithm 6.1 with $m = 1$, we have*

$$I - B_J A_J = (I - T_0)(I - T_1)\cdots(I - T_J) \tag{6.18}$$

where

$$T_0 = P_0, T_k = R_k A_k P_k, \quad 1 \le k \le J.$$

Hence, with such defined operators B_k, the iteration (6.11) is mathematically equivalent to Algorithm 6.6.

Furthermore if B_J^m is obtained from Algorithm 6.1 for general $m \ge 1$, then

$$\|I - B_J^m A_J\|_A \le \|I - B_J A_J\|_A.$$

Based on the above lemma, different proofs may be obtained of the convergence of the backslash cycle multigrid method. In particular, the framework given in §3 can be applied. This new analysis does not depend crucially on the elliptic regularity and hence can be applied more easily to more complicated situations such as the problems with large discontinuous jump coefficients and locally refined meshes (see §6.9) .

We now consider the more general case in which we do not assume any elliptic regularity for the underlying partial differential equations (4.1).

We assume that all the smoothers R_k satisfy

$$(R_k v, v) \le \omega_1 (A_k^{-1} v, v) \quad \forall\, v \in \mathcal{M}_k. \tag{6.19}$$

and, if the R_k are all symmetric,

$$\frac{c_0}{\lambda_k}(v, v) \le (R_k v, v) \le \frac{c_1}{\lambda_k}(v, v) \quad \forall\, v \in \mathcal{M}_k, \tag{6.20}$$

or, in general,

$$\frac{c_0}{\lambda_k}(v, v) \le (\bar{R}_k v, v) \le \frac{c_1}{\lambda_k}(v, v) \quad \forall\, v \in \mathcal{M}_k, \tag{6.21}$$

where $\bar{R}_k$ is the symmetrization of R_k (see eqn. (2.4)).

By Lemma 6.3, the Gauss–Seidel methods satisfy the above assumptions.

The idea is to use the general framework of §3. According to the theory there, we need to estimate two basic parameters, namely $\bar{K}_0$ and $\bar{K}_1$.

By Theorem 4.8 and (6.20), there exists a constant C_0 independent of mesh parameters such that

$$K_0 \le C_0. \tag{6.22}$$

Lemma 6.8 *The $\bar{\epsilon}_{ij}$ defined in (3.14) satisfy, for some $\gamma \in (0, 1)$ independent of mesh parameters,*

$$\bar{\epsilon}_{ij} \lesssim \gamma^{|i-j|/2}. \tag{6.23}$$

Proof Let $i \geq j$. It follows from Lemma 4.5 and (6.20) that

$$(\bar{T}_i v_j, v_j)_A \lesssim \gamma^{i-j} h_i^{-1} \|v_j\|_A \|\bar{T}_i v_j\| \lesssim \gamma^{i-j} (v_j, v_j)_A \quad \forall\, v_j \in \mathcal{V}_j.$$

The desired estimate then follows from the definition of $\bar{\epsilon}_{ij}$. □

By definition, we conclude from the above lemma that there exists a constant C_1 independent of mesh parameters such that

$$\bar{K}_1 \leq C_1. \tag{6.24}$$

With the above results, the following result follows directly from Theorem 3.18.

Theorem 6.9 *Assume that the smoothers R_k satisfy (6.20) and (6.19) with $\omega_1 < 2$; then the backslash cycle (Algorithm 6.1 or Algorithm 6.6) satisfies*

$$\|I - B_J A_J\|_A^2 \leq 1 - \frac{2 - \omega_1}{C}$$

for some positive constant C independent of mesh parameters.

Another convergence analysis using full elliptic regularity If the full elliptic regularity is valid, however, a more straightforward proof can also be obtained in this new framework. We shall first present such a proof.

Theorem 6.10 *For the iteration (6.11) with B_J given by Algorithm 6.1 with one smoothing on each level, then*

$$\|I - B_J A_J\|_A^2 \leq 1 - \frac{c_0}{c_1}.$$

Proof Denote $E_{-1} = I$ and for $0 \leq i \leq J$,

$$E_i = (I - T_i)(I - T_{i-1}) \cdots (I - T_1)(I - T_0).$$

Note that $E_J = (I - B_J A_J)^*$. It follows that, denoting $\tilde{P}_i = P_i - P_{i-1}$,

$$\begin{aligned}
\|v\|_A^2 &= \sum_{i=0}^{J} (\tilde{P}_i v, v)_A \\
&= \sum_{i=0}^{J} (\tilde{P}_i v, E_{i-1} v)_A \quad (\text{since } (I - E_{i-1})v \in \mathcal{M}_{i-1}) \\
&= \sum_{i=0}^{J} (\tilde{P}_i v, A_i P_i E_{i-1} v) \\
&\leq \sqrt{c_1} \sum_{i=0}^{J} \lambda_i^{-1/2} \|\tilde{P}_i v\|_A \|A_i P_i E_{i-1} v\| \quad (\text{by } (6.14))
\end{aligned}$$

$$
\begin{aligned}
&\le \sqrt{\frac{c_1}{c_0}} \sum_{i=0}^{J} \|\tilde{P}_i v\|_A (R_i A_i P_i E_{i-1} v, A_i P_i E_{i-1} v)^{1/2} \quad \text{(by (6.13))} \\
&\le \sqrt{\frac{c_1}{c_0}} \sum_{i=0}^{J} \|\tilde{P}_i v\|_A (T_i E_{i-1} v, E_{i-1} v)_A^{1/2} \\
&\le \sqrt{\frac{c_1}{c_0}} \left(\sum_{i=0}^{J} \|\tilde{P}_i v\|_A^2 \right)^{1/2} \left(\sum_{i=0}^{J} (T_i E_{i-1} v, E_{i-1} v)_A \right)^{1/2} \\
&\le \sqrt{\frac{c_1}{c_0}} \left(\sum_{i=0}^{J} \|\tilde{P}_i v\|_A^2 \right)^{1/2} \left(\sum_{i=0}^{J} (T_i E_{i-1} v, E_{i-1} v)_A \right)^{1/2} \\
&\le \sqrt{\frac{c_1}{c_0}} \|v\|_A \left(\|v\|_A^2 - \|E_J v\|_A^2 \right)^{1/2}.
\end{aligned}
$$

Consequently,

$$
\|E_J v\|_A^2 \le \left(1 - \frac{c_0}{c_1}\right) \|v\|_A^2 \quad \forall\, v \in \mathcal{V}.
$$

The desired estimate then follows easily. □

6.6 Full multigrid cycle

We shall now describe a more efficient multigrid technique, called a full multigrid cycle, originally proposed by Brandt.

On each level of the finite element space $\mathcal{V}_k$, there is a corresponding finite element approximation $u^{(k)} \in \mathcal{V}_k$ such that

$$
A(u^{(k)}, v) = (F, v) \quad \forall v \in \mathcal{V}. \tag{6.25}
$$

Similarly to (4.10), the best error estimate in the H^1 norm is

$$
\|U - u^{(k)}\|_1 = O(h_k^{\alpha}). \tag{6.26}
$$

If $\mu^{(k)} \in \mathbb{R}^{n_k}$ is the nodal value vector of $u^{(k)}$, then

$$
\mathcal{A}_k \mu^{(k)} = b^{(k)} \tag{6.27}
$$

where $b^{(k)} = ((F, \varphi_i^k))$. It can be proved that, with $\mathcal{I}_k^{k+1}$ given by (6.12),

$$
b^{(k)} = (\mathcal{I}_k^{k+1})^t b^{(k+1)} \quad \text{with } b^{(J)} = b.
$$

The full multigrid method is based on the following two observations.

(1) $u^{(k-1)} \in \mathcal{V}_{k-1} \subset \mathcal{V}_k$ is close to $u^{(k)} \in \mathcal{V}_k$ and hence can be used as an initial guess for an iterative scheme for solving $u^{(k)}$.

(2) Each $u^{(k)}$ can be solved within its truncation error shown in (6.26) by a multigrid iterative scheme.

Algorithm 6.7 $\hat{\mu}^{(1)} \leftarrow \mathcal{A}_1^{-1} b^{(1)}$. *For* $k = 2 : J$

(1) $\hat{\mu}^{(k)} \leftarrow \mathcal{I}_{k-1}^{k} \hat{\mu}^{(k-1)}$,

(2) *iterate* $\hat{\mu}^{(k)} \leftarrow \hat{\mu}^{(k)} + \mathcal{B}_k (b^{(k)} - \mathcal{A}_k \hat{\mu}^{(k)})$ *m times.*

The most important fact about the full multigrid method is that it has an optimal computational complexity $O(N)$ to compute the solution within truncation error.

Proposition 6.11 *Assume that that C_0 is a positive constant satisfying (for all k)*

$$\|\mu^{(k)} - \mu^{(k-1)}\|_{\mathcal{A}} \leq C_0 h_k^{\alpha}.$$

Then

$$\|\mu^{(k)} - \hat{\mu}^{(k)}\|_{\mathcal{A}} \leq h_k^{\alpha}$$

if

$$m \geq \frac{\log(2^{\alpha} + C_0)}{|\log \delta|}.$$

Proof By definition, $\|\mu^{(1)} - \hat{\mu}^{(1)}\|_{\mathcal{A}} = 0$. Now assume that

$$\|\mu^{(k-1)} - \hat{\mu}^{(k-1)}\|_{\mathcal{A}} \leq h_{k-1}^{\alpha}.$$

Then

$$\begin{aligned} \|\mu^{(k)} - \hat{\mu}^{(k)}\|_{\mathcal{A}} &\leq \delta^m \|\mu^{(k)} - \hat{\mu}^{(k-1)}\|_{\mathcal{A}} \\ &\leq \delta^m \|\mu^{(k)} - \mu^{(k-1)}\|_{\mathcal{A}} + \delta^m \|\mu^{(k-1)} - \hat{\mu}^{(k-1)}\|_{\mathcal{A}} \\ &\leq \delta^m (C_0 + 2^{\alpha}) h_k^{\alpha} \\ &\leq h_k^{\alpha} \end{aligned}$$

□

6.7 BPX multigrid preconditioners

We shall now describe a parallelized version of the multigrid method studied earlier. This method was first proposed by Bramble, Pasciak and Xu [9] and by Xu [32], and is now often known as the BPX preconditioner in the literature.

6.7.1 *Basic algorithm and theory*

There are different ways of deriving the BPX preconditioners. The method originally resulted from an attempt to parallelize the classical multigrid method. With the current multigrid theoretical technology, the derivation of this method is not so difficult. We shall first derive this preconditioner based on Theorem 4.8 and then, in the next subsection, study the method using the framework of subspace correction.

By Theorem 4.8, we have

$$(Av, v) \eqsim \sum_{k=0}^{J} h_k^{-2} \|(Q_k - Q_{k-1})v\|^2 = (\hat{A}v, v), \quad v \in \mathcal{V},$$

where

$$\hat{A} = \sum_k h_k^{-2}(Q_k - Q_{k-1}).$$

Using Lemma 4.2, we can show that

$$\hat{A}^{-1} = \sum_k h_k^2 (Q_k - Q_{k-1}).$$

Using the fact that $h_k = 2h_{k+1}$, we deduce that

$$\begin{aligned}
(\hat{A}^{-1}v, v) &= \sum_{k=0}^{J} h_k^2((Q_k - Q_{k-1})v, v) \\
&= \sum_{k=0}^{J} h_k^2(Q_k v, v) - \sum_{k=0}^{J-1} h_{k+1}^2(Q_k v, v) \\
&\eqsim h_J^2(v, v) + \sum_{k=0}^{J-1} h_k^2(Q_k v, v) \\
&\eqsim \sum_{k=0}^{J} h_k^2(Q_k v, v) = (\tilde{B}v, v)
\end{aligned}$$

where

$$\tilde{B} = \sum_{k=0}^{J} h_k^2 Q_k.$$

If $R_k : \mathcal{M}_k \mapsto \mathcal{M}_k$ is an SPD operator satisfying

$$(R_k v_k, v_k) \eqsim h_k^2(v_k, v_k) \quad \forall\, v_k \in \mathcal{M}_k \tag{6.28}$$

then, for

$$B = \sum_{k=0}^{J} R_k Q_k \tag{6.29}$$

we have

$$(Bv, v) \eqsim (\tilde{B}v, v) \eqsim (A^{-1}v, v),$$

namely

$$\kappa(BA) \eqsim 1.$$

Theorem 6.12 *Assume that the R_k satisfy (6.28); then the preconditioner (3.6) satisfies*

$$\kappa(BA) \eqsim 1.$$

We note that all the relaxation methods mentioned earlier, such as Richardson, Jacobi and symmetric Gauss–Seidel, satisfy (6.28).

6.7.2 *Subspace correction approach*

In §6.5, the slash cycle multigrid method is interpreted as a successive subspace correction method. Correspondingly, the BPX preconditioner can be interpreted as the relevant PSC (parallel subspace correction) preconditioner. It is possible to use the abstract theory in §3 to derive some estimates for the BPX preconditioner somewhat more refined than that in Theorem 6.12; we leave the details to the interested readers. In §6.9, we shall use this approach to analyze the BPX preconditioner for locally refined meshes.

Implementation Again, in view of (3.27), the algebraic representation of the preconditioner given by (3.27) is

$$\mathcal{B} = \sum_{k=0}^{J} \mathcal{I}_k \mathcal{R}_k \mathcal{I}_k^t, \tag{6.30}$$

where $\mathcal{I}_k \in \mathbb{R}^{n \times n_k}$ is the representation matrix of the nodal basis $\{\varphi_i^k\}$ in $\mathcal{M}_k$ in terms of the nodal basis $\{\varphi_i\}$ of $\mathcal{M}$, i.e.

$$(\varphi_1^k, \cdots, \varphi_{n_k}^k) = (\varphi_1, \cdots, \varphi_n)\mathcal{I}_k.$$

Let $\mathcal{I}_k^{k+1} \in \mathbb{R}^{n_{k+1} \times n_k}$ be as defined in (6.12); then

$$\mathcal{I}_k = \mathcal{I}_{J-1}^J \cdots \mathcal{I}_{k+1}^{k+2} \mathcal{I}_k^{k+1}.$$

This identity is very useful for the efficient implementation of (6.30) on both serial and parallel machines.

If $\mathcal{R}_k$ are given by the Richardson iteration, we have

$$\mathcal{B} = \sum_{k=0}^{J} h_k^{2-d} \mathcal{I}_k \mathcal{I}_k^t. \tag{6.31}$$

From (6.30) or (6.31), we see that the preconditioner depends entirely on the transformation between the nodal bases on multilevel spaces.

For $1 \le l \le J$, let

$$\mathcal{B}_l = \sum_{k=0}^{l} \mathcal{I}_k^l \mathcal{R}_k (\mathcal{I}_k^l)^t.$$

By definition $\mathcal{B} = \mathcal{B}_J$ and

$$\mathcal{B}_l = \mathcal{R}_l + \mathcal{I}_{l-1}^l \mathcal{B}_{l-1} (\mathcal{I}_{l-1}^l)^t.$$

We shall use the above recurrence relation to compute the action of $\mathcal{B}$. Assume that m_l is the number of operations that are needed to compute the action $\mathcal{B}_l \alpha_l$ for $\alpha_l \in \mathbb{R}^{n_l}$. By the identity

$$\mathcal{B}_l \alpha_l = \mathcal{R}_l \alpha_l + \mathcal{I}_{l-1}^l [B_{l-1} (\mathcal{I}_{l-1}^l)^t \alpha_l]$$

we get, for some constant $c_0 > 0$,

$$m_l \leq m_{l-1} + c_0 n_l$$

from which we conclude that

$$m_J \leq m_1 + c_0 \sum_{l=2}^{J} n_l \leq c_1 n$$

for some positive constant c_1. This means that the action of B_J can be carried out within $O(n)$ operations.

Algorithm 6.8. (Computation of $\mathcal{B}\alpha$)

$\alpha_J = \alpha$;
for $l = J : 1$,
  $\alpha_{l-1} = (\mathcal{I}_{l-1}^l)^t \alpha_l$;
end

$\beta_0 = \mathcal{R}_0 \alpha_0$;
for $l = 1 : J$,
  $\beta_l = \mathcal{R}_l \alpha_l + \mathcal{I}_{l-1}^l \beta_{l-1}$;
end
$\mathcal{B}\alpha = \beta_J$.

As discussed above, the number of operations needed in the above algorithm is $O(n)$. We also note that all the vectors α_l for $1 \leq k \leq J$ need to be stored, but the whole storage space for these vectors is also only $O(n)$.

6.8 Hierarchical basis methods

Assume that we are given a nested sequence of multigrid subspaces of $H_0^1(\Omega)$,

$$\mathcal{M}_1 \subset \mathcal{M}_2 \subset \ldots \subset \mathcal{M}_k \subset \ldots \subset \mathcal{M}_J,$$

as described in §4.2. The so-called *hierarchical basis* refers to the special set of nodal basis functions

$$\{\varphi_i^k : x_i^k \in \mathcal{N}_k \setminus \mathcal{N}_{k-1}, k = 0, \cdots, J\}. \tag{6.32}$$

It is easy to see that this set of functions does form a basis of $\mathcal{M}$. For $d \neq 2$, it is often more convenient to use the scaled HB (hierarchical basis) as follows

$$\{h_k^{2-d}\varphi_i^k : x_i^k \in \mathcal{N}_k \setminus \mathcal{N}_{k-1}, k = 0, \cdots, J\}. \tag{6.33}$$

With a proper ordering, we shall denote the scaled HB by $\{\psi_i, i = 1 : N\}$.

The HB in multiple dimensions is formally a direct generalization of the HB in the one dimensional case. But the property for the corresponding stiffness matrix in multiple dimensions is not at all as clear as in one dimension where the stiffness matrix is an identity matrix in some special cases. In this section, we shall show that at least in two dimensions, a hierarchical basis is still very useful.

The hierarchical basis in two dimensions was first analysed by Yserentant in his pioneering paper [43]. The work of Yserentant was apparently motivated by the famous unpublished technical report of Bank and Dupont [1]. Incidently these three authors got together and wrote another important paper [3] on a Gauss–Seidel (or multiplicative) variant of the hierarchical basis method. The presentation of the materials in this section is of course mostly based on the aforementioned papers, and moreover it also adopts the view of subspace correction from Xu [33] (see also Xu [32]).

6.8.1 *Preliminaries*

We shall now discuss multigrid subspaces that are directly related to the HB. Consider the part of the HB functions on level k as follows

$$\{\varphi_i^k : x_i^k \in \mathcal{N}_k \setminus \mathcal{N}_{k-1}\}. \tag{6.34}$$

It is easy to see that the above set of functions spans the subspace

$$\mathcal{V}_k = (I_k - I_{k-1})\mathcal{M} = (I - I_{k-1})\mathcal{M}_k, \quad \text{for } k = 0 : J. \tag{6.35}$$

Here, we recall, $I_{-1} = 0$ and $I_k : \mathcal{M} \mapsto \mathcal{M}_k$ is the nodal value interpolant. The above subspaces obviously give rise to a direct sum decomposition of the space $\mathcal{M}$ as follows:

$$\mathcal{V} = \oplus_{k=0}^J \mathcal{V}_k.$$

In fact, for any $v \in \mathcal{V}$, we have the following unique decomposition:

$$v = \sum_{k=0}^J v_k \quad \text{with} \quad v_k = (I_k - I_{k-1})v.$$

With the subspaces $\mathcal{V}_k$ given by (6.35), the operators A_k are all well conditioned. In fact, by (4.19) and (4.5), we can see that

$$A(v, v) \eqsim h_k^{-2}\|v\| \quad \forall\, v \in \mathcal{V}_k.$$

As a result, the subspace equations can be solved effectively by elementary iterative methods such as Richardson, Jacobi and Gauss–Seidel methods.

6.8.2 *Stiffness matrix in terms of hierarchical basis*

The easiest way of understanding the HB is perhaps, like in one dimension, through the study of the property of the corresponding stiffness matrix. As one may expect, the condition number of the HB stiffness matrix should be smaller than the NB (normal basis) stiffness matrix. This is indeed the case in two and three dimensions.

Theorem 6.13 *Assume that $\hat{\mathcal{A}}$ is the stiffness matrix under the scaled hierarchical basis; then*

$$\kappa(\hat{\mathcal{A}}) \lesssim \kappa_d(h) \tag{6.36}$$

where

$$\kappa_d(h) \lesssim \begin{cases} 1 & \text{if } d = 1; \\ |\log h|^2 & \text{if } d = 2; \\ h^{2-d} & \text{if } d \geq 3. \end{cases} \tag{6.37}$$

In fact, the estimates given in the above theorem can be proven to be sharp. The most interesting case is obviously $d = 2$ for which $\kappa(\hat{\mathcal{A}}) \lesssim |\log h|^2$. Compared with the conditioning of the stiffness matrix under the NB, this is a great improvement. It is also in the case $d = 2$ that the HB is most useful. Indeed for $d = 3$, the $\kappa(\hat{\mathcal{A}}) = O(h^{-1})$ is also one magnitude smaller than the condition number of the NB stiffness matrix, but such an improvement is not attractive as we shall see that a much better approach (such as the BPX preconditioner) is available. There is no doubt that, as far as $\kappa(\hat{\mathcal{A}})$ is concerned, the HB is of no use for $d \geq 4$.

Proof Given $\alpha \in \mathbb{R}^N$, set $v = \sum_{i=1}^N \alpha_i \psi_i$. We can write

$$v = \sum_k v_k \quad \text{with} \quad v_k = (I_k - I_{k-1})v = \sum_{x_i^k \in \mathcal{N}_k \backslash \mathcal{N}_{k-1}} \alpha_i^k \varphi_i^k.$$

It follows that

$$|v_k|_1^2 \eqsim \sum_{x_i^k \in \mathcal{N}_k \backslash \mathcal{N}_{k-1}} v_k^2(x_i^k) = \sum_{x_i^k \in \mathcal{N}_k \backslash \mathcal{N}_{k-1}} (\alpha_i^k)^2 = |\alpha|^2.$$

Thus

$$\begin{aligned} \alpha^t \hat{\mathcal{A}} \alpha &= A(v, v) = \sum_{k,l} A(v_k, v_l) \\ &\lesssim \sum_{k,l} \gamma^{|k-l|} |v_k|_1 |v_l|_1 \quad \text{(by Lemma 4.6)} \\ &\lesssim \sum_k |v_k|_1^2 \eqsim |\alpha|^2. \end{aligned}$$

This implies that $\lambda_{\max}(\hat{\mathcal{A}}) \lesssim 1$. On the other hand

$$\begin{aligned}
|\alpha|^2 &\eqsim \sum_k |v_k|_1^2 = \sum_k |(I_k - I_{k-1})v|_1^2 \\
&\lesssim \sum_k (J-k+1)|v|_1^2 \quad \text{(by (4.19))} \\
&\lesssim \kappa_d(h)|v|_1^2 \lesssim \kappa_d(h)\alpha^t\hat{\mathcal{A}}\alpha
\end{aligned}$$

This proves that $\lambda_{\min}(\hat{\mathcal{A}}) \gtrsim \kappa_d(h)^{-1}$. □

The above proof is essentially the same as in Yserentant [43] (and see also Ong [28] for $d = 3$).

6.8.3 *Subspace correction approach and general case*

Following Xu [33], we shall now study the HB method from the viewpoint of space decomposition and subspace correction.

In view of (3.27), the algebraic representation of the PSC preconditioner is

$$\mathcal{H} = \sum_{k=0}^{J} \mathcal{S}_k \mathcal{R}_k \mathcal{S}_k^t, \tag{6.38}$$

where $\mathcal{S}_k \in \mathbb{R}^{n\times(n_k - n_{k-1})}$ is the representation matrix of the nodal basis $\{\varphi_i^k\}$ in $\mathcal{M}_k$, with $x_i^k \in \mathcal{N}_k \setminus \mathcal{N}_{k-1}$, in terms of the nodal basis $\{\varphi^i\}$ of $\mathcal{M}$.

A special case If $\mathcal{R}_k$ is given by the Richardson iteration $\mathcal{R}_k = h_k^{2-d}\mathcal{I}$, we have

$$\mathcal{H} = \sum_{k=0}^{J} h_k^{2-d} \mathcal{S}_k \mathcal{S}_k^t = \hat{\mathcal{S}}\hat{\mathcal{S}}^t.$$

where

$$\mathcal{S} = (h_1^{1-d/2}\mathcal{S}_1, h_2^{1-d/2}\mathcal{S}_2, \cdots, h_J^{1-d/2}\mathcal{S}_J)$$

is the representation matrix of the HB in terms of NB. Obviously the HB stiffness matrix and NB stiffness matrix are related by $\hat{\mathcal{A}} = \mathcal{S}^t\mathcal{A}\mathcal{S}$. Therefore,

$$\kappa(\mathcal{H}\mathcal{A}) = \kappa(\hat{\mathcal{A}}),$$

and, as a result of Theorem 6.13,

$$\kappa(\mathcal{H}\mathcal{A}) \eqsim \kappa_d(h). \tag{6.39}$$

The above estimate apparently also holds for the more general $\mathcal{H}$ when $\mathcal{R}_k$ is given by either Jacobi or symmetric Gauss–Seidel since either of these iterations satisfies the following spectral equivalence:

$$\mathcal{R}_k \eqsim h_k^{2-d}. \tag{6.40}$$

In fact, the estimate (6.39) also follows easily from the general theory for the PSC preconditioner.

For the SSC iterative method, it is more convenient to choose $\mathcal{R}_k$ to be the symmetric Gauss–Seidel as the other two methods need to be properly scaled to assume that $\omega_1 \in (0,2)$. The resulting algorithm is as follows.

Algorithm 6.9 *Let $\mu^0 \in \mathbb{R}^n$ be given. Assume that $\mu^k \in \mathbb{R}^n$ is obtained. Then μ^{l+1} is defined by*

$$\mu^{l+(k+1)/(J+1)} = \mu^{l+k/(J+1)} + \mathcal{S}_k \mathcal{R}_k \mathcal{S}_k^t (\eta - \mathcal{A}\mu^{l+(i-1)/J})$$

for $k = 0 : J$.

It can be proved later that

$$\|\mu - \mu_l\|_{\mathcal{A}} \leq \left(1 - \frac{c}{\kappa_d(h)}\right)^l \|\mu - \mu_0\|_{\mathcal{A}}. \tag{6.41}$$

6.8.4 *Convergence analysis*

Lemma 6.14 *We assume that R_k is either Richardson, or Jacobi or symmetric Gauss–Seidel iteration; then*

$$\lambda_k^{-2}\|v\|_A^2 \lesssim (R_k A_k v, v)_A \leq \omega_1 (v,v)_A, \quad \forall\, v \in \mathcal{V}_k.$$

where ω_1 is a constant and for symmetric Gauss–Seidel, $\omega_1 = 1$.

Proof The proof for the Richardson or Jacobi method is straightforward. The proof for the symmetric Gauss–Seidel method is almost identical to that of Lemma 6.2 and the detail is left to the reader. □

Lemma 6.15

$$K_0 \lesssim c_d \quad \textit{and} \quad K_1 \lesssim 1,$$

where $c_1 = 1, c_2 = J^2$ and $c_d = 2^{(d-2)J}$ for $d \geq 3$.

Proof For $v \in \mathcal{V}$, it follows from (4.19) that

$$\sum_{k=0}^{J} h_k^{-2}\|v_k\|^2 \lesssim c_d \|v\|_A^2.$$

This gives the estimate of K_0. The estimate of K_1 follows from Lemma 4.6. □

For the SSC iterative method, we apply Theorem 3.18 with Lemma 6.15 and get

Theorem 6.16 *The Algorithm 3.3 with the subspaces $\mathcal{V}_k$ given by (6.35) satisfies*

$$\|E_J\|_A^2 \leq 1 - \frac{2-\omega_1}{Cc_d}$$

provided that the R_k satisfy (6.14) with $\omega_1 < 2$.

Compared with the usual multigrid method, the smoothing in the SSC hierarchical basis method is carried out only on the set of new nodes $\mathcal{N}_k \setminus \mathcal{N}_{k-1}$ on each subspace $\mathcal{M}_k$. The method proposed by Bank, Dupont and Yserentant [3] can be viewed as such an algorithm with R_k given by an appropriate Gauss–Seidel iteration. Numerical examples in [3] show that the hierarchical basis SSC algorithm converges much faster than the corresponding SSC algorithm.

6.8.5 *Relation with BPX preconditioners*

Observing that S_k in (6.38) is a submatrix of $\mathcal{I}_k$ given in (6.30), we then have

$$(\mathcal{H}\alpha, \alpha) \le (\mathcal{B}\alpha, \alpha), \quad \forall\, \alpha \in \mathbb{R}^n.$$

In view of the above inequality, if we take

$$\hat{\mathcal{H}} = \sum_{k=0}^{J-1} h_k^{2-d} \mathcal{S}_k \mathcal{S}_k^t + I,$$

we obtain

$$(\mathcal{H}\alpha, \alpha) \le (\hat{\mathcal{H}}\alpha, \alpha) \le (\mathcal{B}\alpha, \alpha), \quad \forall\, \alpha \in \mathbb{R}^n.$$

Even though $\hat{H}$ appears to be a very slight variation of H, numerical experiments have shown a great improvement over H for $d = 2$. We refer to Xu and Qin [42] for the numerical results.

6.9 Locally refined grids

In practical computations, finite element grids are often locally refined (by using some error estimators or other adaptive strategies). In this subsection, we shall describe optimal multigrid procedures for adaptive grids. Our presentation here is based on [9], [7] and [6].

With appropriate rearrangement and grouping, we may assume that the mesh refinement can be done in the following fashion. We first start with the original domain Ω which is also denoted by Ω_0. We introduce a relatively coarse and quasi-uniform triangulation of Ω_0 with a mesh size h_0 and denote the corresponding finite element space by $\mathcal{V}_0 \subset H_0^1(\Omega_0)$. Let Ω_1 be a subregion where we wish to increase the resolution: we do so by subdividing the elements of the first triangulation to get a new triangulation of Ω_1 with mesh size h_1 in Ω_1 and we introduce an additional finite element space $\mathcal{V}_1 \subset H_0^1(\Omega_1)$. We repeat this process and finally get a collection of subdomains Ω_i together with the corresponding finite element spaces $\mathcal{V}_i$ defined on a triangulation of mesh size h_i for $i = 1, 2, \cdots, J$ for some integer $J > 1$. Throughout, we have

$$\Omega_i \subset \Omega_{i-1}, \quad \mathcal{V}_{i-1} \cap H_0^1(\Omega_i) \subset \mathcal{V}_i \subset H_0^1(\Omega_i), \quad i = 1, 2, \cdots, J.$$

The finite element space on the repeatedly refined mesh can be written as

$$\mathcal{V} = \sum_{i=0}^{J} \mathcal{V}_i.$$

The only restrictions on the mesh domains $\{\Omega_k\}$ are that $\partial\Omega_k$ for $k \geq 1$ consists of edges of mesh triangles in the mesh $\mathcal{T}_{k-1}$ and that there is at least one edge from $\mathcal{T}_{k-1}$ contained in Ω_k.

Let $A : \mathcal{V} \mapsto \mathcal{V}$ be as defined by (4.15). Operators $A_i : \mathcal{V}_i \mapsto \mathcal{V}_i$ and $Q_i, P_i : \mathcal{V} \mapsto \mathcal{V}_i$ can be defined as before. If we choose $R_i : \mathcal{V}_i \mapsto \mathcal{V}_i$ to be some appropriate approximate solvers of the A_i, we then have all the ingredients to define our PSC and SSC algorithms. In this setting, the coarse space $\mathcal{M}_0$ may not be very coarse: we therefore assume that, for the PSC type algorithm, the first subspace solver R_0 is SPD and satisfies

$$(R_0^{-1}v, v) \eqsim (A_0 v, v) \quad \forall\, v \in \mathcal{M}_0, \tag{6.42}$$

and for the SSC type of algorithm, we assume that R_0 satisfies

$$\|(I - R_0 A_0)\|_A \leq \delta_0 \tag{6.43}$$

for some $\delta_1 \in (0, 1)$ independent of mesh parameters.

As for the other subspace solvers R_k for $k > 1$, we assume for clarity that R_k is given by a Gauss–Seidel iteration or properly damped Jacobi iteration. Apparently other reasonable solvers can also be adopted.

We would like to remark that the corresponding PSC and SSC methods in this setting can be viewed as a 'nested' multigrid method associated with the multilevel spaces given by

$$\mathcal{M}_k = \sum_{i=0}^{k} \mathcal{V}_i, \quad 0 \leq k \leq J,$$

but with special coarse space solvers R_k only defined on the subspace $\mathcal{V}_k$, namely the smoothings are only carried out in the refined regions.

To analyze the corresponding PSC and SSC methods, we introduce, for each k, an auxiliary finite element space $\hat{\mathcal{M}}_k$ which is defined on a quasi-uniform triangulation with mesh size h_k and satisfies $\mathcal{M}_k \subset \hat{\mathcal{M}}_k$ and $\hat{\mathcal{M}}_0 \subset \cdots \subset \hat{\mathcal{M}}_J$. It is easy to see that $\hat{\mathcal{M}}_k$ can be well defined. Corresponding to the space $\hat{\mathcal{M}}_k$, let $\hat{Q}_k : \mathcal{V} \mapsto \hat{\mathcal{M}}_k$ be the L^2 projection.

The following result (from [7]) plays a crucial rôle in our analysis for the algorithms discussed in this section.

Lemma 6.17 *Assume that $h_{k-1}/h_k \leq C$. There exists a sequence of lin-*

ear operators $\Pi_k : \mathcal{V} \mapsto \mathcal{M}_k$ *for* $k = 0, 1, 2, \cdots, J$ *with* $\Pi_J = I$ *such that, for any* $v \in \mathcal{V}$, $(\Pi_k - \Pi_{k-1})v \in \mathcal{V}_k$,

$$||(I - \Pi_k)v|| \lesssim ||(I - \hat{Q}_k)v|| \tag{6.44}$$

and

$$||\Pi_k v||_A \lesssim ||v||_A. \tag{6.45}$$

Proof The linear operator Π_k is then defined, for $v \in \mathcal{V}$, by $\Pi_k v = w$, where w is the unique function in $\mathcal{M}_k$ satisfying

$$w = \begin{cases} \hat{Q}_k v & \text{at the nodes of } \mathcal{M}_k \text{ in the interior of } \Omega_{k+1}, \\ v & \text{at the remaining nodes of } \mathcal{M}_k. \end{cases}$$

By this definition, it is clear that $(\Pi_k - \Pi_{k-1})v \in \mathcal{V}_k$. To establish (6.45), we first note that

$$||\hat{Q}_k v - w||^2 \le Ch_k^2 \Sigma' (\hat{Q}_k v(x_i^k) - v(x_i^k))^2 \le C||(I - \hat{Q}_k)v||^2,$$

where the sum $\sum'$ is taken over the nodes x_i^k of $\mathcal{M}_k$ on $\partial\Omega_{k+1}$. Combining the above estimate with (4.11) yields

$$||(I - \Pi_k)v|| \le ||(I - \hat{Q}_k)v|| + ||\hat{Q}_k v - w|| \le ||(I - \hat{Q}_k)v||$$

This proves part of the estimate (6.44). The rest of (6.45) can be estimated similarly by using $||\Pi_k v||_A \le ||(I - \hat{Q}_k)v||_A + ||v - w||_A$. This completes the proof. □

Lemma 6.18 *Let* $\Pi_k : \mathcal{V} \mapsto \mathcal{M}_k$ *be the operator from the previous lemma and* $\Pi_{-1} = 0$; *then, there exists a constant* c_0 *independent of mesh parameters such that*

$$\sum_{k=0}^{J} \lambda_k ||(\Pi_k - \Pi_{k-1})v||_A^2 \le c_0 ||v||_A^2 \quad \forall\, v \in \mathcal{V}. \tag{6.46}$$

Proof By Lemma 6.17 we have, for $k \ge 1$,

$$\begin{aligned} ||(\Pi_k - \Pi_{k-1})v||_A^2 &\le 2(||(I - \Pi_k)v||_A^2 + ||(I - \Pi_{k-1})v||_A^2) \\ &\le 2(||(I - \hat{Q}_{k-1})v||_A^2 + ||(I - \hat{Q}_{k-1})v||_A^2) \end{aligned}$$

Thus, combining with (6.45),

$$\sum_{k=0}^{J} \lambda_k ||(\Pi_k - \Pi_{k-1})v||_A^2 \lesssim ||v||_A^2 + \sum_{k=0}^{J} \lambda_k ||(I - \hat{Q}_k)v||_A^2.$$

The desired estimate then follows by using Theorem 4.9. □

6.9.1 *PSC version*

Let us first consider the preconditioner corresponding to the PSC algorithm:

$$B = \sum_{k=0}^{J} R_k Q_k. \tag{6.47}$$

Thanks to Lemma 6.18, as for the quasi-uniform case, we can use our general framework in §3 to obtain the following theorem.

Theorem 6.19 *If R_0 satisfies (6.42) and R_k are given by Jacobi or symmetric Gauss–Seidel, then the PSC preconditioner (6.47) yields a uniformly bounded condition number*

$$\kappa(BA) \lesssim 1.$$

As a special example, we may choose R_k as follows:

$$R_k v = h_k^{2-d} \sum_{x_i^k \in \mathcal{N}_k} (v, \varphi_i^k)\varphi_i^k. \tag{6.48}$$

As we know, this corresponds to Richardson iteration which is equivalent to Jacobi iteration. For such a choice, we obtain that

$$Bv = R_0 v + \sum_{k=1}^{J} h_k^{2-d} \sum_{x_i^k \in \mathcal{N}_k} (v, \varphi_i^k)\varphi_i^k. \tag{6.49}$$

We notice that (6.49) is exactly the preconditioner given in [9] for locally refined meshes.

The hierarchical basis type algorithms for these composite grids can be obtained by the decomposition with $\mathcal{V}_i = (I_i - I_{i-1})\mathcal{V}$ (here $I_i : \mathcal{V} \mapsto \mathcal{M}_i$ is the nodal value interpolation operator). It is easy to see that the SSC type preconditioner is

$$Hv = R_0 v + \sum_{k=1}^{J} h_k^{2-d} \sum_{x_i^k \in \mathcal{N}_k \setminus \mathcal{N}_{k-1}} (v, \varphi_i^k)\varphi_i^k. \tag{6.50}$$

This preconditioner is equivalent to what is given in [43] for refined meshes. We further point out the corresponding algorithm in [3] for the refined meshes is the SSC algorithm by choosing R_k to be some appropriate Gauss–Seidel iteration.

6.9.2 *SSC version*

The SSC version corresponds to multigrid algorithms with smoothings done in the refined regions. Similarly we have the following convergence theorem.

Theorem 6.20 *If R_0 satisfies (6.43) and R_k are given by properly damped Jacobi or Gauss–Seidel, then the corresponding SSC iteration yields a uni-*

form contraction:

$$\|I - BA\|_A \leq \delta$$

for some $\delta \in [\delta_0, 1)$ *independent of mesh parameters.*

Additional bibliographic comments The multilevel algorithm for finite element or finite difference equations was first developed in the early sixties by the Russian mathematician Fedorenko [19]. In the early seventies, Brandt [12] brought this method to the attention of western countries and extensive research has been done on this method since then. Nowadays it has become one of the most popular and powerful iterative methods.

Multilevel methods for composite grids can be traced back to Brandt [12] or to composite grids in McCormick [24] (see also the references therein). The finite element space on a mesh refined in this way is $\mathcal{V} = \sum_{i=0}^{J} \mathcal{V}_i$. PSC and SSC methods can be naturally obtained with Gauss–Seidel iterations as subspace solvers. The SSC iteration corresponds to a multigrid method with smoothings carried out only on the refined region discussed in Brandt [12]. The PSC preconditioner was first considered in Bramble, Pasciak and Xu [9].

The aforementioned refined grids may not give the minimal degrees of freedom from an approximation point of view, but they are a computationally efficient approach and have the desirable structure for multigrid applications. If more traditional graded meshes are used on each level, proper nested subspaces are then hard to come by and the corresponding nonnested multigrid methods are more complicated (cf. Zhang [45]).

7 Multigrid for unstructured problems

Theories presented in previous sections are based on the assumptions that the multilevel subspaces are nested in the sense the the coarse spaces are subspaces of the finer spaces and that the bilinear forms on each level are all the same. This type of theory can only be effectively applied to problems with good structures (for example, conforming finite elements on structured grids as discussed above).

7.1 Nonnested subspaces and varying bilinear forms

In this section, we shall present a general multigrid theory that only depends on some weak assumptions (see Bramble, Pasciak and Xu [10], Xu [32]). Such a theory has been successfully applied to many situations and some examples of applications will be briefly mentioned near the end of this section.

Assume we are given a Hilbert space H and a hierarchy of real finite dimensional subspaces of H

$$\mathcal{M}_0, \mathcal{M}_1, \mathcal{M}_2, \ldots, \mathcal{M}_J$$

which are related by the so-called prolongation operators $I_k : \mathcal{M}_{k-1} \mapsto \mathcal{M}_k$.

In addition, let $A_k(\cdot,\cdot)$ and $(\cdot,\cdot)_k$ be symmetric positive definite bilinear forms on $\mathcal{M}_k$. We shall develop multigrid algorithms for the solution of the following problem: given $f \in \mathcal{M}_J$, find $u \in \mathcal{M}_J$ satisfying

$$A_J(u,\varphi) = (f,\varphi)_J \quad \forall \varphi \in \mathcal{M}_J.$$

To define the multigrid algorithms, we need to define some auxiliary operators. For $k = 0,\ldots,J$, the operator $A_k : \mathcal{M}_k \mapsto \mathcal{M}_k$ is defined by

$$(A_k w,\varphi)_k = A_k(w,\varphi) \quad \forall w,\varphi \in \mathcal{M}_k.$$

Clearly the operator A_k is symmetric positive definite (in both the $A_k(\cdot,\cdot)$ and $(\cdot,\cdot)_k$ inner products). In terms of the prolongation operator I_k, we have operators $I_k^t : \mathcal{M}_k \mapsto \mathcal{M}_{k-1}$ and $I_k^* : \mathcal{M}_k \mapsto \mathcal{M}_{k-1}$ defined, for all $w \in \mathcal{M}_k, \varphi \in \mathcal{M}_{k-1}$, by

$$(I_k^t w,\varphi)_{k-1} = (w, I_k\varphi)_k \quad A_{k-1}(I_k^* w,\varphi) = A_k(w, I_k\varphi). \qquad (7.1)$$

In other words, the I_k^t and I_k^* are the adjoints of I_k with respect to the inner products $(\cdot,\cdot)_k$ and $A_k(\cdot,\cdot)$ respectively. I_k^t is often called the *restriction* operator, which is another main ingredient of any multigrid algorithm.

Another important component of the multigrid algorithm is the *smoothing*, which will be represented by a sequence of linear operators $R_k : \mathcal{M}_k \mapsto \mathcal{M}_k$ for $1 \le k \le j$ to define the smoothing process. These operators may be symmetric or nonsymmetric with respect to the inner product $(\cdot,\cdot)_k$. If R_k is not symmetric, then we denote by R_k^t its adjoint and set

$$R_k^{(l)} = \begin{cases} R_k & \text{if } l \text{ is odd;} \\ R_k^t & \text{if } l \text{ is even.} \end{cases}$$

With the framework and notation given above, we are now in a position to define a multigrid algorithm, which will be characterized in terms of a sequence of recursively defined operators $B_k : \mathcal{M}_k \mapsto \mathcal{M}_k$. In the following, p, m_k are given positive integers and λ_k is either equal to $\rho(A_k)$ or is an upper bound of $\rho(A_k)$ such that $\lambda_k \eqsim \rho(A_k)$.

Algorithm S

Step 1 $B_0 = A_0^{-1}$.

Step 2 *Assume* B_{k-1} *is defined. Then for* $g \in \mathcal{M}_k (A_k w = g)$, B_k *is defined as follows.*

(1) Pre-smoothing *on* $\mathcal{M}_k$*:*

$$w^0 = 0$$
$$w^l = w^{l-1} + R_k^{(l+m_k)}(g - A_k w^{l-1}), \quad l = 1, 2, \cdots, m_k.$$

(2) Correction *on* $\mathcal{M}_{k-1}$*:* $w^{m_k+1} = w^{m_k} + I_k q^p$ *where* $q^p \in \mathcal{M}_{k-1}$ *is defined as follows, with* $r_k = g - A_k w^{m_k}$.

$$q^0 = 0$$
$$q^l = q^{l-1} + B_{k-1}(I_k^t r_k - A_{k-1} q^{l-1}), \quad l = 1, 2, \cdots, p.$$

(3) Post-smoothing *on* $\mathcal{M}_k$*:*

$$w^l = w^{l-1} + R_k^{(l+m_k+1)}(g - A_k w^{l-1})$$
$$l = m_k + 2, \cdots, 2m_k + 1.$$

Define: $B_k g = w^{2m_k+1}$.

Remark 7.1 Ordinarily, the above multigrid methods can be made more general. For example, in Step 1, B_1 may be defined by an iterative method which solves the equation approximately on $\mathcal{M}_1$. Another generalization is that the number of pre- and post-smoothings are not necessarily the same. Nevertheless we are not going to consider these more general cases here. However in some circumstances it seems crucial to our theory that the number of pre- and post smoothings should be the same, which will guarantee that the multigrid operators B_k are also symmetric (see Lemma 7.2). This is reasonable and important from many viewpoints. The most natural reason would be because the original problems are symmetric themselves.

A great advantage in setting out the algorithm by means of the operators B_k is that we have a very simple recurrence relation for the 'residue' operator $E_k \stackrel{\text{def}}{=} I - B_k A_k$ as given in the following lemma.

Lemma 7.1 *Let* $E_k = I - B_k A_k$ *and* $K_k = I - R_k A_k$*. Then*

$$E_k = (\tilde{K}_k^{m_k})^* \left((I - I_k I_k^*) + I_k E_{k-1}^p I_k^*\right) \tilde{K}_k^{m_k}. \tag{7.2}$$

Furthermore, for any $u, v \in \mathcal{M}_k$

$$A_k(E_k u, v) = A_k((I - I_k I_k^*)\tilde{u}, \tilde{v}) + A_{k-1}(E_{k-1}^p I_k^* \tilde{u}, I_k^* \tilde{v}), \tag{7.3}$$

where $\tilde{u} = \tilde{K}_k^{m_k} u$ *and*

$$\tilde{K}_k^{m_k} = \begin{cases} (K_k^* K_k)^{\frac{m-1}{2}} K_k^*, & \text{if } m \text{ is odd;} \\ (K_k^* K_k)^{\frac{m}{2}}, & \text{if } m \text{ is even.} \end{cases}$$

The verification of the above lemma is straightforward by the definition of the algorithm. The next thing we want to address is that the **Algorithm S** defines a symmetric operator. More specifically, we have

Lemma 7.2 *B_k is symmetric with respect to $(\cdot,\cdot)_k$ and E_k is symmetric with respect to $A_k(\cdot,\cdot)$.*

We observe that in the algorithm stated above, p and m_k are free parameters. With different parameters, we will take account of the three types of algorithm named in the following:

Definition 7.3 *The* **Algorithm S** *is known as the*

(1) *V-cycle if $p = 1$ and $m_k = m \geq 1$ for $k = 0, \cdots, J$,*
(2) *W-cycle if $p = 2$ and $m_k = m \geq 1$ for $k = 0, \cdots, J$,*
(3) *variable V-cycle (VV-cycle) if $p = 1$ and $\gamma_0 m_k \leq m_{k-1} \leq \gamma_1 m_k$ for $k = 0, \cdots, J$, where γ_0 and γ_1 are constants greater than 1.*

The subsequent multigrid convergence theory is largely based on two basic assumptions: one is concerned with the smoothing operators R_k and the other is relevant to the 'regularity' of the underlying problem and the approximation property of the multilevel spaces. More refined convergence estimates depend on more assumptions on prolongation operators and such assumptions will be stated during the presentation of the convergence theory.

The assumption on the smoothing operator R_k which concerns only one level of space is the same as in the nested case.

(A0) $$\|v\|^2 \leq C_0 \lambda_k (\bar{R}_k v, v)$$

where $\bar{R}_k$ is the symmetrization of R_k (see eqn. (2.4)).

A direct consequence of (**A0**) is

(A0′) $$\rho(K_k) < 1,$$

where $\rho(\cdot)$ denotes the spectral radius. This assumption will be used in place of (**A0**) to get some more general (but weaker) results.

The second assumption, usually called the 'regularity and approximation assumption', is that there exists a constant $\beta \in (0, 1]$ and a constant C_1 such that

(A1) $$|A_k((I - I_k I_k^*)v, v)| \leq C_1(\lambda_k{}^{-1}\|A_k v\|_k{}^2)^\beta A_k(v,v)^{1-\beta} \quad \forall v \in \mathcal{M}_k.$$

This is the most crucial assumption in the multigrid theory to be presented in this section. It relates the bilinear forms, different levels of spaces and prolongation operators. In the case of elliptic boundary value problems, its verification is strongly tied to the regularity property of the underlying partial differential equation.

It is not hard to see that (**A1**) implies that

$$A_k(I_k v, I_k v) \leq \tilde{C}_1 A_{k-1}(v,v), \quad \forall\, v \in \mathcal{M}_{k-1}. \tag{7.4}$$

Theorem 7.4 *Under the assumptions* (**A0**) *and* (**A1**), *the Algorithm S has the following convergence properties (with a positive constant M depending only on C_0, C_1 and β):*

1. The W-cycle converges uniformly for sufficiently many smoothings:

$$|||E_k|||_k \le \frac{M}{m^{\beta/2}} \quad \text{if } m^{\beta/2} \ge 2M.$$

2. The variable V-cycle converges uniformly for sufficiently large m_J:

$$|||E_k|||_k \le \frac{c_0}{m^{\beta/2}} \quad \text{if } m_J^{\beta/2} \ge c_0.$$

3. The variable V-cycle gives a uniform preconditioner with any fixed number of smoothings on the finest grid:

$$\kappa(B_J A_J) \le \frac{(M + m^\beta)^2}{m^{2\beta}}.$$

4. If the conditions

(**A2**) $$A_k(I_k v, I_k v) \le A_{k-1}(v,v) \quad A_k(I_k v, I_k v) \le A_{k-1}(v,v)$$

hold for all v in $\mathcal{M}_{k-1}$, then the W-cycle and VV-cycle converge uniformly and the V-cycle converges nearly uniformly with any given number of smoothings:

$$|||E_k|||_k \le \begin{cases} \frac{M}{m_k^\beta + M} & \text{for the variable V-cycle,} \\ \frac{M^\beta}{(m+M)^\beta} & \text{for the W-cycle,} \\ \frac{k^{1/\beta-1} M}{m^\beta + k^{1/\beta-1} M} & \text{for the V-cycle.} \end{cases}$$

5. If the condition

$$A_k(I_k v, I_k v) \le 2A_{k-1}(v,v)$$

holds, the W-cycle converges uniformly with any given number of smoothings.

The proof of the above theorem is not very complicated and may be found in [9] and [32].

7.1.1 *Applications*

The above theory has found many applications. First of all, the development of the theory was motivated by applications to multigrid methods for unstructured grids. Unstructured grids refer to grids that do not have a natural multilevel structure. Most of the grids generated by traditional grid generators may fall into such categories. In this application, one has to coarsen the given grid to obtain a sequence of often nonnested multilevel coarse grids. There have been many coarsening techniques available in two dimensions, see Chan and Smith [15], Bank and Xu [4] and others.

Another important application is to multigrid methods for nonconforming finite elements (especially to fourth order problems). Nonconforming elements often give rise to nonnested multigrid subspaces (even on nested multilevel grids). A major concern in this application is the choice of prolongation operators, which are often obtained by using proper averaging techniques. In most applications, estimates like (A2) or (A3) are hard to satisfy and hence only W-cycle or variable V-cycle can be proved to be convergent with sufficiently many smoothings, or variable V-cyle gives rise to an optimal preconditioner. One interesting exception to the above phenomenon is the work by Chen and Oswald [16] and Chen [17] where they have proved that (A2) or (A3) can be satisfied for some nonconforming P_1 elements in some special situations. For this application and related subjects, we refer to Brenner [14] and the references cited there.

7.2 The auxiliary space method with application to unstructured grids

In this section, an abstract framework of *auxiliary space methods* is proposed and, as an application, an optimal multigrid technique is developed for general unstructured grids. The auxiliary space method is a (nonnested) two level preconditioning technique based on a simple relaxation scheme (smoother) and an auxiliary space (that may be roughly understood as a nonnested coarser space). An optimal multigrid preconditioner is then obtained for a discretized partial differential operator defined on an unstructured grid by using an auxiliary space defined on a more structured grid in which a further *nested* multigrid method can be naturally applied. This new technique makes it possible to apply multigrid methods to general unstructured grids without too much more programming effort than traditional solution methods.

The materials in this section are taken from Xu [38].

7.2.1 *The auxiliary space method*

The auxiliary space method is a general preconditioning approach based on a relaxation scheme and an auxiliary space. As mentioned in the introduction, this method can be interpreted in various ways but it may be best understood as a two level nonnested multigrid preconditioner. A detailed description of this approach will be set out below and two theorems will be given for estimating the condition number of the preconditioned system.

Assume that a linear inner product space $\mathcal{V}$ is given together with a linear operator $A : \mathcal{V} \to \mathcal{V}$ that is SPD with respect to an inner product $(\cdot,\cdot)$. Consider the linear equation (2.1). The main ingredient in the new preconditioning technique is another auxiliary linear inner product space $\mathcal{V}_0$ together with an operator $A_0 : \mathcal{V}_0 \mapsto \mathcal{V}_0$ that is SPD with respect to an inner product $[\cdot,\cdot]$ on $\mathcal{V}_0$. This space, in most applications, may be viewed as some approximation for $\mathcal{V}$. The space $\mathcal{V}_0$ need not be a subspace of $\mathcal{V}$ in general, but it should be, in some sense, simpler than $\mathcal{V}$. The operator A_0 may be viewed as a representation of or approximation to A in the space $\mathcal{V}_0$, and A_0 is assumed to be preconditioned by another SPD operator $B_0 : \mathcal{V}_0 \mapsto \mathcal{V}_0$. In other words, the auxiliary space $\mathcal{V}_0$ is chosen in such a way that the equation given by A_0 can be more easily solved than (2.1).

The auxiliary space $\mathcal{V}_0$ is linked with the original space $\mathcal{V}$ by an operator $\Pi : \mathcal{V}_0 \mapsto \mathcal{V}$. If $\mathcal{V}_0$ is viewed as a 'coarse' space, Π plays the rôle of prolongation in the multigrid method. The 'restriction' operator is given by its adjoint $\Pi^t : \mathcal{V} \mapsto \mathcal{V}_0$ defined by

$$[\Pi^t v, w] = (v, \Pi w) \quad v \in \mathcal{V}, w \in \mathcal{V}_0.$$

Another ingredient is an SPD operator $R : \mathcal{V} \mapsto \mathcal{V}$. The rôle of R is to resolve what cannot be resolved, in preconditioning A by the aforementioned space $\mathcal{V}_0$ and the operators defined on $\mathcal{V}_0$. By the multigrid terminology, R is like a smoother. In most applications, R is given by a simple relaxation scheme such as the Jacobi and Gauss–Seidel methods.

With the ingredients described above, the proposed preconditioner is as follows:

$$B = R + \Pi B_0 \Pi^t. \tag{7.5}$$

In the special case that $\mathcal{V}_0 \subset \mathcal{V}$ and $[\cdot,\cdot] = (\cdot,\cdot)$, Π can obviously be given by the natural inclusion operator and as a result Π^t is nothing but the orthogonal projection $Q_0 : \mathcal{V} \mapsto \mathcal{V}_0$. In this case, the preconditioner (7.5) is reduced to

$$B = R + B_0 Q_0,$$

which is the two-level special case of the general nested multilevel preconditioner in Bramble, Pasciak and Xu [9].

By definition, for any $u, v \in \mathcal{V}_h$,

$$(BAu, v)_A = (RAu, v)_A + [B_0 A_0 \Pi^* u, \Pi^* v]_{A_0} \tag{7.6}$$

where $(\cdot,\cdot)_A = (A\cdot,\cdot)$, $[\cdot,\cdot]_{A_0} = [A_0\cdot,\cdot]$ and $\Pi^* = A_0^{-1}\Pi^t A$ satisfying

$$[\Pi^* v, w]_{A_0} = (v, \Pi w)_A \quad v \in \mathcal{V}, w \in \mathcal{V}_0.$$

Denote $\rho_A = \rho(A)$, the spectral radius of A. Also denote by $\|\cdot\|$ the norm induced by either $(\cdot,\cdot)$ or $[\cdot,\cdot]$. The main abstract result in this section is stated below.

Theorem 7.5 *Assume that there are some nonnegative constants α_0, α_1, λ_0, λ_1, and β_1 such that, for all $v \in \mathcal{V}$ and $w \in \mathcal{V}_0$,*

$$\alpha_0 \rho_A^{-1}(v, v) \le (Rv, v) \le \alpha_1 \rho_A^{-1}(v, v), \tag{7.7}$$

$$\lambda_0 [w, w]_{A_0} \le [B_0 A_0 w, w]_{A_0} \le \lambda_1 [w, w]_{A_0}, \tag{7.8}$$

$$\|\Pi w\|_A^2 \le \beta_1 \|w\|_{A_0}^2, \tag{7.9}$$

and furthermore, assume that there exists a linear operator $P : \mathcal{V} \mapsto \mathcal{V}_0$ and positive constants β_0 and γ_0 such that,

$$\|Pv\|_{A_0}^2 \le \beta_0^{-1}\|v\|_A^2 \tag{7.10}$$

and

$$\|v - \Pi Pv\|^2 \le \gamma_0^{-1}\rho_A^{-1}\|v\|_A^2. \tag{7.11}$$

Then the preconditioner given by (7.5) satisfies

$$\kappa(BA) \le (\alpha_1 + \beta_1\lambda_1)((\alpha_0\gamma_0)^{-1} + (\beta_0\lambda_0)^{-1}). \tag{7.12}$$

In particular, if P is a right inverse of Π, namely $\Pi Pv =$ for $v \in \mathcal{V}$, then

$$\kappa((\Pi B_0 \Pi^t)A) \le \frac{\beta_1}{\beta_0}\frac{\lambda_1}{\lambda_0}. \tag{7.13}$$

Proof One first notes that (7.9) is equivalent to

$$\|\Pi^* v\|_{A_0}^2 \le \beta_1\|v\|_A^2 \quad \forall\, v \in \mathcal{V}.$$

By (7.6) and the assumptions, one has

$$(BAv, v)_A \le \alpha_1\rho_A^{-1}\|Av\|^2 + \lambda_1\|\Pi^* v\|_{A_0}^2 \le (\alpha_1 + \lambda_1\beta_1)\|v\|_A^2.$$

This means that $\lambda_{\max}(BA) \le \alpha_1 + \beta_1\lambda_1$. It follows that

$$\begin{aligned}
(v, v)_A &= (v - \Pi Pv, v)_A + [Pv, \Pi^* v]_{A_0} \\
&\le \|v - \Pi Pv\|\|Av\| + \|Pv\|_{A_0}\|\Pi^* v\|_{A_0} \\
&\le (\gamma_0\rho_A)^{-1/2}\|v\|_A \alpha_0^{-1/2}\rho_A^{1/2}(RAv, Av)^{1/2} \\
&\qquad \text{(by (7.11) and (7.7))} \\
&\quad + \beta_0^{-1/2}\|v\|_A \lambda_0^{-1/2}[B_0 A_0 \Pi^* v, \Pi^* v]_{A_0}^{1/2} \\
&\qquad \text{(by (7.10) and (7.8))} \\
&\le ((\alpha_0\gamma_0)^{-1} + (\beta_0\lambda_0)^{-1})^{1/2}\|v\|_A (BAv, v)_A^{1/2} \quad \text{(by (7.6))}.
\end{aligned}$$

This implies that $\lambda_{\min}(BA) \ge ((\alpha_0\gamma_0)^{-1} + (\beta_0\lambda_0)^{-1})^{-1}$. The estimate (7.12) then follows.

In the particular case that P is a right inverse of Π, one may take $\gamma_0 = \infty$ and $R = 0$; the proof of the corresponding estimate is then clear. □

The last estimate for a special case in the above theorem corresponds to the *fictitious space lemma* of Nepomnyaschikh [25, 26]. A necessary condition for Π to have a right inverse is that $\Pi : \mathcal{V}_0 \mapsto \mathcal{V}$ is surjective. As a consequence, $\dim \mathcal{V}_0 \ge \dim \mathcal{V}$, in other words $\mathcal{V}_0$ has to be at least as rich as $\mathcal{V}$. Furthermore the construction of Π also needs more caution. The introduction of an additional smoother (or relaxation operator) R greatly relaxes the constraints on the choice $\mathcal{V}_0$ and Π in the fictitious space approach; hence the resulting preconditioner is potentially more flexible and more robust.

7.2.2 *A special technique for unstructured grids*

The purpose of this subsection is to construct optimal multigrid preconditioners for a finite element matrix from unstructured grids. The term 'unstructured grids' here loosely mean those grids that do not possess natural or convenient multilevel structures. The main idea is to choose an auxiliary space from a rather structured grid in which a natural nested multilevel structure is available. This idea was briefly discussed in Xu [39].

In this section, the model problem in §3.1 will be used and the conforming piecewise linear finite element spaces over quasi-uniform triangulations $\mathcal{T}_h$ will be considered. Let Ω_h be the mesh domain determined by $\mathcal{T}_h$, namely

$$\bar{\Omega}_h = \cup_{\tau \in \mathcal{T}_h} \bar{\tau}.$$

To avoid unimportant technical difficulty, it is assumed, for all feasible h, that

$$\Omega_h \subset \Omega.$$

The finite element space $\mathcal{V}_h$ consists of continuous piecewise (with respect to $\mathcal{T}_h$) linear functions that vanish on $\bar{\Omega} \setminus \Omega_h$.

7.2.3 *A structured auxiliary finite element space*

Given a uniform square partition of the whole space with mesh size

$$h_0 \equiv 2^{-J} \eqsim h$$

for some integer $J \eqsim |\log h|$, let $\mathcal{T}_0$ be the union of squares ($d = 2$) or cubes ($d = 3$) that are contained in Ω (see Fig. 4(a)). Let Ω_0 denote the mesh domain determined by $\mathcal{T}_0$, namely

$$\bar{\Omega}_0 = \cup_{\tau \in \mathcal{T}_0} \bar{\tau}.$$

By construction, $\Omega_0 \subset \Omega$ and

$$\max_{x \in \bar{\Omega} \setminus \Omega_0} \operatorname{dist}(x, \partial\Omega) \lesssim h.$$

The set of interior nodes of $\mathcal{T}_0$ will be denoted by $\mathcal{N}_0 = \{x_j^0 : 1 \leq j \leq n_0\}$.

An auxiliary finite element space $\mathcal{V}_0$ will be defined to be a space of continuous piecewise bilinear ($d = 2$) or trilinear ($d = 3$) functions that vanish on $\bar{\Omega} \setminus \Omega_0$. Alternatively, if $\mathcal{T}_0$ is a uniform triangulation (see Fig. 4(a)) consisting of triangles ($d = 2$) or tetrahedrals ($d = 3$) from the aforementioned squares ($d = 2$) or cubes ($d = 3$), $\mathcal{V}_0$ may be defined to be a finite element space consisting of continuous piecewise linear functions that vanish on $\bar{\Omega} \setminus \Omega_0$.

The finite element space $\mathcal{V}_0$ will be used as the auxiliary space for the

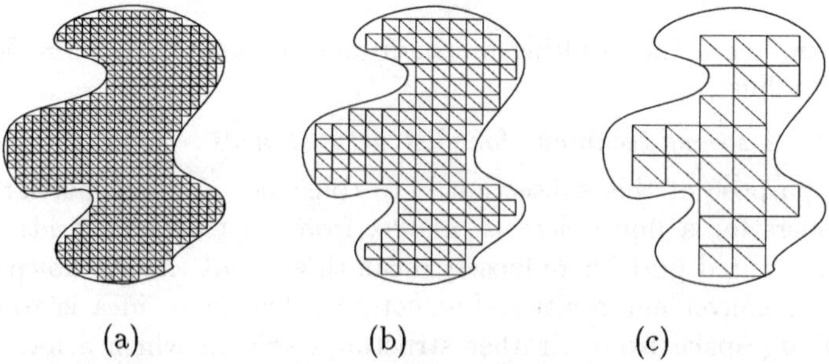

FIG. 4. An example of an auxiliary grid together with two nested coarser grids.

space $\mathcal{V}_h$.

Associated with the spaces $\mathcal{V}_h$ and $\mathcal{V}_0$, the operators: $A_h : \mathcal{V}_h \mapsto \mathcal{V}_h$ and $A_0 : \mathcal{V}_0 \mapsto \mathcal{V}_0$ are defined by

$$(A_h u, v) = a(u, v) \quad \forall\, u, v \in \mathcal{V}_h, \quad (A_0 u_0, v_0) = a(u_0, v_0) \quad \forall\, u_0, v_0 \in \mathcal{V}_0.$$

The operators between $\mathcal{V}_h$ and $\mathcal{V}_0$ will be the standard nodal value interpolants: $\Pi_h : \mathcal{V}_0 \mapsto \mathcal{V}_h$ and $\Pi_0 : \mathcal{V}_h \mapsto \mathcal{V}_0$. The approximation and stability properties of these two operators will be addressed in the following two lemmas.

Lemma 7.6 *For all $v_0 \in \mathcal{V}_0$,*

$$\|v_0 - \Pi_h v_0\| \lesssim h\|v_0\|_1, \quad \|\Pi_h v_0\|_1 \lesssim \|v_0\|_1. \tag{7.14}$$

Lemma 7.7 *For all $v \in \mathcal{V}_h$,*

$$\|v - \Pi_0 v\| \lesssim h\|v\|_1 \text{ and } \|\Pi_0 v\|_1 \lesssim \|v\|_1. \tag{7.15}$$

The proof of the above two lemmas can be found in Xu [38].

7.2.4 *An optimal multigrid preconditioner*

By the abstract approach in §2 and the auxiliary space $\mathcal{V}_0$, a preconditioner B_h for A_h can be obtained as follows

$$B_h = R_h + \Pi_h B_0 \Pi_h^t, \tag{7.16}$$

where $B_0 : \mathcal{V}_0 \mapsto \mathcal{V}_0$ is a given SPD preconditioner of A_0 and $R_h : \mathcal{V}_h \mapsto \mathcal{V}_h$ is an SPD smoother for A satisfying

$$(R_h v, v) \eqsim h^2(v, v).$$

By Theorem 7.5 and the estimates from the previous subsections, there exist some positive constants α, β, γ that are independent of h such that

$$\kappa(B_h A_h) \le (\alpha + \beta\lambda_{\max}(B_0 A_0))(\gamma + \lambda_{\min}^{-1}(B_0 A_0)).$$

The preconditioner B_0 may be obtained by a further multigrid method. Multigrid methods of this type have been discussed by Kornhuber and Yserentant [22]. Setting $\hat{\mathcal{T}}_J = \mathcal{T}_0$ (with $h_J = h_0 = 2^{-J}$) and $\hat{\mathcal{V}}_J = \mathcal{V}_0$, triangulations $\hat{\mathcal{T}}_k$ $(1 \le k \le J)$ (with $h_k = 2^{-k}$) and their corresponding spaces $\hat{\mathcal{V}}_k$ can be defined similarly. As shown in Fig. 4, if Fig. 4(a) corresponds to $\hat{\mathcal{V}}_J = \mathcal{V}_0$, then Fig. 4(b) and Fig. 4 (c) correspond to $\hat{\mathcal{V}}_{J-1}$ and $\hat{\mathcal{V}}_{J-2}$ respectively. Evidently

$$\hat{\mathcal{V}}_1 \subset \hat{\mathcal{V}}_2 \subset \ldots \subset \hat{\mathcal{V}}_J.$$

Consequently, a BPX preconditioner B_0 may be obtained so that [5] $\kappa(B_0A_0) \lesssim 1$ and hence $\kappa(B_hA_h) \lesssim 1$, or a hierarchical basis preconditioner B_0 may be obtained so that $\kappa(B_0A_0) \lesssim |\log h|^2$ and hence $\kappa(BA) \lesssim |\log h|^2$.

Attention is now turned to the issues such as implementation and complexity of the aforementioned preconditioner. Let $\mathcal{A}_h$ and $\mathcal{A}_0$ be the stiffness matrices corresponding to the finite element spaces $\mathcal{V}_h$ and $\mathcal{V}_0$ respectively. Let $\mathcal{I}$ be the matrix representation of the interpolation Π_h, namely

$$(\Pi_h\psi_1, \ldots, \Pi_h\psi_{n_0}) = (\varphi_1, \ldots, \varphi_{n_h})\mathcal{I},$$

where $\{\psi_i : i = 1 : n_0\}$ and $\{\varphi_j : j = 1 : n_h\}$ are the nodal bases for $\mathcal{V}_h$ and $\mathcal{V}_0$ respectively.

The precondition matrix for the stiffness matrix $\mathcal{A}$ can be written as (cf. [33])

$$\mathcal{B}_h = \mathcal{R}_h + \mathcal{I}\mathcal{B}_0\mathcal{I}^t$$

where R_h represents Richardson, Jacobi or Gauss–Seidel iteration.

By definition, $\mathcal{I} = (\alpha_{ij}) \in \mathbb{R}^{n_h \times n_0}$ with $\alpha_{ij} = \psi_j(x_i^h)$. Obviously $\mathcal{I}$ is a sparse matrix with $O(n_h)$ nonzeros. The evaluation of α_{ij} depends on the location of x_i^h relative to the partition $\mathcal{T}_0$. Because of the regularity of $\mathcal{T}_0$, each x_i^h can be located in $\mathcal{T}_0$ with $O(1)$ operations by, for example, comparing the magnitude of the coordinates of x_i^h. Therefore $\mathcal{I}$ and $\mathcal{I}^t$ can both be obtained with $O(n_h)$ operations. For a more direct way of computing the action of $\mathcal{I}$, for example if $\xi \in \mathbb{R}^{n_0}$ and $\eta = \mathcal{I}\xi$, then $\eta_i = w(x_i^h)$ with $w = \sum_{j=1}^{n_0} \xi_j\psi_j$. Again $w(x_i^h)$ can easily be obtained as long as the location of x_i^h is known in $\mathcal{T}_0$.

In summary, when B_0 is given by a BPX preconditioner, the resulting preconditioner which may be called the *BPX preconditioner for unstructured grids* has the following features: 1. one action of B requires only $O(n_h)$ operations; 2. the condition number of BA is uniformly bounded independently of h in both two and three dimensions; 3. it can be applied to unstructured grids.

---

[5] Although the corresponding estimate was not optimal in [22] for their more general considerations, it can easily be proved to be optimal in the current context by the technique in [33].

Remark 7.2 In practical computations, the auxiliary grid $\mathcal{T}_0$ can be more flexible than given above. For example, one does not have to take the elements that are exactly contained in Ω.

Remark 7.3 For simplicity, the details for unstructured grids are only given for Dirichlet boundary value problems. Applications to more general cases are also possible. For Neumann boundary conditions, for example, it is not sufficient that the auxiliary grid only consists of those regular elements that are contained in Ω. Instead, the auxiliary grid consists of all those regular elements that *intersect* with Ω. The application of the technique to locally refined meshes is a little more complicated. Again the idea is to use a structured refined mesh to define the auxiliary space. Locally refined meshes with multilevel structures were discussed in Brandt [12], McCormick [24] (see also the references therein) Bramble, Pasciak and Xu [9] and Bramble and Pasciak [6].

The main spirit of this section is that, with the help of an additional smoother, quite a rough auxiliary space can be used to construct an optimal preconditioner for a discretized partial differential operator. For an elliptic partial differential equation of order $2m$, for example, the auxiliary space $\mathcal{V}_0$ for a given finite element space $\mathcal{V}$ defined on a grid of size h only needs to satisfy the following approximation property:

$$\inf_{w\in\mathcal{V}_0} \|v-w\|_0 \lesssim h^m \|v\|_{m,h} \quad \forall\, v\in\mathcal{V} \tag{7.17}$$

where $\|\cdot\|_0$ is the L^2 norm and $\|\cdot\|_{m,h}$ is a (discretized) H^m norm.

The approximation property (7.17) is a very weak one and is certainly much weaker than the approximation property that $\mathcal{V}$ (as any reasonable finite element space) should have. One important point to address is that the rôle of $\mathcal{V}_0$ or the rôle of a coarse grid in a multigrid algorithm is to resolve the spectrum of the discretized differential operator and there is no reason that $\mathcal{V}_0$ should be comparable with $\mathcal{V}$ as far as their approximation properties are concerned. Roughly speaking, the spectral property of a discretized partial differential operator is mainly determined by the original partial differential operator rather than the underlying discretization space. Hence in order to capture the spectrum of a discretized operator, the auxiliary space only need have an approximation property like (7.17) that is related to the order of the original differential operator but is not, in a certain sense, strongly related to the discretization space ($\mathcal{V}$).

The weak approximation property (7.17) makes it possible to use a simple and structured auxiliary space for preconditioning a complicated and unstructured problem. As the main application of this general idea, the main concrete conclusion of this section is that a finite element space defined on an unstructured grid can be well preconditioned by combining a simple relaxation scheme and a structured grid. As a consequence, it is possible to solve a finite element equation on a general unstructured grid by a multigrid approach with an optimal computational complexity.

8 Nonsymmetric and/or indefinite linear problems

In this section, we shall study a class of iterative methods for solving nonsymmetric or indefinite equations that are governed by some SPD systems. The main idea is to use two grids of different sizes in which the coarse grid is used to resolve the lower frequencies while the fine grid is used to resolve high frequencies on which the leading SPD operator really dominates and hence an SPD preconditioner can be used. The development of this type of two-grid method can be found in Xu and Cai [41], Xu [35, 34, 36, 40, 37].

8.1 Model problems

In this section, we shall discuss finite element discretizations for nonsymmetric and/or indefinite linear partial differential equations. These results, mostly well known, lay down the groundwork for the further analysis of nonlinear problems.

Linear elliptic partial differential operators Let α, β, γ (with the ranges in $\mathbb{R}^{2\times 2}$, $\mathbb{R}^2$ and $\mathbb{R}^1$ respectively) be smooth functions on $\overline{\Omega}$ satisfying, for some positive constant α_0,

$$\xi^T \alpha(x)\xi \geq \alpha_0|\xi|^2 \quad \forall\, \xi \in \mathbb{R}^2.$$

We shall study the following two linear operators: [6]

$$\mathcal{L}\, v = -\mathrm{div}(\alpha(x)\nabla v) \quad \text{and} \quad \hat{\mathcal{L}}\, v = \mathcal{L}\, v + \beta(x)\cdot\nabla v + \gamma(x)v. \tag{8.1}$$

Obviously $\mathcal{L} : H_0^1(\Omega) \to H^{-1}(\Omega)$ is an isomorphism. Our basic assumption is that $\hat{\mathcal{L}} : H_0^1(\Omega) \to H^{-1}(\Omega)$ is also an isomorphism. (A simple sufficient condition for this assumption to be satisfied is that $\gamma(x) \geq 0$.) An application of the open mapping theorem yields

$$||v||_{H^1(\Omega)} \lesssim ||\hat{\mathcal{L}}v||_{H^{-1}(\Omega)} \quad \forall\, v \in H_0^1(\Omega). \tag{8.2}$$

It is easy to see that if $\hat{\mathcal{L}}$ satisfies the above assumption, so does its formal adjoint:

$$\hat{\mathcal{L}}^* u = -\mathrm{div}(\alpha(x)\nabla u + \beta(x)u) + \gamma(x)u.$$

Namely $\hat{\mathcal{L}}^* : H_0^1(\Omega) \to H^{-1}(\Omega)$ is also isomorphic and satisfies (8.2).

Corresponding to $\mathcal{L}$ and $\hat{\mathcal{L}}$, we define two bilinear forms, for $u, v \in$

---

[6] Nonsymmetric and/or indefinite operators or bilinear forms will in general be denoted by symbols with 'hats'.

$H_0^1(\Omega)$, as follows:

$$A(u,v)=\int_\Omega \alpha(x)\nabla u\cdot\nabla v\,dx,\quad \hat{A}(u,v)=A(u,v)+\int_\Omega((\beta\cdot\nabla u)v+\gamma(x)uv)\,dx. \tag{8.3}$$

We shall often use the following well known regularity result (using (8.2)).

Lemma 8.1 *If $u\in H_0^1(\Omega)$ and $\hat{\mathcal{L}}u\in L^2(\Omega)$, then $u\in H^2(\Omega)$ and*

$$||u||_{H^2(\Omega)}\le C||\hat{\mathcal{L}}u||$$

for some positive constant C depending on the coefficients of $\hat{\mathcal{L}}$ and the domain Ω.

Finite element discretizations We assume that Ω is partitioned by a quasi-uniform triangulation $T_h=\{\tau_i\}$. By this we mean that the τ_i are simplices of size h with $h\in(0,1)$ and $\bar{\Omega}=\cup_i\bar{\tau}_i$ and there exist constants C_0 and C_1 not depending on h such that each element τ_i is contained in (contains) a ball of radius C_1h (respectively C_0h).

For a given triangulation T_h, a finite element space $\mathcal{V}_h\subset\mathcal{V}\equiv H_0^1(\Omega)$ is defined by

$$\mathcal{V}_h=\{v\in C(\bar{\Omega}):v|_\tau\in\mathcal{V}_\tau^r\quad\forall\,\tau\in T_h, v|_{\partial\Omega}=0\},$$

where $\mathcal{V}_\tau^r$ is the space of polynomials of degree not greater than a positive integer r. For a given $v\in C(\bar{\Omega})$, $v_I\in\mathcal{V}_h$ will denote the standard nodal value interpolation of v.

It is well known (cf. [18]) that $\mathcal{V}_h$ satisfies the following approximation property

$$\inf_{\chi\in\mathcal{V}_h}\{||v-\chi||_{L^p(\Omega)}+h||v-\chi||_{W_p^1(\Omega)}\}\lesssim h^k|v|_{W_p^k(\Omega)}, \tag{8.4}$$

for all $v\in W_p^k(\Omega)\cap H_0^1(\Omega), 2\le k\le r+1$ and $1\le p\le\infty$.

Let $P_h:\mathcal{V}\longrightarrow\mathcal{V}_h$ be the standard Galerkin projection defined by

$$A(P_hv,\chi)=A(v,\chi)\quad\forall\,\chi\in\mathcal{V}_h. \tag{8.5}$$

Using Lemma 8.1 and a standard duality argument, we have

$$||v-P_hv||\lesssim h||v||_{H^1(\Omega)}\quad\forall\,v\in\mathcal{V}. \tag{8.6}$$

For the nonsymmetric and/or indefinite problems, the following lemma (based on Schatz [30]) is of fundamental importance.

Lemma 8.2 *If $h\ll 1$, then*

$$||v_h||_{H^1(\Omega)}\lesssim\sup_{\varphi\in\mathcal{V}_h}\frac{\hat{A}(v_h,\varphi)}{||\varphi||_{H^1(\Omega)}}\quad\textit{and}\quad||v_h||_{H^1(\Omega)}\lesssim\sup_{\varphi\in\mathcal{V}_h}\frac{\hat{A}(\varphi,v_h)}{||\varphi||_{H^1(\Omega)}}\;\forall\,v_h\in\mathcal{V}_h. \tag{8.7}$$

The same results are also valid for $\varepsilon \ll 1$ if $\hat{A}$ in (8.7) is replaced by $\hat{A}_\varepsilon$ defined by

$$\hat{A}_\epsilon(u,v) = \int_\Omega (\alpha_\epsilon(x)\nabla u \cdot \nabla v + (\beta_\epsilon \cdot \nabla u)v + \gamma_\epsilon(x)uv)\, dx$$

with the functions $\alpha_\epsilon, \beta_\epsilon, \gamma_\epsilon \in L_\infty(\Omega)$ satisfying

$$||\alpha - \alpha_\varepsilon||_{L^\infty(\Omega)} + ||\beta - \beta_\varepsilon||_{L^\infty(\Omega)} + ||\gamma - \gamma_\varepsilon||_{L^\infty(\Omega)} = \delta_\varepsilon$$

where $\delta_\epsilon = o(1)$ as $\varepsilon \to 0$.

Proof Since $\hat{\mathcal{L}} : H_0^1(\Omega) \to H^{-1}(\Omega)$ is an isomorphism, we have

$$||v_h||_{H^1(\Omega)} \lesssim \sup_{w\in\mathcal{V}} \frac{\hat{A}(v_h, w)}{||w||_{H^1(\Omega)}}.$$

Note that, by definition and (8.6),

$$\begin{aligned} \hat{A}(v_h, P_h w) &= \hat{A}(v_h, w) - \hat{A}(v_h, w - P_h w) \\ &= \hat{A}(v_h, w) + (A - \hat{A})(v_h, w - P_h w) \\ &\geq \hat{A}(v_h, w) - c||v_h||_{H^1(\Omega)}||w - P_h w|| \\ &\geq \hat{A}(v_h, w) - c_1 h||v_h||_{H^1(\Omega)}||w||_{H^1(\Omega)}. \end{aligned}$$

The proof of the first estimate in (8.7) then follows by using the fact that $||P_h w||_{H^1(\Omega)} \lesssim ||w||_{H^1(\Omega)}$. The proof of the second estimate is similar.

For the form $\hat{A}_\epsilon(\cdot,\cdot)$, it follows from the assumption that

$$\hat{A}_\epsilon(v_h, \varphi) \geq \hat{A}(v_h, \varphi) - c\delta_\epsilon||v_h||_{H^1(\Omega)}||\varphi||_{H^1(\Omega)}.$$

The desired result then follows easily if $\epsilon \ll 1$. □

Now, define $\hat{P}_h : \mathcal{V} \longrightarrow \mathcal{V}_h$ by

$$\hat{A}(\hat{P}_h v, \chi) = \hat{A}(v, \chi) \quad \forall\, \chi \in \mathcal{V}_h. \tag{8.8}$$

Following (8.7) and Lemma 8.1, we have

Lemma 8.3 *If $h \ll 1$, then $\hat{P}_h$ is well defined and*

$$||u - \hat{P}_h u|| + h||u - \hat{P}_h u||_{H^1(\Omega)} \lesssim h \inf_{\chi\in\mathcal{V}_h} ||u - \chi||_{H^1(\Omega)} \quad \forall\, u \in \mathcal{V}.$$

The following results show that P_h and $\hat{P}_h$ are 'super-close' in the $H^1(\Omega)$ and $W_\infty^1(\Omega)$ norms.

Lemma 8.4 *Assume that P_h and $\hat{P}_h$ are defined by (8.5) and (8.8) respectively; then*

$$||P_h u - \hat{P}_h u||_{H^1(\Omega)} \lesssim ||u - \hat{P}_h u||.$$

Proof By definition

$$A(P_h u - \hat{P}_h u, \chi) = (A - \hat{A})(u - \hat{P}_h u, \chi) \quad \forall\, \chi \in \mathcal{V}_h.$$

The desired estimates then follow by taking $\chi = P_h u - \hat{P}_h u$. □

We end this section by stating some basic error estimates for $\hat{P}_h$.

Lemma 8.5 *The projection $\hat{P}_h$ admits the following estimate*

$$\begin{aligned} \|u - \hat{P}_h u\| &\lesssim h^{r+1}\|u\|_{H^{r+1}(\Omega)}, \\ \|u - \hat{P}_h u\|_{H^1(\Omega)} &\lesssim h^{r+1}\|u\|_{H^{r+1}(\Omega)}. \end{aligned}$$

8.2 Two-grid discretizations

In this section, we shall present a number of algorithms for non-SPD problems based on two finite element spaces. The idea is to reduce a non-SPD problem to an SPD problem by solving a non-SPD problem on a much smaller space.

The basic mechanisms in our approach are two quasi-uniform triangulations of Ω, T_H and T_h, with two different mesh sizes H and h ($H > h$), and the corresponding finite element spaces $\mathcal{V}_H$ and $\mathcal{V}_h$ which will be called coarse and fine space respectively. In the applications given below, we shall always assume that

$$H = O(h^{\lambda}), \quad \text{for some } 0 < \lambda < 1. \tag{8.9}$$

With the bilinear form $\hat{A}$ defined in (8.3), for $h \ll 1$, let $u_h \in \mathcal{V}_h$ be the unique solution of

$$\hat{A}(u_h, \chi) = (f, \chi) \quad \forall\, \chi \in \mathcal{V}_h$$

and denote the bilinear form of the lower order terms of the operator $\hat{\mathcal{L}}$ (in (8.1)) by

$$N(v, \chi) = (\hat{A} - A)(v, \chi) = (\beta \cdot \nabla v, \chi) + (\gamma v, \chi).$$

Let us now present our first two-grid algorithm.

Algorithm 8.1

1. *Find $u_H \in \mathcal{V}_H$ such that* $\hat{A}(u_H, \varphi) = (f, \varphi) \quad \forall\, \varphi \in \mathcal{V}_H$.
2. *Find $u^h \in \mathcal{V}_h$ such that* $A(u^h, \chi) + N(u_H, \chi) = (f, \chi) \quad \forall\, \chi \in \mathcal{V}_h$.

We note that the linear system in the second step of the above algorithm is SPD.

Theorem 8.6 *Assume $u^h \in \mathcal{V}_h$ is the solution obtained by Algorithm 8.1 for $H \ll 1$. Then*

$$\|u_h - u^h\|_{H^1(\Omega)} \lesssim H^{r+1}\|u\|_{H^{r+1}(\Omega)}$$

and

$$\|u - u^h\|_{H^1(\Omega)} \lesssim (h^r + H^{r+1})\|u\|_{H^{r+1}(\Omega)}$$

provided that $u \in H^{r+1}(\Omega)$.

Proof A direct calculation and an application of Lemma 8.5 shows that

$$\begin{aligned} A(u_h - u^h, \chi) &= -N((I - \hat{P}_H)\, u_h, \chi) \\ &\lesssim \|(I - \hat{P}_H)u_h\| \, \|\chi\|_{H^1(\Omega)} \\ &\lesssim (H\|u - u_h\|_{H^1(\Omega)} + \|(I - \hat{P}_H)u\|) \, \|\chi\|_{H^1(\Omega)} \\ &\lesssim H^{r+1} \, \|u\|_{H^{r+1}(\Omega)} \, \|\chi\|_{H^1(\Omega)}. \end{aligned}$$

The desired result then follows. □

Remark 8.1 If $\beta(x) = 0$ and $r \geq 2$, we have

$$\|P_h u - u^h\|_{H^1(\Omega)} \lesssim \|u - u_H\|_{H^{-1}(\Omega)} \lesssim H^{r+2}\|u\|_{H^{r+2}(\Omega)}$$

and

$$\|u - u^h\|_{H^1(\Omega)} \lesssim (h^r + H^{r+2})\|u\|_{H^{r+2}(\Omega)}.$$

Algorithm 8.1 can be applied in a successive fashion.

Algorithm 8.2 *Let* $u_h^0 = 0$*; assume that* $u_h^k \in \mathcal{V}_h$ *has been obtained;* $u_h^{k+1} \in \mathcal{V}_h$ *is defined as follows.*

1. *Find* $e_H \in \mathcal{V}_H$ *such that* $\hat{A}(e_H + u_h^k, \varphi) = (f, \varphi) \quad \forall\, \varphi \in \mathcal{V}_H$.
2. *Find* $u^h \in \mathcal{V}_h$ *such that* $A(u_h^{k+1}, \chi) + N(u_h^k + e_H, \chi) = (f, \chi) \quad \forall\, \chi \in \mathcal{V}_h$.

As is well known, most linear iterative methods for solving algebraic systems can be obtained by an appropriate matrix (or operator) splitting. For the nonsymmetric system under consideration, the most natural splitting would lead to the following iterative method

$$A(u_h^{k+1}, \chi) + N(u_h^k, \chi) = (f, \chi) \quad \forall\, \chi \in \mathcal{V}_h.$$

This iterative scheme, however, is not convergent in general. The Algorithm 8.2 may be considered as a modification of this 'natural' iterative scheme with recourse to an additional coarse space.

Theorem 8.7 *Assume* $u_h^k \in \mathcal{V}_h$ *is the solution obtained by Algorithm 8.2 for* $k \geq 1$*; then*

$$\|u_h - u_h^k\|_{H^1(\Omega)} \lesssim H^{k+r}\|u\|_{H^{r+1}(\Omega)},$$

and

$$\|u - u_h^k\|_{H^1(\Omega)} \lesssim (h^r + H^{k+r})\|u\|_{H^{r+1}(\Omega)}$$

Proof By definition and Lemma 8.5,

$$\begin{aligned} A(u_h - u_h^k, \chi) &= N((I - \hat{P}_H)(u_h^{k-1} - u_h), \chi) \\ &\leq \|(I - \hat{P}_H)(u_h^{k-1} - u_h)\| \, \|\chi\|_{H^1(\Omega)} \\ &\lesssim H \, \|u_h^{k-1} - u_h\|_{H^1(\Omega)} \, \|\chi\|_{H^1(\Omega)}. \end{aligned}$$

This implies

$$\|u_h - u_h^k\|_{H^1(\Omega)} \lesssim H\|u_h - u_h^{k-1}\|_{H^1(\Omega)}.$$

Applying the above estimate successively and then using Theorem 8.6 yields

$$\|u_h - u_h^k\|_{H^1(\Omega)} \lesssim H^{k-1}\|u_h - u_h^1\|_{H^1(\Omega)} \lesssim H^{k+r}\|u\|_{H^{r+1}(\Omega)}.$$

□

Before ending this section, we present an algorithm for symmetric and indefinite problems (namely $\beta(x) = 0$ in (8.1)). This algorithm is based on the finite element space

$$\hat{\mathcal{V}}_h = (I - \hat{P}_H)\mathcal{V}_h.$$

Algorithm 8.3

1. Find $u_H \in \mathcal{V}_H$ *such that* $\hat{A}(u_H, \varphi) = (f, \varphi) \quad \forall \, \varphi \in \mathcal{V}_H.$
2. Find $e_h \in \hat{\mathcal{V}}_h$ *such that* $A(e_h, \chi) = (f, \chi) \quad \forall \, \chi \in \hat{\mathcal{V}}_h.$
3. $u^h = u_H + e_h.$

We note that the system in the second step of Algorithm 8.3 is SPD. But since it is on the space $\hat{\mathcal{V}}_h$, this system may not be solved very easily. Nevertheless this algorithm is of some theoretical interest. In fact, as shown in the next theorem,

$$\|u - u^h\|_{H^1(\Omega)} \lesssim (h + H^3)\|u\|_{H^2(\Omega)}$$

if linear finite elements are used.

Theorem 8.8 *Assume* $u^h \in \mathcal{V}_h$ *is obtained by Algorithm 8.3; then*

$$\|u - u^h\|_{H^1(\Omega)} \lesssim (h^r + H^{r+2})\|u\|_{H^{r+1}(\Omega)}.$$

Proof As $\hat{A}$ is symmetric, so is $\hat{P}_H$. Thus

$$\hat{A}((I - \hat{P}_H)u, \chi) = (f, \chi) \quad \forall \, \chi \in \hat{\mathcal{V}}_h.$$

Therefore

$$\begin{aligned} A(u_h - (u_H + e_h), \chi) &= -(\gamma(u - u_H), \chi) \lesssim H\|u - u_H\|\|\chi\|_{H^1(\Omega)} \\ &\lesssim H^{r+2}\|u\|_{H^{r+1}(\Omega)}\|\chi\|_{H^1(\Omega)}. \end{aligned}$$

where we have used the fact that $\|\chi\| \lesssim H\|\chi\|_{H^1(\Omega)}$ for $\chi \in \hat{\mathcal{V}}_h$. The desired result follows by taking $\chi = u_h - (u_H + e_h) \in \hat{\mathcal{V}}_h$. □

8.3 Iteration and precondition

The algorithms discussed above are based on exact solvers for the SPD problems. We shall now discuss algorithms based on inexact SPD solvers. For generality and clarity, we begin our discussion in an abstract setting.

We assume that $\mathcal{V}$ is a given linear vector space equipped with an inner product $(\cdot,\cdot)$. Let $L(\mathcal{V})$ denote the space of all linear operators from $\mathcal{V}$ to itself. We are interested in solving the equation

$$\hat{A}u = f, \tag{8.10}$$

for a given $f \in \mathcal{V}$. Here $\hat{A} \in L(\mathcal{V})$ is a given invertible operator satisfying

$$\hat{A} = A + N,$$

and $A \in L(\mathcal{V})$ is SPD in the sense that

$$(Au, v) = (u, Av) \quad \forall\, u, v \in \mathcal{V} \quad \text{and} \quad (Av, v) > 0 \quad \text{if} \quad v \neq 0;$$

the perturbation operator $N \in L(\mathcal{V})$ is not SPD in general.

As A is SPD, $(\cdot,\cdot)_A = (A\cdot,\cdot)$ defines an inner product on $\mathcal{V}$ and induces a norm on $\mathcal{V}$, denoted by $\|\cdot\|_A$. Given $G \in L(\mathcal{V})$, we define its A-norm by

$$\|G\|_A = \sup_{v\in\mathcal{V}} \frac{\|Gv\|_A}{\|v\|_A}.$$

The construction of an iterative algorithm for (8.10) often amounts to the construction of a $\hat{B} \in L(\mathcal{V})$ which behaves like $\hat{A}^{-1}$. One approach is to use $\hat{B}$ to obtain a linear iterative scheme as follows

$$u^{k+1} = u^k + \hat{B}(f - \hat{A}u^k), \tag{8.11}$$

for $k = 0, 1, 2, \cdots$, and any $u^0 \in \mathcal{V}$. Obviously a sufficient condition for the convergence of scheme (8.11) is

$$\eta = \|I - \hat{B}\hat{A}\|_A < 1,$$

and in this case

$$\|u - u^k\|_A \leq \eta^k \|u\|_A.$$

Another approach is to use $\hat{B}$ as a preconditioner for (8.10) in conjunction with GMRES type methods (cf. [29]). Unlike the conjugate gradient method for SPD problems, the GMRES method may not be convergent without proper preconditioning. A preconditioner for the GMRES method is not only to speed up the convergence but more importantly to guarantee the convergence as well. More precisely, if there are two constants

$\alpha_0, \alpha_1 > 0$ such that

$$(\hat{B}\hat{A}v, v)_A \geq \alpha_0(v, v)_A, \quad \|\hat{B}\hat{A}v\|_A \leq \alpha_1\|v\|_A, \quad \forall\, v \in \mathcal{V},$$

then the GMRES method applied to the preconditioned system

$$\hat{B}\hat{A}u = \hat{B}f$$

with the inner product $(\cdot,\cdot)_A$ converges at the rate $1 - \alpha_0^2/\alpha_1^2$ (cf. [29]).

Now we assume that a subspace $\mathcal{V}_0 \subset \mathcal{V}$ is given; we define an operator $\hat{A}_0 : \mathcal{V}_0 \mapsto \mathcal{V}_0$, and three projections $Q_0, P_0, \hat{P}_0 : \mathcal{V} \mapsto \mathcal{V}_0$ by, for all $u_0, v_0 \in \mathcal{V}_0$,

$$(\hat{A}_0 u_0, v_0) = (\hat{A}u_0, v_0),$$

and for all $u \in \mathcal{V}, v_0 \in \mathcal{V}_0$,

$$(AP_0u, v_0) = (Au, v_0), \quad (\hat{A}\hat{P}_0u, v_0) = (\hat{A}u, v_0), \quad (Q_0u, v_0) = (u, v_0).$$

It is clear that $\hat{A}_0$, P_0 and Q_0 are well defined. We shall assume that $\hat{A}_0$ is invertible, which implies that $\hat{P}_0$ is also well defined. By the definitions of $\hat{P}_0, \hat{A}_0$ and Q_0,

$$\hat{A}_0\hat{P}_0 = Q_0\hat{A}.$$

It follows that, for a given $f \in \mathcal{V}$,

$$\hat{u}_0 = \hat{A}_0^{-1}Q_0f \quad \text{if and only if} \quad (\hat{A}\hat{u}_0, v_0) = (f, v_0), \quad \forall\, v_0 \in \mathcal{V}_0.$$

Many estimates in this paper will be established in terms of the parameter

$$\delta_0 = \sup_{u,v\in\mathcal{V}} \frac{(N(I - \hat{P}_0)u, v)}{\|u\|_A\|v\|_A}. \tag{8.12}$$

The assumption that we shall make later is that δ_0 can be sufficiently small if the subspace $\mathcal{V}_0$ is properly chosen.

In the study of preconditioners, we need to use another parameter defined by

$$\bar{\delta} = \sup_{u,v\in\mathcal{V}} \frac{(Nu, v)}{\|u\|_A\|v\|_A}.$$

It is easy to see that

$$\|A^{-1}N\|_A \leq \bar{\delta}. \tag{8.13}$$

Observe that $\bar{\delta} = \delta_0$ if $\mathcal{V}_0 = \{0\}$. Without loss of generality, we assume that $\delta_0 \leq \bar{\delta}$.

Lemma 8.9 *For any* $u \in \mathcal{V}$,

$$\|(\hat{P}_0 - P_0)u\|_A \leq \delta_0\|u\|_A, \quad \|u - \hat{P}_0u\|_A \leq (1 + \delta_0)\|u\|_A. \tag{8.14}$$

Proof It follows from the definitions of $\hat{P}_0$ and P_0 that

$$(A(\hat{P}_0 - P_0)u, v_0) = (N(I - \hat{P}_0)u, v_0), \quad \forall\, u \in \mathcal{V}, v_0 \in \mathcal{V}_0,$$

which, with $v_0 = (\hat{P}_0 - P_0)u$, implies the first inequality in (8.14). The second estimate obviously follows from the first one. □

Linear iterative algorithms We now present the main algorithm proposed in Xu [35]. The algorithm depends on a given solver for A, represented by a $B \in L(\mathcal{V})$, satisfying

$$\|I - BA\|_A < 1.$$

Algorithm 8.4 *Given $u^0 \in \mathcal{V}$, assume u^k is defined for $k \geq 0$; then*

1. *solve (exactly) the equation on $\mathcal{V}_0$:*
$$\hat{A}_0 \hat{u}_0 = Q_0(f - \hat{A}u^k);$$
2. *set $g = f - \hat{A}(u^k + \hat{u}_0)$, for $i = 0, 1, \cdots, p$ and $v^0 = 0$,*
$$v^{i+1} = v^i + B(g - Av^i);$$
3. $u^{k+1} = u^k + \hat{u}_0 + v^p$.

Like in the classical multigrid method, the first step of the above algorithm plays the rôle of correction on the small subspace $\mathcal{V}_0$; the second step plays the rôle of smoothing (by the SPD operator A).

Let us derive the error equation of the above algorithm. Without loss of generality, we assume that $p = 1$. Note that $f = \hat{A}u$, and it follows that

$$\hat{u}_0 = \hat{P}_0(u - u^k) \quad \text{and} \quad v^1 = B\hat{A}(I - \hat{P}_0)(u - u^k).$$

Thus

$$u - u^{k+1} = (I - B\hat{A})(I - \hat{P}_0)(u - u^k).$$

Obviously Algorithm 8.4 is identical to (8.11) if $\hat{B}$ satisfies

$$I - \hat{B}\hat{A} = (I - B\hat{A})(I - \hat{P}_0). \tag{8.15}$$

Theorem 8.10 *Assume that $\hat{B}$ is given by (8.15); then*

$$\|I - \hat{B}\hat{A}\|_A \leq \eta,$$

where

$$\eta = \rho^p + 3\delta_0, \quad \rho = \|I - BA\|_A. \tag{8.16}$$

Consequently

$$\|u - u^k\|_A \leq (\rho^p + 3\delta_0)^k \|u - u^0\|_A,$$

where the u^k are defined by Algorithm 8.4 and u is the solution of (8.10). Therefore the Algorithm 8.4 is convergent if δ_0 is small enough to ensure that $3\delta_0 < 1 - \rho^p$.

Proof Without loss of generality, we assume that $p = 1$. Given $u \in \mathcal{V}$, denote $u_0 = \hat{P}_0 u, v = A^{-1}\hat{A}(u - u_0)$ and $w = u - u_0$. We shall first show that

$$\|w - v\|_A \le \delta_0 \|u\|_A, \quad \|v\|_A \le (1 + 2\delta_0)\|u\|_A. \tag{8.17}$$

In fact

$$\begin{aligned}\|w - v\|_A^2 &= (A(w - v), w - v) = ((A - \hat{A})(u - u_0), w - v) \\ &= -(N(u - u_0), w - v) \le \delta_0 \|u\|_A \|w - v\|_A.\end{aligned}$$

The first estimate in (8.17) then follows. To see the second estimate in (8.17), by Lemma 8.9

$$\begin{aligned}\|v\|_A^2 &= (Av, v) = (\hat{A}(u - u_0), v) = (A(u - u_0), v) + (N(u - u_0), v) \\ &\le (1 + \delta_0)\|u\|_A\|v\|_A + \delta_0\|u\|_A\|v\|_A \le (1 + 2\delta_0)\|u\|_A\|v\|_A.\end{aligned}$$

Therefore (8.17) is justified. Thanks to (8.17), the rest of the proof is easy:

$$\begin{aligned}&\|(I - B\hat{A})(I - \hat{P}_0)u\|_A = \|w - B(Av)\|_A \\ &\le \|w - v\|_A + \|v - B(Av)\|_A \le \delta_0\|u\|_A + \rho\|v\|_A \\ &\le (\delta_0 + \rho(1 + 2\delta_0))\,\|u\|_A \le (\rho + 3\delta_0)\,\|u\|_A\end{aligned}$$

as desired. □

Preconditioners for GMRES type methods Based on the theory just developed, a number of preconditioners can be derived in a straightforward fashion.

First, as a direct consequence of Theorem 2.1, we have

Theorem 8.11 *Suppose*

$$\hat{B} = (I - B\hat{A})\hat{A}_0^{-1}Q_0 + B; \tag{8.18}$$

then, for all $v \in \mathcal{V}$,

$$(\hat{B}\hat{A}v, v)_A \ge (1 - \eta)(v, v)_A, \quad \|\hat{B}\hat{A}v\|_A \le (1 + \eta)\|v\|_A.$$

The proof of the above theorem is straightforward and hence is omitted. We shall now derive the theory developed in [41].

Theorem 8.12 *Let*

$$\hat{B} = \omega\hat{A}_0^{-1}Q_0 + B. \tag{8.19}$$

Then, for η given by (8.16) and for all $v \in \mathcal{V}$,

$$(\hat{B}\hat{A}v, v)_A \ge \frac{1}{2}(1 - \eta)(v, v)_A, \quad \|\hat{B}\hat{A}v\|_A \le (\omega + 2)(1 + \bar{\delta})\|v\|_A, \tag{8.20}$$

provided that ω is sufficiently large and δ_0 is sufficiently small, e.g.

$$\omega \geq \frac{(1+2\bar{\delta})^2}{1-\eta}, \quad \delta_0 \leq \frac{1}{4}\frac{1-\eta}{\omega+1+2\bar{\delta}}. \tag{8.21}$$

Proof Obviously

$$\begin{aligned}\hat{B}\hat{A} &= \omega\hat{P}_0 + B\hat{A} = (\omega - 1 + B\hat{A})\hat{P}_0 + \hat{P}_0 + B\hat{A}(I-\hat{P}_0)\\ &= (\omega - 1 + B\hat{A})P_0 + (\omega - 1 + B\hat{A})(\hat{P}_0 - P_0) + \hat{P}_0 + B\hat{A}(I-\hat{P}_0)\end{aligned}$$

By (8.13) and the fact that $||I - BA||_A < 1$, it is easy to show that

$$||I - B\hat{A}||_A \leq 1 + 2\bar{\delta}.$$

Hence, by (8.14)

$$((\omega - 1 + B\hat{A})(\hat{P}_0 - P_0)v, v)_A \leq (\omega + 1 + 2\bar{\delta})\delta_0||v||_A^2.$$

An application of the Cauchy–Schwarz inequality gives

$$\begin{aligned}((I - B\hat{A})P_0 v, v)_A &\leq ||I - B\hat{A}||_A||P_0 v||_A||v||_A\\ &\leq \frac{1}{1-\eta}(1+2\bar{\delta})^2||P_0 v||_A^2 + \frac{1-\eta}{4}||v||_A^2.\end{aligned}$$

Combining the above two estimates with Theorem 8.11 yields

$$(\hat{B}\hat{A}v, v)_A \geq (\omega - \frac{(1+2\bar{\delta})^2}{1-\eta})||P_0 v||_A^2 + \left(\frac{3(1-\eta)}{4} - (\omega + 1 + 2\bar{\delta})\delta_0\right)||v||_A^2.$$

The first estimate in (8.20) then follows if (8.21) holds. The rest of the proof is straightforward. □

We are now in a position to derive the main result in [41].

Theorem 8.13 *Assume that $\bar{B}$ is a SPD preconditioner for A and*

$$\hat{B} = \omega\hat{A}_0^{-1}Q_0 + \bar{B}. \tag{8.22}$$

Then, for all $v \in \mathcal{V}$,

$$(\hat{B}\hat{A}v, v)_A \geq \frac{\lambda_0 + \lambda_1}{4}(\frac{2\lambda_0}{\lambda_1 + \lambda_0} - 3\delta_0)A(v, v),$$

and

$$||\hat{B}\hat{A}v||_A \leq (\omega + 2)(1 + \bar{\delta})\frac{\lambda_0 + \lambda_1}{2}||v||_A,$$

provided that ω is sufficiently large and δ_0 is sufficiently small. Here

$$\lambda_0 = \lambda_{\min}(BA), \lambda_1 = \lambda_{\max}(BA).$$

Proof Let $B = \frac{2}{\lambda_0+\lambda_1}\bar{B}$. Then

$$\rho = \|I - BA\|_A \le \frac{\lambda_1 - \lambda_0}{\lambda_1 + \lambda_0} < 1.$$

The desired result can be derived from Theorem 8.12. □

Subspace correction method The algorithms we have studied above are based on a given iterative algorithm for the SPD problem. In this section, we shall discuss a special class of iterative methods for SPD problems and discuss the corresponding Algorithm 8.4 and its modifications.

Suppose that $\mathcal{V}_0$ used in the definition of Algorithm 8.4 coincides with that in the decomposition (3.1). Then, if Algorithm 8.4 is applied with Algorithm 3.3, the subspace problems on $\mathcal{V}_0$ are solved twice in each iteration, once for A_0 and once for $\hat{A}_0$. We shall remove the solver for A_0 from Algorithm 3.3 and modify Algorithm 8.4 as follows.

Algorithm 8.5 *Given $u^0 \in \mathcal{V}$, assume that u^k is defined for $k \ge 1$; then we define $u^{k+1} = \hat{u}^k + v^J$ where*

$$\hat{u}^k = u^k + \hat{A}_0^{-1} Q_0 (f - \hat{A} u^k)$$

and, for $i = 1, \cdots, J$,

$$v^i = v^{i-1} + R_i Q_i (g - A v^{i-1})$$

with $g = f - \hat{A}\hat{u}^k$ and $v^0 = 0$.

The error equation of the above algorithm is

$$u - u^{k+1} = (I - \tilde{B}\hat{A})(I - \hat{P}_0)(u - u^k)$$

where

$$I - \tilde{B}A = (I - T_J)(I - T_{J-1}) \cdots (I - T_1). \tag{8.23}$$

Theorem 8.14 *Assume that $\omega_1 < 2$. Then Algorithm 8.5 converges if δ_0, given by (8.12), is sufficiently small. Furthermore the error operator $\tilde{E} = (I - \tilde{B}\hat{A})(I - \hat{P}_0)$ satisfies*

$$\|\tilde{E}\|_A \le \eta$$

where

$$\eta = 5\delta_0 + \sqrt{1 - \frac{2 - \omega_1}{K_0(1 + \omega_1(K_1 - 1))^2}}. \tag{8.24}$$

Proof Define B by

$$I - BA = (I - \tilde{B}A)(I - P_0). \tag{8.25}$$

Then by the estimate (3.24) in Thereom 3.18,

$$\rho = \rho(I_B A) \le \sqrt{1 - \frac{2-\omega_1}{K_0(1+\omega_1(K_1-1))^2}}. \tag{8.26}$$

A direct manipulation yields

$$\begin{aligned}(I - \tilde{B}\hat{A})(I - \hat{P}_0) &= (I - B\hat{A})(I - \hat{P}_0)\\ &+ \quad (I - \tilde{B}A)(P_0 - \hat{P}_0) + (B - \tilde{B})N(I - \hat{P}_0).\end{aligned}$$

Thus

$$\begin{aligned}\|(I - \tilde{B}\hat{A})(I - \hat{P}_0)u\|_A &\le \|(I - B\hat{A})(I - \hat{P}_0)u\|_A\\ &+ \quad \|(I - \tilde{B}A)(P_0 - \hat{P}_0)u\|_A + \|(B - \tilde{B})N(I - \hat{P}_0)u\|_A\\ &\equiv \quad I_1 + I_2 + I_3.\end{aligned}$$

The estimate of I_1 is given by Theorem 8.10

$$I_1 \le (\rho + 3\delta_0)\|u\|_A$$

where ρ is given by (8.26). By the assumption on R_i,

$$\|I - T_i\|_A \le 1,$$

which implies that $\|I - \tilde{B}A\|_A \le 1$. Hence, by (8.14),

$$I_2 \le \|(P_0 - \hat{P}_0)u\|_A \le \delta_0\|u\|_A.$$

It remains to estimate I_3. We first note that, by (8.25),

$$(B - \tilde{B})A = (I - \tilde{B}A)P_0.$$

Thus

$$\|(B - \tilde{B})A\|_A = \|I - \tilde{B}A\|_A\|P_0\|_A \le 1.$$

Let 't' and '$*$' denote the transpositions with respect to the inner products $(\cdot,\cdot)$ and $(A\cdot,\cdot)$ respectively; then

$$\|(B - \tilde{B})^t A\|_A = \|[(B - \tilde{B})A]^*\|_A = \|(B - \tilde{B})A\|_A \le 1.$$

Consequently

$$\begin{aligned}\|(B - \tilde{B})N(I - \hat{P}_0)u\|_A^2 &= ((B - \tilde{B})N(I - \hat{P}_0)u, A(B - \tilde{B})N(I - \hat{P}_0)u)\\ &\le \delta_0\|u\|_A\|(B - \tilde{B})^t A(B - \tilde{B})N(I - \hat{P}_0)u\|_A\\ &\le \delta_0\|u\|_A\|(B - \tilde{B})^t A\|_A\|(B - \tilde{B})N(I - \hat{P}_0)u\|_A\\ &\le \delta_0\|u\|_A\|(B - \tilde{B})N(I - \hat{P}_0)u\|_A.\end{aligned}$$

Hence

$$I_3 = \|(B - \tilde{B})N(I - \hat{P}_0)u\|_A \le \delta_0\|u\|_A.$$

The desired estimate then follows. □

With the subspace correction methods for the SPD problem, we shall now discuss the corresponding preconditioners studied in Section 3.2.

Theorem 8.15 *For $\tilde{B}$ given by (8.23), we have*

$$\hat{B} = (I - \tilde{B}\hat{A})\hat{A}_0^{-1}Q_0 + \tilde{B}. \tag{8.27}$$

Then, for η given by (8.24) and for all $v \in \mathcal{V}$,

$$(\hat{B}\hat{A}v, v)_A \geq (1-\eta)(v,v)_A, \quad \|\hat{B}\hat{A}v\|_A \leq (1+\eta)\|v\|_A.$$

Because of Theorem 8.14, the proof of this theorem or the next one is identical to that of Theorem 8.11 or Theorem 8.12.

Theorem 8.16 *For $\tilde{B}$ given by (8.23), define*

$$\hat{B} = \omega\hat{A}_0^{-1}Q_0 + \tilde{B}. \tag{8.28}$$

Then, for all $v \in \mathcal{V}$,

$$(\hat{B}\hat{A}v, v)_A \geq \frac{1}{2}(1-\eta)A(v,v),$$

and

$$\|\hat{B}\hat{A}v\|_A \leq (\omega+2)(1+\bar{\delta})\|v\|_A,$$

provided that ω is sufficiently large and δ_0 is sufficiently small.

Note that preconditioner (8.28) may also be applied in the SPD case.

Theorem 8.17 *Suppose*

$$\hat{B} = \omega\hat{A}_0^{-1}Q_0 + \sum_{i=1}^{J} R_i Q_i. \tag{8.29}$$

Then, for all $v \in \mathcal{V}$,

$$(\hat{B}\hat{A}v, v)_A \geq \frac{\lambda_0+\lambda_1}{4}\left(\frac{2\lambda_0}{\lambda_1+\lambda_0} - 4\delta_0\right)(v,v)_A,$$

and

$$\|\hat{B}\hat{A}v\|_A \leq (\omega+2)(1+\bar{\delta})\frac{\lambda_0+\lambda_1}{2}\|v\|_A,$$

provided that ω is sufficiently large and δ_0 is sufficiently small.

Proof Using the obvious identity

$$\hat{B}\hat{A} = \hat{P}_0 - P_0 + (\omega-1)\hat{P}_0 + \bar{B}\hat{A},$$

the desired result then follows by (8.14) and Theorem 8.13. □

Bibliography

1. R. Bank and T. Dupont. *Analysis of a two-level scheme for solving finite element equations.* Technical Report Report CNA-159, Center for Numerical Analysis, University of Texas at Austin (1980).
2. R. E. Bank and T. Dupont. *An optimal order process for solving elliptic finite element equations.* Math. Comp. **36**, 35–51 (1981).
3. R. E. Bank, T. Dupont, and H. Yserentant. *The hierarchical basis multigrid method.* Numer. Math. **52**, 427–458 (1988).
4. R.E. Bank and J. Xu. *An algorithm for coarsening unstructured meshes.* Numer. Math. **73**, 1–36 (1996).
5. J. H. Bramble. *Multigrid Methods*, volume 294 of *Pitman Research Notes in Mathematical Sciences.* Longman Scientific & Technical, Essex, England (1993).
6. J. H. Bramble and J. E. Pasciak. *New estimates for multigrid algorithms including the V-cycle.* Math. Comp. **60**, 447–471 (1993).
7. J. H. Bramble, J. E. Pasciak, J. Wang, and J. Xu. *Convergence estimates for multigrid algorithms without regularity assumptions.* Math. Comp. **57**, 23–45 (1991).
8. J. H. Bramble, J. E. Pasciak, J. Wang, and J. Xu. *Convergence estimates for product iterative methods with applications to domain decomposition.* Math. Comp. **57**, 1–21 (1991).
9. J. H. Bramble, J. E. Pasciak, and J. Xu. *Parallel multilevel preconditioners.* Math. Comp. **55**, 1–22 (1990).
10. J. H. Bramble, J. E. Pasciak, and J. Xu. *The analysis of multigrid algorithms with nonnested spaces or noninherited quadratic forms.* Math. Comp. **56**, 1–34 (1991).
11. J. H. Bramble and J. Xu. *Some estimates for a weighted l^2 projection.* Math. Comp. **56**, 463–476 (1991).
12. A. Brandt. *Multi-level adaptive solutions to boundary-value problems.* Math. Comp. **31**, 333–390 (1977).
13. A. Brandt. *Multigrid techniques: 1984 guide with applications to fluid dynamics.* GMD–Studien Nr. 85. Gesellschaft für Mathematik und Datenverarbeitung, St. Augustin (1984).
14. S. C. Brenner. *Multigrid methods for nonconforming finite elements.* In J. Mandel and S. F. McCormick, editors, *Preliminary Proc. of the 4th Copper Mountain Conference on Multigrid Methods*, volume 1, pages 135–149, Denver (1989). Computational Mathematics Group, Univ. of Colorado.
15. T. F. Chan and Barry Smith. *Domain decomposition and multigrid methods for elliptic problems on unstructured meshes.* In David Keyes and Jinchao Xu, editors, *Domain Decomposition Methods in Science*

and Engineering, Proceedings of the Seventh International Conference on Domain Decomposition, October 27–30, 1993, The Pennsylvania State University. American Mathematical Society, Providence (1994).

16. Z. Chen and P. Oswald. *Multigrid and multilevel methods for nonconforming rotated q_1 elements.* Preprint (1996).
17. Z. Chen. *On the convergence of nonconforming multigrid methods for second-order elliptic problems.* Preprint (1996).
18. P. G. Ciarlet. *The Finite Element Method for Elliptic Problems.* North–Holland, Amsterdam, New York (1978).
19. R. P. Fedorenko. *A relaxation method for solving elliptic difference equations.* Z. Vycisl. Mat. i. Mat. Fiz. **1**, 922–927 (1961). Also in USSR Comput. Math. and Math. Phys. **1**, 1092–1096 (1962).
20. W. Hackbusch. *Multigrid Methods and Applications*, volume 4 of *Computational Mathematics.* Springer-Verlag, Berlin (1985).
21. W. Hackbusch. *Iterative Solution of Large Sparse Systems of Equations.* Springer-Verlag, Berlin (1993).
22. R. Kornhuber and H. Yserentant. *Multilevel methods for elliptic problems on domains not resolved by the coarse grid.* In *Domain Decomposition Methods in Scientific and Engineering Computing: Proceedings of the Seventh International Conference on Domain Decomposition*, volume 180 of *Contemporary Mathematics*, pages 49–60, Providence, Rhode Island (1994). American Mathematical Society.
23. S. F. McCormick. *Multigrid Methods*, volume 3 of *Frontiers in Applied Mathematics.* SIAM Books, Philadelphia (1987).
24. S. F. McCormick. *Multilevel Adaptive Methods for Partial Differential Equations*, volume 6 of *Frontiers in Applied Mathematics.* SIAM Books, Philadelphia (1989).
25. S. V. Nepomnyaschikh. *Mesh theorems of traces, normalizations of function traces and their inversion.* Sov. J. Numer. Anal. Math. Modeling **6**, 151–168 (1991).
26. S. V. Nepomnyaschikh. *Decomposition and fictitious domains methods for elliptic boundary value problems.* In D. E. Keyes, T. F. Chan, G. A. Meurant, J. S. Scroggs, and R. G. Voigt, editors, *Fifth International Symposium on Domain Decomposition Methods for Partial Differential Equations*, pages 62–72, Philadelphia (1992). SIAM.
27. R. A. Nicolaides. *On a geometrical aspect of SOR and the theory of consistent ordering for positive definite matrices.* Numer. Math. **23**, 99–104 (1974).
28. M. E. G. Ong. *The 3D linear hierarchical basis preconditioner and its shared memory parallel implementation.* In *Vector and Parallel Computing: issues in applied research and development*, pages 273–

283. Ellis Horwood Ltd. (J. Wiley & Sons), Chichester (1989).

29. Y. Saad and M. H. Schultz. *GMRES: A generalized minimal residual algorithm for solving nonsymmetric linear systems.* SIAM J. Sci. Stat. Comput. **7**, 856–869 (1986).
30. Alfred H. Schatz. *An observation concerning Ritz–Galerkin methods with indefinite bilinear forms.* Math. Comp. **28**, 959–962 (1974).
31. P. Wesseling. *An Introduction to Multigrid Methods.* John Wiley & Sons, Chichester (1992).
32. J. Xu. *Theory of Multilevel Methods.* PhD thesis, Cornell University (1989).
33. J. Xu. *Iterative methods by space decomposition and subspace correction: A unifying approach.* SIAM Review **34**, 581–613 (1992).
34. J. Xu. *Iterative methods by SPD and small subspace solvers for nonsymmetric or indefinite problems.* In D. E. Keyes, T. F. Chan, G. A. Meurant, J. S. Scroggs, and R. G. Voigt, editors, *Fifth International Symposium on Domain Decomposition Methods for Partial Differential Equations*, pages 106–118, Philadelphia (1992). SIAM.
35. J. Xu. *A new class of iterative methods for nonselfadjoint or indefinite problems.* SIAM J. Numer. Anal. **29**, 303–319 (1992).
36. J. Xu. *A novel two-grid method for semilinear elliptic equations.* SIAM J. Sci. Comput. **15**, 231–237 (1994).
37. J. Xu. *Some two-grid finite element methods.* In *Domain Decomposition Methods in Science and Engineering: The Sixth International Conference on Domain Decomposition*, volume 157 of *Contemporary Mathematics*, pages 79–87, Providence, Rhode Island (1994). American Mathematical Society.
38. J. Xu. *An auxiliary space preconditioning technique with application to unstructured grids.* To appear in Numer. Math.
39. J. Xu. *Multigrid and domain decomposition methods.* In J. Xu W. Cai, J. Shu and Z. Shi, editors, *Numerical Analysis in Applied Sciences.* Science Press (1995).
40. J. Xu. *Two-grid discretization techniques for linear and nonlinear elliptic PDEs.* To appear in SIAM J. Numer. Anal.
41. J. Xu and X.-C. Cai. *A preconditioned GMRES method for nonsymmetric or indefinite problems.* Math. Comp. *59*, 311–319 (1992).
42. J. Xu and J. Qin. *Some remarks on a multigrid preconditioner.* SIAM J. Sci. Comput. **15**, 172–184 (1994).
43. H. Yserentant. *On the multi-level splitting of finite element spaces.* Numer. Math. **49**, 379–412 (1986).
44. H. Yserentant. *Old and new convergence proofs for multigrid methods.* Acta Numerica. Cambridge University Press (1992).

45. S.Y. Zhang *Optimal order non-nested multigrid methods for solving finite element equations I–III.* Part I in Math. Comp. **55**, 23–36 (1990); Part II in Math. Comp. **55**, 439–450 (1990); Part III in Math. Comp. **64**, 23–49 (1995).